MATERIALS CONCEPTS FOR SOLAR CELLS

Second Edition

MATERIALS CONCEPTS FOR SOLAR CELLS

Second Edition

Thomas Dittrich

Helmholtz Center Berlin for Materials and Energy, Germany

World Scientific

NEW JERSEY · LONDON · SINGAPORE · BEIJING · SHANGHAI · HONG KONG · TAIPEI · CHENNAI · TOKYO

Published by

World Scientific Publishing Europe Ltd.

57 Shelton Street, Covent Garden, London WC2H 9HE

Head office: 5 Toh Tuck Link, Singapore 596224

USA office: 27 Warren Street, Suite 401-402, Hackensack, NJ 07601

Library of Congress Cataloging-in-Publication Data
Names: Dittrich, Thomas, author.
Title: Materials concepts for solar cells / Thomas Dittrich
     (Helmholtz Center Berlin for Materials and Energy, Germany).
Description: Second edition. | Singapore ; Hackensack, NJ : World Scientific, [2017]
Identifiers: LCCN 2017030469| ISBN 9781786344489 (hc ; alk. paper) |
     ISBN 1786344483 (hc ; alk. paper)
Subjects: LCSH: Solar cells--Materials. | Photovoltaic cells--Materials.
Classification: LCC TK2960 .D57 2017 | DDC 621.31/2440284--dc23
LC record available at https://lccn.loc.gov/2017030469

British Library Cataloguing-in-Publication Data
A catalogue record for this book is available from the British Library.

For any available supplementary material, please visit
http://www.worldscientific.com/worldscibooks/10.1142/Q0131#t=suppl

Desk Editors: Suraj Kumar/Jennifer Brough/Koe Shi Ying

Typeset by Stallion Press
Email: enquiries@stallionpress.com

Printed in Singapore

Preface

This textbook results from lecture courses I have been giving to students who are interested in principles of solar energy conversion and in solar cell materials. Students listening to my courses usually study in the fields of engineering, materials science or natural science at universities around the world. I have had, and I continue to have, the great pleasure to teach not only in Europe but also in Asia, Africa and Latin America. A broad understanding of materials and combinations of materials for photovoltaic solar energy conversion helps implement renewable energy worldwide. The textbook gives a comprehensive introduction to materials concepts for solar cells including basic principles and materials specific concepts.

The success in the development and application of solar cells is closely related to countless improvements of materials and to the development of new materials for solar cells. Quality criteria of solar cell materials are consequences of basic principles of solar cells. Materials and technological concepts follow from the ways in which different photovoltaic absorbers can be realized. Materials concepts of solar cells are based (i) on the growth of large crystals (wafer-based technology), (ii) on the growth of sophisticated layer systems on substrate crystals (epitaxy-based technology), (iii) on the deposition of absorber layers on foreign substrates (thin-film photovoltaics) and (iv) on the combination of very different materials on a nanometer scale (nanocomposite solar cells). Interfaces between materials for charge separation, electric contacts to external leads and passivation are considered for all kinds of solar cells.

I would like to express my appreciation to all of my students, but especially those from the Free University Berlin for their interest and questions, to my former teacher Fred Koch for inspiration, to Günther Seliger

v

for the opportunity to teach at GPE (Global Production Engineering) at the Technical University Berlin, to friends at the Kasetsart University in Bangkok and other universities around the world for supporting my teaching projects, to Martha Lux-Steiner at the Helmholtz Centre Berlin for giving me the opportunity to teach alongside my scientific work, to Bernhard Reinhold for discussing aspects of this textbook, to Brian Edlefsen Lasch and Steffen Fengler for critical reading of the manuscript, and to Catharina Weijman from Imperial College Press for excellent collaboration.

In the last five years, several breakthroughs appeared in photovoltaics and in research on solar cells and solar cell materials. For example, the high potential of crystalline silicon with charge-selective hetero-junctions and alkaline treatments of thin-film absorbers based on chalcopyrite enabled new records and research activities were boosted by the class of hybrid organic–inorganic metal halide perovskites, a promising newcomer in the field. These and other aspects have been considered in the second edition of the book.

Contents

Symbols and Abbreviations

Those Beginning with Latin Letters

A, A$^-$	acceptor, ionized acceptor
A	surface area
A	ampere, unit of the current
a	distance between two grid fingers
A/cm^2	unit of the current density
A_{cr}	area of a cross section
a_0	Bohr radius of the hydrogen atom (0.051 nm)
a_0	interatomic distance
A_{cell}	area of the solar cell
$a_{e1h1,b}$	radius of an exciton with the lowest energy in bulk semiconductor
a_{lc}	lattice constant
A_{fc}	area of the front contact
ALD	atomic layer deposition
AM	air mass
AM0	air mass zero
AM1.5d	air mass 1.5 for direct irradiation
AM1.5G	air mass 1.5 for global irradiation $(P_{sun}(AM1.5G) \equiv 1000\,W/cm^2)$
As	unit of the charge (ampere-second or Coulomb)
a-Si:H	hydrogenated amorphous silicon
a-SiC:H	hydrogenated amorphous silicon carbide
AU	astronomical unit

B	width
B	radiative recombination rate constant
B_{ideal}	radiative recombination rate constant of an ideal absorber
C	heat capacitance
C_A	Auger recombination rate constant
$C_{A,e}$	Auger recombination rate constant of electrons
$C_{A,h}$	Auger recombination rate constant of holes
c	velocity of light ($2.99 \cdot 10^8$ m/s)
C_{i0}	initial concentration of impurity atoms
C_{is}	concentration of impurity atoms in the growing crystal
C_m	concentration of impurity atoms in the melt
c-Si	crystalline silicon
CSS	close space sublimation
Cu_{In}^{2-}	copper replacing an In atom at a lattice side of a chalcopyrite
$CuInS_2$	copper indium disulfide
$Cu(In,Ga)Se_2$	copper indium gallium diselenide
CVD	chemical vapor deposition
D	diffusion coefficient
D, D^+	donor, ionized donor
d	thickness
d_{abs}	thickness of the absorber
D_e	density of states for free electrons in the conduction band
dE	interval of energy
d_{em}	thickness of the emitter layer
$d(h\nu)$	interval of photon energies
D_h	density of states for free holes in the valence band
Di	diffusion constant of an impurity
d_k	thickness of cuvette
d_{local}	local thickness
D_n	diffusion coefficient of free electrons
D_p	diffusion coefficient of free holes

dbs	dangling bonds
$DOS_{st,a}$	density of donor-like surface states
$DOS_{st,d}$	density of acceptor-like surface states
dR	resistance of a very thin slice
D_{th}	thermal diffusion constant
dx	interval of distances
DSSC	dye-sensitized solar cell
E	energy (in units of eV)
e, e^-	free electron
E_A	energy of an acceptor
E_A	activation energy
E_{Ai}	activation energy of a diffusing impurity
$E_A^{SRH,pn}$	activation energy for Shockley–Read–Hall recombination in a *pn*-junction
E_{ab}	energy of an anti-bonding state
$E_{abs}(r_{QD})$	absorption edge of a semiconductor quantum dot
E_b	energy of a bonding state
E_C	energy of the conduction band edge, conduction band edge
$E_{C(n)}$	conduction band edge in an *n*-type doped region
$E_{C(p)}$	conduction band edge in a *p*-type doped region
$E_{C,abs}$	conduction band edge of the photovoltaic absorber
$E_{C,pass}$	conduction band edge of the passivation layer
E_{Cs}	conduction band edge at the surface of a photovoltaic absorber
E_D	energy of a donor
E_{e1h1}	lowest energy of an exciton
$E_{e1h1,b}$	lowest energy of an exciton in a bulk semiconductor
E_F	Fermi-energy
E_{F0}	Fermi-energy in thermal equilibrium
E_{Fm}	Fermi-energy of a metal
E_{Fn}	Fermi-energy of free electrons
$E_{Fn(n)}$	Fermi-energy of free electrons in an *n*-type doped region
$E_{Fn(p)}$	Fermi-energy of free electrons in a *p*-type doped region

$E_{F(n)}^0$	Fermi-energy in thermal equilibrium in an n-type doped region
E_{Fp}	Fermi-energy of free holes
$E_{Fp(n)}$	Fermi-energy of free holes in an n-type doped region
$E_{Fp(p)}$	Fermi-energy of free holes in a p-type doped region
$E_{F(p)}^0$	Fermi-energy in thermal equilibrium in a p-type doped region
E_{Fn}-E_{Fp}	Fermi-level splitting
E_{Fs}	Fermi-energy at the surface of a photovoltaic absorber
E_g	energy of the forbidden band gap
E_{g0}	energy of the forbidden band gap at the temperature of 0 K
E_{HL}	energy of the HOMO–LUMO gap of an organic semiconductor
$E_{HL}^{absorption}$	energy of the HOMO–LUMO gap for the absorption of an organic semiconductor
$E_{HL}^{transport}$	energy of the HOMO–LUMO gap for the transport in an organic semiconductor
E_{HOMO}	lowest energy of an unoccupied state in the HOMO band of an organic semiconductor
E_i	intrinsic Fermi-level
$E_{ion}^{D,A}$	ionization energies of donor and acceptor states
E_{ion}^H	ion ionization energy of the hydrogen atom (13.56 eV)
$E_{kin,e}$	kinetic energy of a free electron
$E_{kin,h}$	kinetic energy of a free hole
E_{LUMO}	lowest energy of an occupied state in the LUMO band of an organic semiconductor
E_{og}	energy of the optical band gap
E_{ph}	photon energy
E_{redox}	potential of a redox reaction
$E_{S/S+}$	energy of the ground state and of the ionized state of a dye molecule
E_S^*	energy of the excited state of a dye molecule
E_t	energy characterizing exponential absorption tails
E_t	energy of a trap state in the forbidden band gap
E_V	energy of the valence band edge, valence band edge
$E_{V(n)}$	valence band edge in an n-type doped region

$E_{V(p)}$	valence band edge in a p-type doped region
$E_{V,abs}$	valence band edge of the photovoltaic absorber
$E_{V,pass}$	valence band edge of the passivation layer
E_{Vs}	valence band edge at the surface of a photovoltaic absorber
E_{vac}	energy of the vacuum
eV	electron Volt, unit of energy
f	occupation probability of an electron
FF	fill factor
FTO	fluorine-doped tin oxide
FZ	float zone
G	generation rate of free charge carriers
G_0	thermal generation rate of an ideal absorber
G_e	emission rate of free electrons
G_{fn}	geometry factor for the diode saturation current density of electron diffusion
G_{fp}	geometry factor for the diode saturation current density of hole diffusion
G_h	emission rate of free holes
G_{sun}	photo-generation rate of an ideal absorber
GaAs	gallium arsenide
GaP	gallium phosphide
Ge	germanium
H	height
h	Planck constant ($6.626 \cdot 10^{-34}$ Js)
$\hbar$	Planck constant in terms of $h/(2\pi)$
h, h^+	free hole
H_n	thickness of the n-type doped region
H_p	thickness of the p-type doped region
HF	hydrofluoric acid
HIT	hetero-junction with intrinsic thin layer
HOMO	highest occupied molecular orbital
I	current, current density
I_0	diode saturation current, diode saturation current density

I_{00} pre-factor of the diode saturation current density

$I_{0,SRH}$ diode saturation current density related to Shockley–Read–Hall recombination

I_D current across a diode

I_{mp} current, current density in the maximum power point

I_{ph} photo current, photo current density

I_S ionization energy of a semiconductor (in units of eV)

I_{SC} short-circuit current (in units of A), short circuit current density (in units of A/cm^2)

I_{SC} short-circuit current density in the diode equation

I_{SC}^* short-circuit current density following from the $I–V$ characteristics

I_{SC}^{abs} short-circuit current density under consideration of transmission losses

I_{SC}^{max} maximum short-circuit current density

I_{SRH} diode current related to Shockley–Read–Hall recombination

$I_{tunn,e}$ tunneling current density for electrons

$I_{tunn,h}$ tunneling current density for holes

ILGAR®   ion layer gas reaction

In_{Cu}^{2+} indium replacing a copper atom at a lattice side of a chalcopyrite

InP   indium phosphide

ITO   indium tin oxide

$I–V$ current–voltage (characteristics)

J_S solar constant (1356 W/m^2)

k wave vector

k heat conductivity

k_0 segregation coefficient in equilibrium

k_B Boltzmann constant ($1.38 \cdot 10^{-23}$ J/K)

k_e momentum of a free electron

k_h momentum of a free hole

k_s segregation coefficient

KCN   potassium cyanide

kWh   kilowatt hour, unit of the electrical energy

L	length
L	diffusion length
L_{drift}	drift length
L_f	length of a grid finger
L_n	diffusion length of electrons
L_p	diffusion length of holes
L_{th}	thermal diffusion length
LPE	liquid phase epitaxy
LUMO	lowest unoccupied molecular orbital
m_e	electron mass in free space ($9.1 \cdot 10^{-31}$ kg)
m_e^*	effective mass of free electrons in the conduction band of a semiconductor
m_h^*	effective mass of free holes in the valence band of a semiconductor
MBE	molecular beam epitaxy
mc-Si	multi-crystalline silicon
MOCVD	metal organic chemical vapor deposition
MOVPE	metal organic vapor phase deposition
MOSFET	metal oxide semiconductor field-effect transistor
N	number
N	density of particles (units of cm^{-3})
N_C	effective density of states in the conduction band
N_{db}	density of dangling bonds
N_e	number of photo-generated electrons
N_{i0}	initial number of impurity atoms
N_{im}	number of impurity atoms
N_{ph}	number of incident photons
N_{st}	density of surface defects
$N_{st,a}$	density of acceptor-like surface defects
$N_{st,d}$	density of donor-like surface defects
N_t	density of defects in the forbidden band gap
N_V	effective density of states in the valence band
n	ideality factor in the diode equation
n	density of free electrons (unit in cm^{-3})

n_i	intrinsic density of free charge carriers
n^+ (n^{++})	density of free electrons in a highly (very highly) n-type doped semiconductor
n_0	density of free electrons in thermal equilibrium
n_n	density of free electrons in an n-type doped region
n_n^0	density of free electrons in an n-type doped region in thermal equilibrium
n_p	density of free electron in a p-type doped region
n_p^0	density of free electron in a p-type doped region in thermal equilibrium
n_s	density of free electrons at the surface of a photovoltaic absorber
n_t	density of occupied traps
n_{air}	refractive index of air
n_{AR}	refractive index of an antireflection coating layer
n_S	refractive index of a photovoltaic absorber
n_{TCO}	refractive index of a TCO layer
nwSC	nanowire solar cell
OVC	ordered vacancy compound
OSC	organic solar cell
P	electric power (in units of W)
P_e	sun's power received on earth
P_i	phosphorus atom at an interstitial lattice side
P_{inst}	installed power (in units of W_p)
P_{light}	light intensity (in units of W/cm^2)
P_{loss}	electric power losses due to resistive heating
P_{mp}	electric power in the maximum power point
P_{Si}	phosphorus atom at a lattice side in a silicon crystal
P_{Si}^+	ionized phosphorus atom at a lattice side in a silicon crystal
P_{sun}	areal density of the power of the sun received on earth (in units of W/m^2)
$\langle P_{sun} \rangle$	average areal density of the power of the sun received on earth
P_{sun}^{total}	total power emitted by the sun

PECVD	plasma-enhanced chemical vapor deposition
PERL	passivated emitter rear locally diffused
PV	photovoltaic
PVD	physical vapor deposition
p	density of free holes (in units of cm^{-3})
p^+ (p^{++})	density of free holes in a highly (very highly) p-type doped semiconductor
p_0	density of free holes in thermal equilibrium
p_n	density of free holes in an n-type doped region
p_n^0	density of free holes in an n-type doped region in thermal equilibrium
p_p	density of free holes in a p-type doped region
p_p^0	density of free holes in a p-type doped region in thermal equilibrium
p_s	density of holes at the surface of a photovoltaic absorber
Q_{is}	total amount of a diffusing species
q	elementary charge ($1.602 \cdot 10^{-19}$ As)
Q_{SC}	charge in the space charge region
$Q_{st,a}$	charge in acceptor-like surface states
$Q_{st,d}$	charge in donor-like surface states
qc-Si	quasi-mono-crystalline silicon
QD	quantum dot
QD-SC	quantum dot-based solar cell
QE	quantum efficiency
R	recombination rate of free charge carriers (in units $s^{-1}cm^{-3}$)
R	reflection coefficient
R	resistance (in units of Ω)
R_{Auger}	Auger recombination rate (in units of $s^{-1}cm^3$)
R_{base}^{c-Si}	resistance of the base of a c-Si solar cell
R_C	contact resistance
$R_{C,t}$	tunneling resistance
$R_{C,tunn,e}$	tunneling resistance of free electrons
$R_{C,tunn,h}$	tunneling resistance of free holes

R_e	radius of the earth (in units of km)
R_e	capture rate of free electrons (in units of $s^{-1} cm^3$)
R_{em}^{c-Si}	emitter resistance of a c-Si solar cell
R_{fc}^{c-Si}	resistance of the front contact of a c-Si solar cell
R_h	capture rate of free holes
R_{if}	distance between initial and finite states
R_L	load resistance
R_{min}	minimum reflection coefficient
R_p	parallel resistance, shunt resistance
$R_{p,tol}$	tolerable parallel resistance, tolerable shunt resistance
r_{QD}	radius of a semiconductor quantum dot (in units of nm)
R_{rad}	radiative recombination rate
R_s	series resistance
$R_s^{contact}$	series resistance of contacts
$R_s^{contact-pn}$	series resistance of contacts between two *pn*-junctions
R_{se}	capture rate for electrons at surface states
R_{sh}	capture rate for holes at surface states
R_{SRH}	Shockley–Read–Hall recombination rate
$R_{s,tol}$	tolerable series resistance
R_{sun}	radius of the sun
R_{TCO}	resistance of a TCO layer
$R_{\square}$	sheet resistance
RD	reflection difference
RF	radio frequency
RHEED	reflection high-energy electron diffraction
S	empirical parameter for describing the dependence of the barrier height
Si_i	silicon atom at an interstitial lattice side
Si_{Si}	silicon atoms on the lattice side of a silicon crystal
SR	spectral response (in units of A/W)
SRH	Shockley–Read–Hall recombination
STC	standard test conditions
T	absolute temperature (in units of K)
t	time (in units of s)

T_0	ambient temperature for the operation of a solar cell
$t_{tr,e}$	transit time of electrons
$t_{tr,h}$	transit time of holes
T_{tr}	transmission probability of a barrier
$T_{tr,e}$	transmission probability of a barrier for electrons
$T_{tr,h}$	transmission probability of a barrier for holes
T_{otr}	optical transmission coefficient
T_S	temperature at the surface of the sun
T_{SC}	temperature of a solar cell under operation
TCO	transparent conducting oxide
TMG	trimethyl gallium
U	voltage, potential (in units of V)
U_D	diffusion potential
U_n	potential drop across the depletion region at an n-type doped side
U_p	potential drop across the depletion region at a p-type doped side
U_{pot}	potential energy (in units of eV)
UHV	ultra-high vacuum
V	volume (in units of cm^3)
V	volt, unit of the electrical potential
V_0	initial volume
v_{cg}	velocity of crystal growth
v_e	electron drift velocity
V_m	volume of a melt
V_{mp}	potential in the maximum power point
V_{OC}	open-circuit voltage
V_s	volume of a growing crystal
V_{Si}	vacancy at a silicon lattice side
v_{th}	thermal velocity
V_{Cu}	copper vacancy (in chalcopyrites, $CuInSe_2$)
V_{Cu}^-	charged copper vacancy (in chalcopyrites, $CuInSe_2$)
VLS	vapor–liquid–solid
vtaSC	very thin absorber solar cell

W_{el}	electric energy (in units of kWh)
$W_{ideal\ gas}$	kinetic energy of an ideal gas (in units of eV)
W_p	watt peak, unit of the installed power
X	concentration factor
X_{max}	maximum concentration factor
X_{path}	factor by which an optical path can be increased
x	parameter for describing the stoichiometry in compounds and alloys
x	depth
x_n	width of the space charge region in an n-type doped semiconductor
x_p	width of the space charge region in a p-type doped semiconductor
x_{scr}	width of the space charge region
x_{scr}^{n+}	width of the space charge region in a highly n-type doped semiconductor
x_{scr}^{p+}	width of the space charge region in a highly p-type doped semiconductor
ZnO	zinc oxide

Those Beginning with Greek Letters

α	absorption coefficient
α	coefficient in the temperature dependence of the band gap
α^{-1}	absorption length
α_{tunn}	reciprocal tunneling length
β	specific temperature in the temperature dependence of the band gap
β_e	emission rate constants for free electrons
β_h	emission rate constants for free holes
Γ	phonon, quantum of lattice and atom vibrations
γ	photon, quantum of light

Δ	drop of the potential energy across a dipole layer at a metal–semiconductor junction
Δa_{lc}	difference of lattice constants
ΔG	photo-generation rate
ΔE_C	offset of the conduction band edges
ΔE_{dipole}	drop of the potential energy across a dipole layer at a hetero-junction
ΔE_{HOMO}	width of the HOMO band
ΔE_{LUMO}	width of the LUMO band
ΔE_V	offset of the valence band edges
ΔI	current difference
Δp	density of photo-generated holes
Δk	uncertainty of momentum
Δn_p	density of photo-generated electrons in a p-type doped region
Δp	density of photo-generated holes
Δp_n	density of photo-generated holes in an n-type doped region
ΔQ_{pol}	polarization charge
ΔR	recombination rate of photo-generated free charge carriers
ΔU	voltage difference
$\Delta \Phi_B$	barrier lowering due to image charge at a metal–semiconductor contact
Δx	uncertainty of localization
ε	electric field (in unit of V/cm)
ε_0	dielectric constant of vacuum ($8.854 \cdot 10.12$ As/(Vm))
ε_{mol}	molar extinction coefficient
ε_{in}	relative dielectric constant in a quantum dot
ε_{out}	relative dielectric constant in the surrounding of a quantum dot
ε_r	relative dielectric constant
η	solar energy conversion efficiency
η_C	efficiency of the Carnot process
η_{multi}	solar energy conversion efficiency of multi-junction solar cells

η_{record} solar energy conversion efficiency of record solar cells

η_{SQ} solar energy conversion efficiency in the Shockley–Queisser
                limit

η_{th} efficiency at the thermodynamic limit

$\eta_{ultimate}$ ultimate solar energy conversion efficiency

θ angle of incidence

θ_C critical angle

λ wavelength (in units of nm)

λ_B de Broglie wavelength

μ mobility of free charge carriers (in units of cm^2/Vs)

μ_e electron drift mobility

μ_{eh} effective mass of an exciton (in units of m_e)

μ_{n+} mobility of free electrons in the highly doped emitter

μ_p mobility of free holes

ν photon frequency

ρ density (in units of g/cm^3)

ρ density of charge (in units of $1/cm^3$)

ρ specific electrical resistance (in units of Ωcm)

ρ_e specific emitter resistance

σ electrical conductivity (in units of S/cm)

σ_e capture cross section for free electrons

σ_h capture cross section for free holes

σ_p conductivity of a p-type semiconductor

σ_S Stefan–Boltzmann constant ($5.67 \cdot 10^{-8}\,Wm^{-2}K^{-4}$)

σ_{se} capture cross section for free electrons at surface states

σ_{sh} capture cross section for free holes at surface states

τ time constant (in units of s)

τ life time

τ_0 minimum Shockley–Read–Hall recombination lifetime

τ_{Auger} Auger recombination lifetime

$\tau_{e,SRH}^{min}$ minimum Shockley–Read–Hall recombination lifetime of
                free electrons

$\tau_{e,SRH}^{min}$	minimum Shockley–Read–Hall recombination lifetime of free holes
τ_{min}	minimum lifetime
τ_{rad}	radiative recombination lifetime
τ_{se}	effective surface recombination lifetime of free electrons
τ_{sh}	effective surface recombination lifetime of free holes
τ_{total}	total lifetime
Φ_0	photon flux of blackbody radiation
Φ_B	barrier height at a metal–semiconductor contact (in unit of eV)
Φ_m	metal work function (in unit of eV)
Φ_s	work function of a semiconductor (in unit of eV)
Φ_{ph}	photon flux
Φ_{sun}	photon flux from the sun (in unit of photon/(cm^{2}s))
Φ_{sun}^{abs}	part of the photon flux from the sun absorbed by the absorber of the solar cell
Φ_{sun}^{refl}	part of the photon flux from the sun reflected from the absorber of the solar cell
Φ_{sun}^{trans}	part of the photon flux transmitted by the absorber of the solar cell
φ	electric potential
φ_0	surface band bending
χ_s	electron affinity of a semiconductor
Ψ	wave function
ω_{tunn}	tunneling rate for Miller–Abrahams tunneling (in units of s^{-1})
$\omega_{0,tunn}$	maximum tunneling rate
Ω	ohm (unit of the electrical resistance)
Ω_S	space angle

Part I

Basics of Solar Cells and Materials Demands

1

Basic Characteristics and Characterization of Solar Cells

Solar cells convert power of sunlight into electric power. As an introduction, therefore, Chapter 1 is devoted to a brief characterization of sunlight and basic electric parameters of solar cells. The power of sun is given in terms of the solar constant, the power spectrum and power losses in earth atmosphere expressed by the so-called air mass. The basic characteristics of a solar cell are the short-circuit current (I_{SC}), the open-circuit voltage (V_{OC}), the fill factor (FF) and the solar energy conversion efficiency (η). The influence of both the diode saturation current density and of I_{SC} on V_{OC}, FF and η is analyzed for ideal solar cells. The importance of concentrated sunlight for increasing η is shown. Tolerable series and parallel resistances are introduced as an evaluation criterion for resistive losses in real solar cells. The influence of the series resistance (R_s) and parallel resistance (R_p) on I_{SC}, V_{OC}, FF and η is investigated. The specific role of R_s and R_p is discussed in detail for the dependence of η on I_{SC}. Concepts are described for measuring the basic characteristics of solar cells and their dependencies on light intensity, temperature and light spectra. Attention is paid to principle work with various kinds of load resistances, to the function of a pyranometer, of a sun simulator and to the measurement of the quantum efficiency of solar cells.

1.1 Solar Radiation and Two Fundamental Functions of a Solar Cell

The sun is a hot sphere radiating energy in form of light or photons into space. The absolute temperature (T) of the outer photo sphere of the sun

(T of the sun, T_S) is about 5800 K. The radius of the sun (R_{sun}) is 6.96 · 10^5 km. The power of thermal radiation and therefore the total power emitted by the sun (P_{sun}^{total}) can be calculated by using the Stefan–Boltzmann equation and the surface area of the sun.

$$P_{sun}^{total} = \sigma_S \cdot T_S^4 \cdot 4\pi \cdot R_{sun}^2 \tag{1.1}$$

The Stefan–Boltzmann constant (σ_S) is 5.67 · 10^{-8} Wm^{-2}K^{-4}.

The sun's power received on earth (P_e) is proportional to the cross-section of the earth and to the reciprocal area of a sphere with the radius equal to one astronomical unit (AU), the distance between the sun and the earth. This is shown schematically in Figure 1.1. The shortest and longest distances between the sun and the earth, the so-called perihelion and aphelion, are equal to 1.47 and 1.52 · 10^8 km, respectively. The radius of the earth (R_e) is about 6400 km.

$$P_e = P_{sun}^{total} \cdot \frac{\pi \cdot R_e^2}{4\pi \cdot (AU)^2} \tag{1.2}$$

The solar constant (J_s) is defined as the power of the sun (P_{sun}) received on earth over 1 m^2.

$$J_s \equiv \frac{P_{sun}^{total}}{4\pi \cdot (AU)^2} \tag{1.3}$$

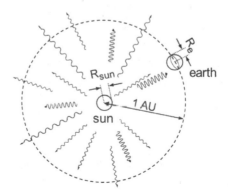

Figure 1.1. Sun emitting photons and earth receiving a proportion of photons emitted by the sun. The radius of the sun, the radius of earth and the distance between sun and earth are denoted by R_{sun}, R_e and AU, respectively.

In reality, the solar constant is not a constant since the distance between the earth and the sun and the temperature distribution at the surface of the sun are not constant. A solar constant of 1356 W/m^2 will be taken into account in the following.

A photon is the smallest portion or quantum of light, the energy of which is proportional to the frequency of the light (v). The factor between the energy of a photon (photon energy, E_{ph}) and the frequency is called the Planck constant ($h = 6.626 \cdot 10^{-34}$ Js).

$$E_{ph} = h \cdot v \tag{1.4}$$

The power spectrum of the sun, i.e. the dependence of power emitted within an interval of photon energies (E_{ph}, $E_{ph} + dE_{ph}$), can be approximated by the radiation of a blackbody with the temperature T_S which is given by the Planck equation.

$$\frac{dJ_S(E_{ph})}{dE_{ph}} = \frac{8\pi}{h^3 c^3} \cdot \frac{E_{ph}^3}{\exp\left(\frac{E_{ph}}{k_B \cdot T_S}\right) - 1} \tag{1.5}$$

The velocity of light (c) is equal to $2.99 \cdot 10^8$ m/s and the Boltzmann constant (k_B) is given as $1.38 \cdot 10^{-23}$ J/K. The spectrum of blackbody radiation for 5800 K is shown in Figure 1.2.

Spectra are usually measured in units of wavelength (λ). The λ of light is proportional to the reciprocal frequency while the proportionality factor is the velocity of light.

$$\lambda = \frac{c}{v} \tag{1.6}$$

The E_{ph} is given in units of eV, which means that the energy is divided by the elementary charge ($q = 1.6 \cdot 10^{-19}$ As). Therefore, the product of the photon energy and the wavelength is given by

$$E_{ph} \cdot \lambda = \frac{h \cdot c}{q} = 1240 \, eV \cdot nm \tag{1.7}$$

The λ can be easily transformed into the E_{ph} by using Equation (1.7). The E_{ph} and the λ in the maximum of the corresponding spectra of blackbody radiation are obtained in tasks T1.3 and T1.4 (see end of this chapter), respectively, for a blackbody with T of 5800 K.

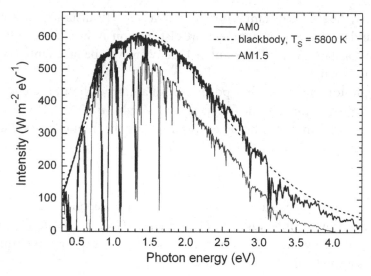

Figure 1.2. Spectrum of sunlight (sun spectrum) outside the earth's atmosphere (air mass (AM) 0, thick solid line), on earth for a zenith angle of 48.2° (AM1.5, thin solid line) and of a blackbody with a temperature of 5800 K (dashed line).

The power received over a certain area on earth depends on geographical location, on the rotation of the earth (day–night cycle), on the inclination angle of the earth's axis (sun in summer and winter) and on the given distance between sun and earth. Further, light is scattered and absorbed by molecules and particles in the earth's atmosphere. The average power of sunlight received on earth, for example, is reduced by a factor of π as a result of the day–night cycle.

The optical path of sunlight through the earth's atmosphere normalized to the thickness of the earth's atmosphere defines the so-called air mass (AM). The power of the sun corresponds at zenith to AM1 and at a zenith angle of 48.2° to AM1.5. It is agreed worldwide that P_{sun} is equal to $1 \, kW/m^2$ at AM1.5G (IEC, 2008) (G denotes global radiation including direct light and scattered light, see also the AM1.5 spectrum (the sun spectrum can be downloaded, for example from (RReDc, 2013)) in Figure 1.2; where direct radiation is less than the global radiation by a factor of about 1.1). The average power of the sun on earth ($\langle P_{sun} \rangle$) is, for example, $0.12 \, kW/m^2$ in Germany and between 0.2 and $0.27 \, kW/m^2$ in most regions in the so-called sunbelt (i.e. those between about the -30 and $+30°$ of latitude).

A solar cell converts P_{sun} into electric power (P), i.e. the product of electric current (I) and electric potential or voltage (U).

$$P = I \cdot U \tag{1.8}$$

With respect to Equation (1.8), the two fundamental functions of a solar cell are (i) the photocurrent generation and (ii) the generation of a photovoltage.

Photocurrent generation means the creation of mobile photogenerated charge carriers by absorbing light and their collection at external contacts. A potential difference between external contacts is necessary for the collection of photogenerated charge carriers. It is not important for the measurement of a photocurrent whether the potential difference between external contacts originates from an external or internal voltage source. Charge separation at a charge-selective contact inside the solar cell is the origin for an internal voltage source and leads to the generation of a photovoltage.

The current or photocurrent is often normalized to the area of the solar cell under consideration. Current and current density will not be distinguished in the following since their meaning follows directly from the unit used (A for current and A/cm^2 for current density) or from the context.

The electric power of solar cells and photovoltaic (PV) modules is on the order of 1 mW to 300 W. PV power plants can be installed for the kW–MW range, and even higher. The extreme scalability of solar cells and PV power plants over many orders of magnitude makes the application of PV solar energy conversion very flexible. This is unique in comparison to any other technology of electricity production.

The installed power of a PV power plant (P_{inst}) is defined as the maximum power which can be given by the power plant at P_{sun} (AM1.5) during the first year after installation. P_{inst} is proportional to the area covered by the PV modules (A) and their solar energy conversion efficiency (η, Equation (1.15)). P_{inst} is given in W_p (Watt peak).

$$P_{inst} = A \cdot \eta \cdot P_{sun}(AM1.5) \tag{1.9}$$

The electric energy (W_{el}) which can be produced with a PV power plant is the product of A, η, $\langle P_{sun} \rangle$ and the time period of operation (t).

$$W_{el} = A \cdot \eta \cdot \langle P_{sun} \rangle \cdot t \tag{1.10}$$

PV power plants can be described by the ratio between W_{el} and P_{inst}.

$$\frac{W_{el}}{P_{inst}} = \frac{\langle P_{sun} \rangle}{P_{sun}(AM1.5)} \cdot t = \left[\frac{kWh}{kW_p} \right] \tag{1.11}$$

The η of PV modules decreases during operation time at a so-called degradation rate (k_{deg}) of between about 0.3% and 1.5% per year (Jordan *et al.*, 2016). The degradation time of a PV module (Δt_η) can be defined as the time after which η reduced to a certain value (see also task T1.2).

The energy payback time of a PV module (t_{EPB}) is defined as the operation time of the PV module resulting in W_{el} equal to the energy which was treated for the production of the corresponding PV module. Low values of k_{deg} and t_{EPB} are important for high energy pay back and therefore for sustainable PV solar energy conversion (see also task T1.3). Changing environmental parameters, such as T and humidity, cause stress in materials and combinations of materials. Therefore, degradation rates depend on the stability of materials and combinations of materials under stress induced, for example, by different thermal expansion coefficients of materials.

1.2 Basic Characteristics of a Solar Cell

An illuminated solar cell can provide a certain photovoltage at a given photocurrent. A combination of values of photocurrent and photovoltage at which a solar cell can be operated is called a working point. A particular working point of a solar cell is fixed with a load resistance (R_L) due to the Ohm's law.

$$R_L = \frac{U}{I} \tag{1.12}$$

In terms of Ohm's law, the photovoltage is very low at very low R_L, and the photocurrent is very low at very high R_L. The short- and open-circuit operation conditions of a solar cell are defined as a R_L which is equal to zero or which is infinitely high, respectively. The values of the photocurrent and of the photovoltage at short- and open-circuit conditions are called short-circuit current (I_{SC}) and open-circuit voltage (V_{OC}), respectively. The electric power is equal to zero at short- and open-circuit operation of a solar cell.

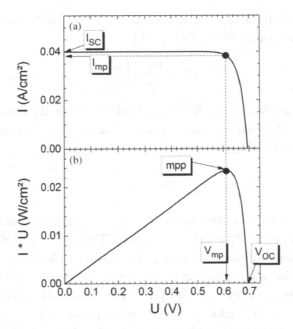

Figure 1.3. Example of a current–voltage characteristic (a) and of the corresponding power–voltage characteristic (b) of a solar cell under illumination. The short-circuit current density, the open-circuit voltage, the maximum power point and the voltage and current density at the maximum power point are denoted by I_{SC}, V_{OC}, mpp, V_{mp} and I_{mp}, respectively.

A current–voltage characteristic (I–V characteristic) of a solar cell is a plot of all possible working points in a considered range. Figure 1.3 shows schematically the I–V characteristic of a solar cell under illumination.

There is one combination of current and voltage at which the power of the solar cell has its maximum (I_{mp} and V_{mp}, respectively). This point on the I–V characteristic of an illuminated solar cell is called the maximum power point (mpp).

$$P_{mpp} = I_{mp} \cdot U_{mp} \tag{1.13}$$

The values of I_{SC} and V_{OC} can be measured easily. Therefore, it is convenient to characterize the maximum power of a solar cell with I_{SC}, V_{OC} and an additional parameter instead of characterizing the solar cell with I_{mp} and U_{mp}. The additional parameter sets the product of I_{mp} and U_{mp} in relation to the product of I_{SC} and V_{OC}. This parameter describes the amount by which the I_{SC}–V_{OC} rectangle is filled by the I_{mp}–U_{mp} rectangle and is therefore called the fill factor (FF).

$$FF = \frac{I_{mp} \cdot U_{mp}}{I_{SC} \cdot V_{OC}} \tag{1.14}$$

The η of a solar cell is defined as the ratio between the power extracted at the mpp of the solar cell and the power of the sunlight at which the solar cell is illuminated (P_{sun}).

$$\eta = FF\frac{I_{SC} \cdot V_{OC}}{P_{sun}} \tag{1.15}$$

The solar energy conversion efficiency is a decisive parameter for costs and sustainability of PV energy production. The higher the value of η, the lower the amount of material and area needed for a PV power plant with a given P_{inst}. A lot of effort has been directed in the past and will continue to be in the future (i) to the investigation of fundamental limitations of η for different types of solar cells, (ii) to materials science for finding out suitable materials and materials combinations for high values of η and (iii) to the development of technologies allowing the realization of maximum values of η in mass production and with a minimum of resources.

1.3 Ideal Solar Cell Under Illumination

1.3.1 *Diode equation of the ideal solar cell*

The two fundamental functions of a solar cell are expressed by two different elements in the equivalent circuit of the ideal solar cell (Figure 1.4). The first element is a current source driven by illumination, i.e. the photocurrent generator. The second element has to fulfill the condition of charge separation, i.e. the current can pass through the element in one

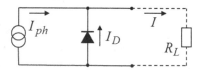

Figure 1.4. Equivalent circuit of an ideal solar cell containing a photocurrent generator and a diode for charge separation and connected to a R_L.

direction but not in the other. This is the characteristic property of a diode, the second element in the equivalent circuit of an ideal solar cell.

A current across a diode (I_D) is described by the diode equation

$$I_D = I_0 \cdot \left[\exp\left(\frac{q \cdot U}{k_B \cdot T} \right) - 1 \right] \qquad (1.16)$$

The diode equation contains the proportionality factor which is called the diode saturation current or diode saturation current density (I_0). The diode saturation current is a specific characteristic of each solar cell depending on the absorber and contact materials as well on their geometry. The diode saturation current is multiplied with an exponential of the voltage while the voltage is multiplied with the q and divided by the product of k_B and T. The diode current increases exponentially with increasing positive voltage (forward direction or forward bias). At zero voltage, the diode current should be equal to zero. It is for this reason that 1 has to be subtracted from the exponential term in the diode equation. At negative voltage, the diode current decreases to the very low negative I_0 (reverse direction or reverse bias).

Any charge-selective contact can be described by a diode equation. The diode equation will be derived for the pn-junction in Chapter 4 and for the ideal solar cell in the Shockley–Queisser limit in Chapter 6.

The resistance of an ideal diode is extremely low in forward direction and very large in reverse direction. The photocurrent flows through the R_L, i.e. the photocurrent is not shunted by the diode. Therefore the photocurrent is a reverse current and has to be subtracted from the diode current. This results in the diode equation of the ideal solar cell.

$$I = I_0 \cdot \left[\exp\left(\frac{q \cdot U}{k_B \cdot T} \right) - 1 \right] - I_{ph} \qquad (1.17)$$

Figure 1.5 shows I–V characteristics of an ideal solar cell with $I_0 = 10^{-13}$ A/cm^2 in the dark and under illumination ($I_{ph} = 0.04$ A/cm^2).

Equation (1.17) is used for the analysis of I–V characteristics of illuminated solar cells by setting I_{ph} equal to I_{SC}. Typical values of the I_{SC} are of the order of tens of mA/cm^2 at AM1.5.

The I–V characteristics of solar cells under illumination are often plotted in the first quadrant for convenience. For this purpose, the

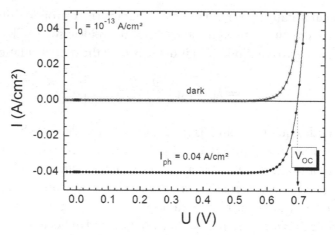

Figure 1.5. Current–voltage characteristics of an ideal solar cell with $I_0 = 10^{-13}$ A/cm^2 in the dark (stars) and under illumination at $I_{ph} = 0.04$ A/cm^2 (circles).

negative current is plotted in the I–V characteristic of a solar cell. As a convention, the negative I_{ph} corresponds to I_{SC}.

$$-I = I_0 \cdot \left[\exp\left(\frac{q \cdot U}{k_B \cdot T} \right) - 1 \right] - I_{SC} \tag{1.18}$$

1.3.2 Open-circuit voltage and diode saturation current

The V_{OC} of an ideal solar cell can be obtained from Equation (1.18) by setting the current to zero.

$$V_{OC} = \frac{k_B \cdot T}{q} \cdot \ln\left(\frac{I_{SC}}{I_0} + 1 \right) \approx \frac{k_B \cdot T}{q} \cdot \ln\left(\frac{I_{SC}}{I_0} \right) \tag{1.19}$$

The value of I_0 can change over many orders of magnitude depending on the materials and on interfaces between materials forming the diode. The importance of I_0 for the performance of a solar cell follows from the limitation of V_{OC} by I_0. The minimization of I_0 is crucial for the optimization of the performance of solar cells in general.

The value of V_{OC} increases with increasing I_{SC}. The dependence of V_{OC} on I_{SC} is shown in Figure 1.6(a) for three different values of I_0.

It is useful to remember that $k_B \cdot T/q$ amounts to 0.026 V at room temperature and that the natural logarithm of 10 is equal to 2.3. Therefore,

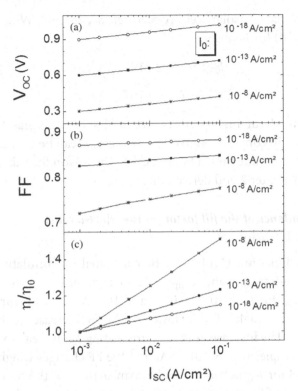

Figure 1.6. Dependence of V_{OC} (a), FF (b) and the normalized η (c) on I_{SC} for an ideal solar cell with $I_0 = 10^{-18}, 10^{-13}$ and 10^{-8} A/cm^2 (squares, circles and stars, respectively).

for an ideal solar cell, V_{OC} increases at room temperature by 60 mV if I_{SC} increases by one order of magnitude.

$$\left[\frac{\Delta V_{OC}}{\text{decade of } I_{SC}} \right]_{300\,K} = 60\,\text{mV per decade} \tag{1.20}$$

The order of magnitude of I_0 can be estimated for crystalline silicon (c-Si) and gallium arsenide (GaAs) record solar cells if assuming them to be ideal.

$$I_0 = \frac{I_{SC}}{\exp\left(\frac{q \cdot V_{OC}}{k_B \cdot T}\right)} \tag{1.21}$$

The values of I_{SC} and V_{OC} are 41.8 mA/cm^2 and 0.74 V, respectively, for a c-Si record solar cell and 29.68 mA/cm^2 and 1.122 V, respectively, for a GaAs record solar cell (Green *et al.*, 2016). With respect to Equation (1.18) for I_0 one gets

$$I_0(\text{c-Si}) \approx 2 \cdot 10^{-14} \, \text{A/cm}^2 \tag{1.22$'$}$$

$$I_0(\text{GaAs}) \approx 6 \cdot 10^{-21} \, \text{A/cm}^2 \tag{1.22$''$}$$

Therefore, I_0 of high-performance c-Si solar cells has to be about 10^{-13} A/cm^2 . In comparison to c-Si solar cells, I_0 can be reduced by about 6 orders of magnitude for GaAs solar cells. The reason for this difference is explored in Chapter 2 and derived in Chapter 4.

1.3.3 *Dependence of the fill factor on the short-circuit current density*

The FF (see Equation (1.14)) can be calculated by simulating Equation (1.18) and finding I_{mp} and U_{mp}. As an example, the FF is plotted as a function of I_{SC} for $I_0 = 10^{-18}, 10^{-13}$ and 10^{-9} A/cm^2 in Figure 1.6(b). At fixed I_{SC} the FF is higher for the lower I_0. The FF increases with increasing I_{SC} at fixed I_0. The increase of FF with increasing I_{SC} is weak for low values of I_0. For example, at $I_0 = 10^{-18}$ A/cm^2 the FF changes roughly between 0.87 and 0.88 for I_{SC} between 1 and 100 mA/cm^2. On the other hand, the increase of FF with increasing I_{SC} is stronger for lower values of I_0. For example, at $I_0 = 10^{-8}$ A/cm^2 the FF changes roughly between 0.72 and 0.78 for I_{SC} between 1 and 100 mA/cm^2.

Due to the influence of I_0, the FFs of solar cells made from similar materials and materials combinations are practically identical at a given I_{SC} if a similar architecture and fabrication technology were used for all compared solar cells. Therefore, besides the relatively weak change of the FF with I_{SC}, FF obtains specific values depending on the type of a solar cell and its production process. For example, FF is lower for thin-film solar cells (see Chapter 9) than for c-Si solar cells (see Chapter 7).

1.3.4 *Dependence of the efficiency of an ideal solar cell on the light intensity*

The solar energy conversion efficiency depends on P_{sun} since FF, I_{SC} and V_{OC} depend on light intensity. The ratio of two solar energy conversion efficiencies (η/η_0), where the index 0 is related to the lower value of P_{sun}, are considered in the following:

$$\frac{\eta}{\eta_0} = \frac{FF}{FF_0} \cdot \frac{I_{SC}}{I_{SC,0}} \cdot \frac{V_{OC}}{V_{OC,0}} \cdot \frac{P_{sun,0}}{P_{sun}}$$ (1.23)

Equation (1.23) can be simplified if assuming that I_{SC} increases linearly with increasing light intensity, i.e. I_{SC} is proportional to P_{sun}. With respect to Equation (1.19), therefore, one gets

$$\frac{\eta}{\eta_0} = \frac{FF}{FF_0} \cdot \frac{T}{T_0} \cdot \frac{\ln \frac{I_{SC}}{I_0}}{\ln \frac{I_{SC,0}}{I_{0,0}}}$$ (1.24)

Equation (1.24) can be further simplified if heating due to increased P_{sun} is neglected and if a constant I_0 is assumed. Of course, one has to be careful with both of these assumptions since the T of solar cells generally increases with increasing P_{sun} and since I_0 can depend on P_{sun}. After a transformation, the following equation is obtained:

$$\frac{\eta}{\eta_0} = \frac{FF}{FF_0} \cdot \left[1 + \frac{X}{\ln \frac{I_{SC,0}}{I_0}} \right]$$ (1.25)

X is the so-called concentration factor.

$$X = \frac{P_{sun}}{P_{sun,0}}$$ (1.26)

The ratio of the FFs is larger than 1, as determined in Section 1.3.3, and X and the natural logarithm of the ratio of $I_{SC,0}$ and I_0 are positive. Therefore, η of an ideal solar cell is larger at the higher light intensity. This is the reason why the absolute largest solar energy conversion efficiencies are reached under concentrated sunlight.

The solar energy conversion efficiency of an ideal solar cell normalized to η at $I_{SC} = 1 \, \text{mA/cm}^2$ is plotted in Figure 1.6(c) as a function of I_{SC} for $I_0 = 10^{-18}$, 10^{-13} and $10^{-9} \, \text{A/cm}^2$. If I_{SC} increases by two orders of magnitude, the relative increases of η is 13, 20 and 50% for $I_0 = 10^{-18}$, 10^{-13} and $10^{-9} \, \text{A/cm}^2$, respectively. However, it has to be taken into account that the assumptions of (i) linear I_{SC} response, (ii) absence of additional heating and (iii) independence of I_0 on P_{sun} may not be fulfilled for real solar cells so that Equation (1.25) can no longer be applied.

1.4 Real Solar Cells: Consideration of Series and Parallel Resistances

1.4.1 *Resistive losses and tolerable series and parallel resistances*

In reality, any solar cell has losses due to ohmic resistances. For example, contact resistances or material resistances are responsible for series resistances and local shunts or other imperfections cause parallel resistances in a solar cell. One common series resistance (R_s) and one common parallel resistance (R_p), which is also called shunt resistance, are considered in the equivalent circuit of a real solar cell (Figure 1.7).

The current flowing through the R_s is equal to the current flowing through R_L. Therefore, the potential across the R_L is reduced by the voltage drop across the R_s. Further, the current flowing through the R_L is reduced by the current flowing through the R_p. The voltage drop across R_s and the current shunted by R_p are taken into account in the following diode equation:

$$-I = I_0 \cdot \left[\exp\left(\frac{q \cdot (U - I \cdot R_s)}{k_B \cdot T} \right) - 1 \right] + \frac{U - I \cdot R_s}{R_p} - I_{SC} \qquad (1.27)$$

The voltage drop across R_s is assumed to be significantly smaller than U_{mp}. Then, the power losses due to resistive heating resistive power losses of R_s and R_p are given by

$$P_{loss}(R_s) = R_s \cdot I_{mp}^2 \qquad (1.28')$$

$$P_{loss}(R_p) \approx \frac{U_{mp}^2}{R_p} \qquad (1.28'')$$

It is useful at this juncture to introduce so-called tolerable series and parallel resistances ($R_{s,tol}$ and $R_{p,tol}$, respectively) in accordance to the power losses that can be tolerated in comparison to the power in the mpp. It makes

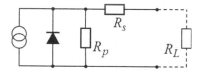

Figure 1.7. Equivalent circuit of a real solar cell containing a photocurrent generator, an ideal diode, a shunt resistance (R_p) and a series resistance (R_s) and being connected with a load resistance (R_L).

sense to tolerate a power loss caused by resistive heating of about 1% of the power in the mpp of a solar cell with very high η. The R_L in the mpp is of the same order of magnitude as the ratio of V_{OC} and I_{SC} (R_L^*). As an approximation rule, one can write the following conditions for $R_{s,tol}$ and $R_{p,tol}$:

$$R_{s,tol} < \frac{1}{100} \cdot \frac{V_{OC}}{I_{SC}} = \frac{R_L^*}{100} \qquad (1.29')$$

$$R_{p,tol} > 100 \cdot \frac{V_{OC}}{I_{SC}} = 100 \cdot R_L^* \qquad (1.29'')$$

Equations (1.29) describe rather harsh conditions. For example, a solar cell based on a c-Si wafer (for details about c-Si solar cells, see in Chapter 7) with an area of about 100 cm² has an I_{SC} of about 3–4 A and a V_{OC} of about 0.6–0.7 V. This means that the overall R_s or R_p of this solar cell should be less than about 2 mΩ or larger than 20 Ω, respectively.

A major source for R_s in c-Si solar cells are the bus bars collecting the charge carriers from the whole solar cell. The bus bars are mainly based on silver. The resistance of a bus bar can be calculated if the specific conductivity (σ, 6.2×10^7 S/m for silver), the length (L), the width (B) and the height (H) of the bus bar are known.

$$R = \frac{L}{\sigma \cdot B \cdot H} \qquad (1.30)$$

The resistance is about 3 mΩ for a typical bus bar with a length of 10 cm, a width of 5 mm and a height of 0.1 mm which is larger than the 2 mΩ demanded. Crystalline silicon solar cells have usually several bus bars depending on the area of the solar cell and on the resistance of the emitter (see Chapter 7).

Current and voltage can be related to area fractions of a solar cell, to solar cells connected in PV modules, to PV modules connected in strings or to strings of PV modules connected in large PV power plants. The conditions for the area of the cross section of bus bars in c-Si solar cells or of copper cables connecting strings of PV modules are equivalent and can be obtained by using Equations (1.29') and (1.30). For this purpose, the series and parallel connection of solar cells has to be taken into account. In a series connection, the current remains constant whereas the individual values of the voltage are added.

$$U = \sum_i U_i \tag{1.31'}$$

$$I = I_i \tag{1.31''}$$

For parallel connection, the voltage remains unchanged and the individual values of the current are added.

$$U = U_i \tag{1.32'}$$

$$I = \sum_i I_i \tag{1.32''}$$

Series connection is favored if an increase of the area of the cross section of connecting cables and therefore of weight or amount of material is unwanted. Binning of solar cells or PV modules in narrow ranges is very important for series connection since the solar cell or the PV module with the lowest photocurrent limits the photocurrent of all other connected solar cells or PV modules. As a consequence the area fraction in an isolated solar cell, the solar cell or the PV module in a string of PV modules with the lowest η will limit the efficiency of the whole system where they are part, i.e. of the complete solar cell, of the PV module or of the PV power plant. This is a reason, for example, why only solar cells with an area of at least $1\,cm^2$ are considered for world records (AM1.5) and why η decreases with increasing area of solar cells, PV modules or PV power plants.

1.4.2 *Influence of series and parallel resistances on the basic characteristics of a solar cell*

The R_s and R_p can heavily influence the $I–V$ characteristics of a solar cell. The influence of R_s and R_p on $I–V$ characteristics will be analyzed separately in the following for a solar cell with fixed I_0 ($10^{-13}\,A/cm^2$) and fixed I_{SC} ($0.04\,A/cm^2$), which is close to the record c-Si solar cells illuminated at P_{sun}(AM1.5). Under these conditions, V_{OC} is close to 0.7 V and R_L^* is about 17 Ωcm^2 in case of the ideal solar cell (R_{Lmpp}^{ideal}). As remark, the I_{SC} in Equation (1.27) and the short-circuit current obtained at zero potential (I_{SC}^*, see also Figure 1.8(a)) can be rather different depending on R_s and thus have to be distinguished.

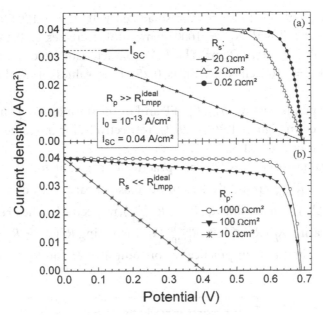

Figure 1.8. *I–V* characteristics for a solar cell with negligible R_p (a) or R_s (b) for different values of R_s (a) and R_p (b) and $I_0 = 10^{-13}$ A/cm^2 and $I_{SC} = 0.04$ A/cm^2.

Figure 1.8(a) shows the $I–V$ characteristics for solar cells for which R_p can be neglected and for which the R_s are about three or one orders of magnitude lower than $R_{\mathrm{Lmpp}}^{\mathrm{ideal}}$ or very close to $R_{\mathrm{Lmpp}}^{\mathrm{ideal}}$. The values of I_{SC} and V_{OC} are practically identical for the $I–V$ characteristics with $R_s = 0.02\ \Omega\mathrm{cm}^2$ and $R_s = 2\ \Omega\mathrm{cm}^2$, while FF decreases significantly for the $I–V$ characteristic with $R_s = 2\ \Omega\mathrm{cm}^2$. The value of V_{OC} remains constant with further increase of R_s since R_s has no influence if there is no current flowing. However, I_{SC}^* decreases to about 0.032 A/cm^2, which is less than I_{SC}, and the $I–V$ characteristic becomes a straight line for $R_s = 20\ \Omega\mathrm{cm}^2$. The FF is 0.25 for a linear $I–V$ characteristic. An FF of 0.25 corresponds to the minimum possible FF of a solar cell which can be described with an equivalent circuit as shown in Figure 1.7. High values of R_s limit I_{SC} and FF.

Figure 1.8(b) shows the $I–V$ characteristics for solar cells in which R_s can be neglected and for which the R_p are about two or one orders of magnitude larger than $R_{\mathrm{Lmpp}}^{\mathrm{ideal}}$ or very close to $R_{\mathrm{Lmpp}}^{\mathrm{ideal}}$. The values of I_{SC} are identical for all R_p since the shunt resistance can be neglected under the I_{SC} condition corresponding to an infinitely low R_L. The value of V_{OC} is a little bit lower for the solar cell with $R_p = 100\,\Omega\text{cm}^2$ than for the solar cell with $R_p = 1000\,\Omega\text{cm}^2$ and decreases to $0.4\,\text{V}$ for the solar cell with $R_p = 10\,\Omega\text{cm}^2$. The I–V characteristic becomes a straight line for $R_p = 10\,\Omega\text{cm}^2$, i.e. FF is 0.25. Low values of R_p limit V_{OC} and FF.

The dependencies of V_{OC}, I_{SC}^*, FF and of the normalized η on R_s or R_p are presented in Figures 1.9(a)–1.9(d), respectively, for the same values of I_0 and I_{SC} as in Figure 1.8.

The values of V_{OC} or I_{SC} are practically constant for $R_p > R_{\text{Lmpp}}^{\text{ideal}}$ or for $R_s < R_{\text{Lmpp}}^{\text{ideal}}$, respectively, and decrease linearly with decreasing R_p $(R_p < R_{\text{Lmpp}}^{\text{ideal}})$ or increasing R_s $(R_s > R_{\text{Lmpp}}^{\text{ideal}})$, respectively. The FFs decrease with decreasing R_p $(R_p \leq 100 \cdot R_{\text{Lmpp}}^{\text{ideal}})$ or increasing R_s $(R_s \geq R_{\text{Lmpp}}^{\text{ideal}}/100)$, respectively, and remain practically constant at 0.25 for $R_p < R_{\text{Lmpp}}^{\text{ideal}}$ or

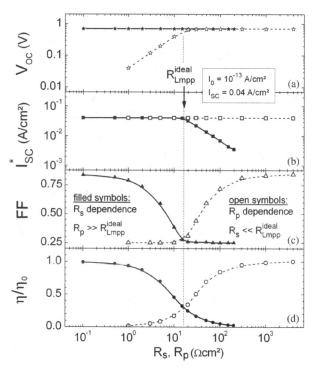

Figure 1.9. Dependence of the V_{OC} (a), the I_{SC}^* density (b), the FF (c) and the normalized η (d) on the R_s (filled symbols) and R_p (open symbols) for solar cells with $I_0 = 10^{-13}\,\text{A/cm}^2$, $I_{SC} = 0.04\,\text{A/cm}^2$.

for $R_s > R_{Lmpp}^{ideal}$, respectively. Therefore, η is limited by FF in the ranges $R_{Lmpp}^{ideal} \leq R_p \leq 100 \cdot R_{Lmpp}^{ideal}$ and $R_L \geq R_s \geq R_{Lmpp}^{ideal}/100$ and by V_{OC} and I_{SC} in the ranges of $R_p < R_{Lmpp}^{ideal}$ and $R_s > R_{Lmpp}^{ideal}$, respectively.

The qualitative influence of R_p and R_s on the basic characteristics of a solar cell is generally similar to the dependencies shown in Figures 1.8 and 1.9 where R_s and R_p should be scaled with respect to I_0 and I_{SC}.

1.4.3 *Influence of series and parallel resistances on the intensity dependence of the basic characteristics of a solar cell*

The influence of R_s and R_p on the I–V characteristics of a solar cell can change strongly with changing light intensity. For this reason R_s and R_p have a tremendous influence on the intensity dependence of the basic characteristics of a solar cell. The I_{SC} in the diode equation increases linearly with increasing P_{sun}.

The influence of I_{SC} on the I–V characteristics will be analyzed separately in the following for a solar cell with fixed I_0 (10^{-13} A/cm^2) and given values of R_s ($10\,\Omega$cm^2) and R_p ($100\,\Omega$cm^2), where the respective values of R_p ($R_s = 10\,\Omega$cm^2) and R_s ($R_p = 100\,\Omega$cm^2) are neglected (Figures 1.10(a) and 1.10(b), respectively).

As shown in Figure 1.10(a) the FF decreases with increasing I_{SC} for a given R_s. The value of I_{SC}^* is proportional to I_{SC} at values for which $I_{SC} < V_{OC}/R_s$. The FF reaches 0.25 when I_{SC} becomes larger than V_{OC}/R_s. In this region I_{SC}^* becomes proportional to V_{OC}. The FF decreases with decreasing I_{SC} for a given R_p (Figure 1.10(b)). The logarithm of the value of V_{OC} is proportional to I_{SC} at values for which $I_{SC} > V_{OC}/R_p$. The FF reaches 0.25 when I_{SC} becomes lower than V_{OC}/R_p. In this region V_{OC} decreases linearly with decreasing I_{SC}.

The voltage drop across the R_s increases with increasing current, which leads to a decrease of the η. This means that R_s limits the η at high values of I_{SC}. On the other hand, the influence of R_p increases with decreasing I_{SC}, R_p limits the η at low values of I_{SC}. The intensity dependencies of V_{OC}, I_{SC}^* and FF can be summarized as the dependencies of the normalized η on I_{SC}.

Figure 1.11 shows the dependence of the normalized η of solar cells with $I_0 = 10^{-13}$ A/cm^2 for R_s of 0.01, 1 and 100 Ωcm^2 (a) and for R_p of 1, 100 and 10,000 Ωcm^2 (b). The range of I_{SC} is chosen in such a way

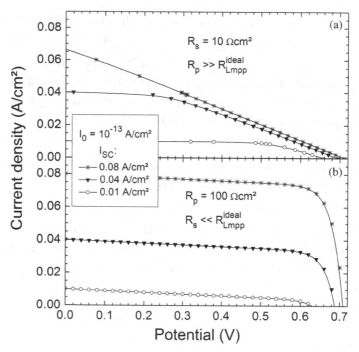

Figure 1.10.   *I–V* characteristics for solar cells with negligible R_p (a) or R_s (b) for $I_0 = 10^{-13}$ A/cm^2 and $I_{SC} = 0.08, 0.04$ and 0.01 A/cm^2 (stars, triangles and circles, respectively) for $R_s = 10\ \Omega$cm^2 (a) and $R_p = 100\ \Omega$cm^2 (b). Incidentally, I_{SC} corresponds to the value in the diode equation.

that the influence of concentrated sunlight or of weak illumination can be analyzed. The illumination target range of the application of a given type of solar cell has significant technological and economic consequences. Concentration factors of up to 1000 are important for highly efficient concentrator solar cells. At very high concentration factors R_s should be significantly lower than 0.01 Ωcm^2.

The R_p of c-Si solar cells fabricated in conventional mass production is only of the order of several hundred Ωcm^2, which is not favorable for highly efficient operation at reduced light intensity. More energy can be produced over the year with a PV power plant based on solar cells with higher R_p due to reduced losses in the morning and evening hours. For example, high values of R_p can be reached with amorphous silicon solar cells (see Chapter 9).

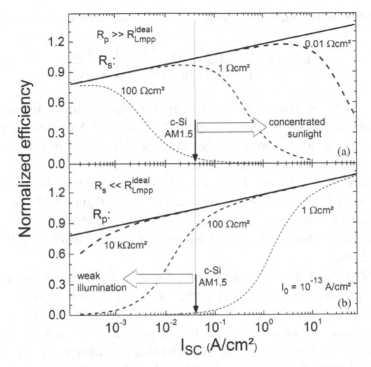

Figure 1.11. Dependencies of the normalized solar energy conversion efficiency on I_{SC} for a solar cell with $I_0 = 10^{-13}$ A/cm^2 at different values of R_s (a) (R_p neglected) and R_p (b) (R_s neglected). The thin and thick arrows mark I_{SC} of an ideal c-Si solar cell at AM1.5 and the directions towards concentrated sunlight and weak illumination, respectively.

Light intensities reduced by up to about 2 orders of magnitude in comparison to AM1.5 can make sense for low-power indoor applications of solar cells and can be relevant for standby functions of electronic devices. The highest values of R_p can be achieved with dye sensitized solar cells (see Chapter 10).

1.5 Characterization of Solar Cells

1.5.1 I_{SC}–V_{OC} characteristics and ideality factor

The simplest measurements on solar cells are performed with just one multimeter by measuring the V_{OC} as a function of the I_{SC}. Regarding to Equations (1.19) and (1.27), considering a low R_s and taking into account an ideality factor (n), the following I_{SC}–V_{OC} characteristic is obtained:

$$I_{SC} = I_0 \cdot \left[\exp\left(\frac{q \cdot V_{OC}}{n \cdot k_B \cdot T} \right) - 1 \right] + \frac{V_{OC}}{R_p} \qquad (1.33)$$

It has to be taken into account that a high accuracy for current measurement over 4–6 orders of magnitude is required. An intensity control of the light is not needed since I_{SC} and V_{OC} are correlated directly and measured at identical light intensities.

Equation (1.33) has I_0, n and R_p as parameters which can be determined from the I_{SC}–V_{OC} characteristic (see Figure 1.12). An I_{SC}–V_{OC} characteristic can be separated into two parts that are dominated by the exponential term and by the ohmic term at high and low values of V_{OC}, respectively.

The slope of the exponential term is usually significantly larger than 60 mV per decade, i.e. a higher potential than for an ideal solar cell is needed to increase the current by one order of magnitude. The ratio of the measured slope and 60 mV per decade is called the ideality factor and is equal to one for an ideal solar cell. The ideality factor is usually on the order of 1.1–1.3 for c-Si solar cells. The value of I_0 is also found from the

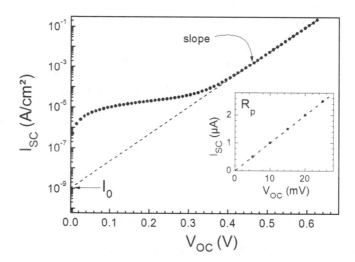

Figure 1.12. Dependence of the I_{SC} on the V_{OC} for a given solar cell in a logarithmic scale and in a linear scale (zoom, inset).

exponential part of the I_{SC}–V_{OC} characteristic. The value of R_p is found from the slope of the ohmic part of the I_{SC}–V_{OC} characteristic.

1.5.2 Temperature-dependent diode saturation current and activation energy

The value of I_0 can be obtained from I_{SC}–V_{OC} characteristics measured at different T. The temperature dependence of I_0 is usually plotted in a so-called Arrhenius plot. Arrhenius plots generally present a T-dependent parameter on a logarithmic scale as a function of T in a reciprocal scale. Figure 1.13 shows an example of a typical Arrhenius plot for I_0.

Arrhenius plots allow for the extraction of the relevant parameter controlling the T dependence of I_0. The point is that the T dependence of I_0 usually follows a straight line in the Arrhenius plot, which means that I_0 depends exponentially on the T. The control parameter is the thermal activation energy (E_A) which can be extracted from the slope of the T-dependent I_0 in the Arrhenius plot. The T-independent pre-factor I_{00} can be obtained from the approximation towards the inverse T equal to 0.

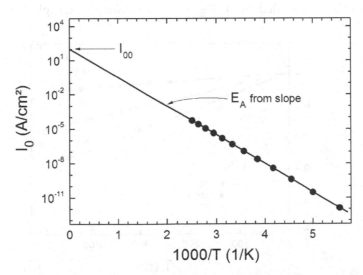

Figure 1.13. Arrhenius plot of the I_0 for a given solar cell. The logarithm of the I_0 should be multiplied with the temperature dependent ideality factor in case of a non-ideal solar cell.

The value of I_{00} amounts to about $100\,\mathrm{A/cm^2}$ for the example shown in Figure 1.13.

$$I_o = I_{00} \cdot \exp\left(-\frac{E_A}{k_B \cdot T}\right) \qquad (1.34)$$

The E_A is given in the unit of eV and depends on fundamental processes in the solar cell. For c-Si solar cells, the E_A is about 1.1 eV. For non-ideal solar cells, the ideality factor can also depend on T. In this case the product of the logarithm of I_0 and the T-dependent ideality factor has to be plotted in the Arrhenius plot.

The thermal activation of I_0 determines the T dependence of V_{OC}. For an ideal solar cell the T dependence of V_{OC} is given by the following equation:

$$V_{OC} = \frac{E_A}{q} - \frac{k_B \cdot T}{q} \cdot \ln\left(\frac{I_{00}}{I_{SC}}\right) \qquad (1.35)$$

The V_{OC} decreases linearly with increasing T following Equation (1.35), while the slope scales with the natural logarithm of the ratio between I_{00} and I_{SC}. The maximum V_{OC} can theoretically be reached at 0 K and is equal to the E_A divided by q. Figure 1.14 shows an example for the T

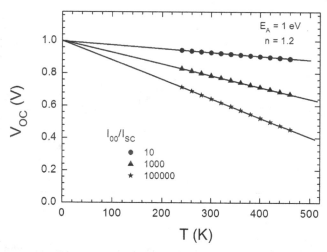

Figure 1.14. Temperature dependence of the V_{OC} for a solar cell with given activation energy and ideality factor measured at different I_{00}/I_{SC} ratios.

dependence of V_{OC} at different ratios of I_{00}/I_{SC}. As can be seen, the higher I_{SC}, the lower the T dependence of V_{OC}.

The solar energy conversion efficiency of solar cells decreases with increasing T due to the thermal activation of I_0. Solar cells are certified for standard test conditions, i.e. at AM1.5 and 25°C. However, the T of solar cells can easily exceed 25°C and increase to values on the order of 60°C, and even higher under operation conditions (see task T1.5 in Section 1.7). This should be taken into account when calculating the expected energy production of a projected PV power plant. Further, there are solar cells with stronger or weaker T dependence of V_{OC} depending on materials, materials combinations and technology. For this reason, the temperature coefficient of PV modules is usually certified as well.

1.5.3 *Measurement of I–V characteristics with loads*

The measurement of complete $I-V$ characteristics of illuminated solar cells is needed to calculate the FF and therefore the η. In addition, the R_s and R_p can be obtained from complete $I-V$ characteristics. The ratio between the current and voltage is different at each point at an $I-V$ characteristic, and therefore variable R_L for probing different working points are necessary. Resistors, electronic loads or capacitors (Figure 1.15) can be applied for $I-V$ measurements depending on requirements for accuracy, costs of measurement equipment, flexibility and speed of measurement.

The use of a set of known resistances and a multimeter for voltage measurement is the easiest and cheapest way to measure an $I-V$ characteristic of an illuminated solar cell. The resistance and the multimeter are connected in parallel with the illuminated solar cell (Figure 1.15(a)). The current is found by Ohm's law from the values of the resistance and the measured voltage.

The values of suitable R_L can be estimated by the following procedure. First, V_{OC} and I_{SC} are measured and the ratio of V_{OC} and I_{SC}, R_L^*, is calculated. The value of R_L^* is close to the R_L in the mpp. Values of resistances less than or greater than R_L^* by roughly one order of magnitude or more are chosen to get points on the $I-V$ characteristic towards I_{SC} or towards V_{OC}, respectively, where more values of R_L have to be chosen around the mpp in order to increase the accuracy for the measurement of the FF. The described method works very well for solar cells and mini-modules

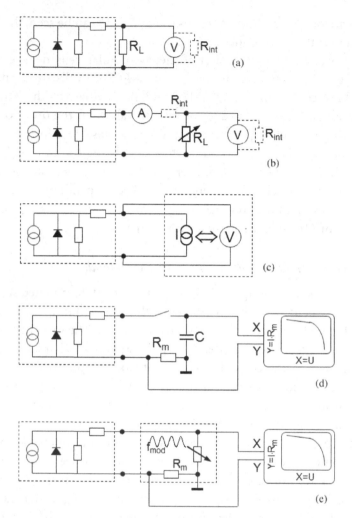

Figure 1.15. Arrangement for the measurement of $I-V$ characteristics of solar cells for (a) a known set of resistances and a multimeter, (b) a potentiometer and two multimeters, (c) an electronic load or constant current source, (d) a switch and capacitor with oscilloscope and (e) a periodically variable R_L and oscilloscope.

with relatively low power when the R_L in the mpp is about 10 Ω or larger. The point is that contact and cable resistances can become an important source of error for low R_L. A fixed correction resistance can be introduced into the analysis if the contact resistance and the resistances of cables are constant. Furthermore, the power range of a given resistor has to be considered since the resistance increases under heating at high power. The accuracy of the described method depends on the accuracy of the measurements of the voltage and of the resistance.

The measurement procedures of I–V characteristics and of the R_p and R_s are depicted in Figures 1.16(a)–1.16(c), respectively. The value of R_p is the negative slope of the I–V characteristic near I_{SC} where the diode current can be neglected.

$$R_p = - \left. \frac{\Delta U}{\Delta I} \right|_{U \to 0} \qquad (1.36)$$

The R_s has a strong influence on the I–V characteristic towards V_{OC}. Two I–V characteristics at different intensities should be measured for

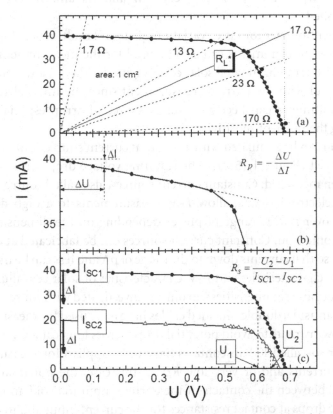

Figure 1.16. Procedure for the measurement of I–V characteristics (a) and determination of the R_p (b) and R_s (c).

obtaining R_s from a simple procedure. The two intensities have to be chosen in such a way that the respective I_{SC} (I_{SC1} and I_{SC2}) differ by a factor of roughly two. The differences between I_{SC1} and I_{SC2} and an identical current difference ΔI are determined (I_1 and I_2). It is recommended to choose a value of ΔI between about half of I_{SC1} and 90% of I_{SC2} for the measurement of R_s. The potentials U_1 and U_2 corresponding to I_1 and I_2 are found on the respective I–V characteristics (Figure 1.16(a)). The R_s is then given by

$$R_s = \frac{U_2 - U_1}{I_{SC1} - I_{SC2}} \tag{1.37}$$

The described procedure of measuring R_s works well for conventional solar cells when the FF is not very low and R_p and R_s can be well distinguished.

A potentiometer for faster variation of R_L and multimeters for current and voltage measurements can also be used for the measurement of I–V characteristics of illuminated solar cells (Figure 1.15(b)). However, one has to be careful with the current measurement since the internal resistance of the multimeter can become a series source of error, especially when changing the range of sensitivity.

A load can be simulated with a constant current source connected with the solar cell (Figure 1.15(c)). The resulting voltage drop across the solar cell is then measured. Constant current sources, also called source measure units or electronic loads, allow I–V measurements to a high degree of accuracy over a wide range of power depending on the dimensioning of the electronic load. Constant current sources can be fabricated at excellent precision so that currents down to the range of pA and less can be measured. Dark I–V characteristics and I–V characteristics of solar cells illuminated over a wide range of intensities can only be investigated with source measure units. As an aside, double-shielded cables have to be used for measurements at very low currents to avoid potential drops across long cables. Four-point probes or so-called Kelvin contacts have to be applied for measurements at high currents and/or very low R_L. In this case, the potential drop is measured between the contacts of the current input to avoid an influence of voltage drops at contact resistances for the current input and to avoid an influence of an inhomogeneous current flow.

Fast and reliable measurements of I_{SC}, V_{OC} and FF are needed for in-line control and binning of solar cells. For example, the tact cycle of a production line for c-Si solar cells is of the order of only 1 s or less. Time for handling, contacting and sorting of a solar cell is needed within one tact cycle in addition to the measurement time. This means that a c-Si solar cell has to be characterized within a time even less than 0.1–0.3 s. Very fast measurements of I–V characteristics become possible with a load changing its resistance automatically and rapidly during illumination with a light flash. The principle of such a flasher is given in Figure 1.15(d). The heart of a flasher is a capacitor (or an electronically simulated capacitor). The resistance of a capacitor varies over a huge range during charging. At the beginning of charging, the resistance of a capacitor is extremely low so that the maximum current which can be provided by the solar cell can flow. Therefore, the current flowing through the capacitor is equal to I_{SC} at the beginning of charging. At the end of charging, the resistance of the capacitor is extremely high since the charged capacitor behaves like an insulator. The maximum voltage across the charged capacitor is equal to the maximum voltage which can be provided by the solar cell, i.e. V_{OC}. The resistance of the capacitor increases continuously during the charging process so that the exact and complete I–V characteristic of the solar cell can be monitored with an oscilloscope. In the simplest case, the voltage drop across the capacitor is applied to the input of the x-channel of the oscilloscope and the current is converted with a low measurement resistance to a voltage signal and applied to the y-channel of the same oscilloscope. It is worth noting a switch connecting and disconnecting the solar cell can be applied instead of using a light pulse for flash characterization.

Rapid but not very fast measurements of I–V characteristics at high accuracy are required when an external parameter such as light intensity is continuously changed during the measurement of I–V characteristics. For such purposes the channel resistance of a field-effect transistor can be varied periodically with a frequency generator at a moderate frequency (Figure 1.15(e)). The I–V characteristic is visualized with an oscilloscope. The voltage drop across the periodically variable load is applied to the x-channel of the oscilloscope and the current transformed to a voltage signal is applied to the y-channel of the oscilloscope.

Periodically variable loads can be realized at relatively small size and low cost, and allow highly accurate and fast measurements and have an excellent performance, which makes them especially advantageous for teaching.

1.5.4 *Measurement of the solar energy conversion efficiency with a pyranometer*

For the measurement of the solar energy conversion efficiency of a solar cell the $I–V$ characteristic has to be measured along with the P_{sun} at the exact moment at which the $I–V$-characteristic is obtained. The P_{sun} is measured with a pyranometer in the unit W/m^2 (Figure 1.17). The angles of incidence must be identical for the solar cell and for the pyranometer. The area of the solar cell has to be measured with a ruler for normalizing the power in the mpp of the solar cell to its area.

In the pyranometer, the blackbody absorbs the light of the complete sun spectrum. The decisive advantage of a blackbody is that the sunlight is absorbed over the whole relevant spectral range at the same sensitivity.

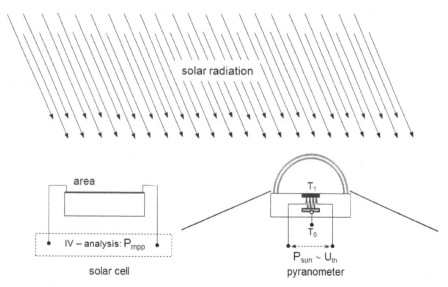

Figure 1.17. Procedure of an outdoor measurement of the solar energy conversion efficiency of a solar cell with a pyranometer.

In the blackbody, the radiation energy of the sunlight is converted into heat leading to an increase of the temperature (T_1). The T_1 is measured with a thermopile consisting of tens of thermocouples in series. A thermocouple is an intimate contact between two different metals. The contact potential between two different metals depends on the T. Therefore a thermocouple converts the T_1 into a voltage signal. The signal of the thermopile is measured as the difference of the series of contact potentials between the heated blackbody and a body at the ambient or reference temperature T_0. The voltage signal of the thermopile at the output of the pyranometer is calibrated with respect to P_{sun}. It shall be noted that the calibration of a pyranometer demands calibration standards and is a rather critical requirement. Pyranometers of high performance have an absolute accuracy of 1% at AM1.5, i.e. $10\,\text{W/m}^2$.

The blackbody in the pyranometer has to be protected from variable heat transfer, i.e. from convection of surrounding air and from changes in the heat conductivity of surrounding materials. For this purpose, the blackbody is surrounded by a double-walled glass dome and embedded into heat-insulating materials. In addition, the inner part of a pyranometer is kept dry, for example, with silica gel, in order to avoid variable heat exchange due to penetrating water molecules. Furthermore, the part of the pyranometer with the body kept at the T_0 is shaded in order to avoid changes of T_0 during the measurement.

The advantage of the measurement of η with a pyranometer is that any solar cell, PV module or string of PV modules can be characterized under given conditions at a certain moment. Therefore, accurate performance monitoring of PV power plants is coupled with real-time monitoring of P_{sun} using pyranometers.

There is a low-cost version of this which involves the measurement of P_{sun} with a photodiode. For his purpose, the dependence of I_{ph} of the photodiode on P_{sun} has to be calibrated with a pyranometer. However, in contrast to a blackbody, a silicon photodiode does not integrate over the whole relevant sun spectrum with identical sensitivity. But the sun spectrum can change depending on geography, weather conditions and time of day. Therefore, the absolute accuracy of measurements with photodiodes is in principle much lower than for measurements with a pyranometer.

1.5.5 *Measurement of the solar energy conversion efficiency with a sun simulator*

The measurement of η with a pyranometer has a great disadvantage which is related to the sun. The intensity and the spectrum of sunlight change dramatically over the day and depends on various conditions. This makes the direct comparison of the η of solar cells measured at different places and at different times problematic. A sun simulator overcomes the problem of comparability of measurements. The standard test conditions (STC; AM1.5 with 1000 W/m^2 and T of the solar cell 25°C) are the common standard for the characterization of the η of solar cells and PV modules (IEC, 2008).

A sun simulator is an artificial light source with an intensity spectrum very close to that of the sun at AM1.5. The artificial reproduction of the sun spectrum at AM1.5 demands appropriate light sources, combinations of light sources and filtering of light. For example, the sun spectrum at AM0 can be approximated with a blackbody with a T of 5800 K. Related blackbodies are not available on earth.

A halogen lamp can be described as a blackbody with a T of about 3000 K (Figure 1.18). The maximum intensity of a halogen lamp is at a wavelength of about 1000 nm. The simulation of the sun spectrum with

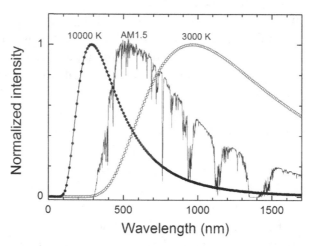

Figure 1.18. Normalized spectra of the intensity of a blackbody at 3000 K and at 10,000 K in comparison with the normalized AM1.5 spectrum.

a halogen lamp is impossible due to the very low intensity in the spectral range between blue and ultra violet light.

A xenon arc lamp is a light source with a T in the arc of about 10,000 K which is much higher than the T_S. The maximum intensity of a xenon arc lamp is at a wavelength of about 300 nm. In addition, a xenon arc lamp contains numerous spectral lines of high intensity, especially in the near infrared range. The simulation of the sun spectrum with a xenon arc lamp is nearly impossible due to the low intensity in the near infrared range.

The sun spectrum can be simulated with a combination of a halogen lamp providing enough intensity in the infrared and near infrared range and a xenon arc lamp providing high intensity in the blue and violet spectral range. Figure 1.19 shows the principle set-up of a sun simulator.

The light of the halogen and xenon arc lamps passes through a half mirror to a plane mirror from where the light is reflected to the collimator. The light passes the collimator and illuminates the solar cell. There are additional filters in the sun simulator to get rid of the intense infrared lines of the xenon arc lamp and to absorb light in the regions of decreased

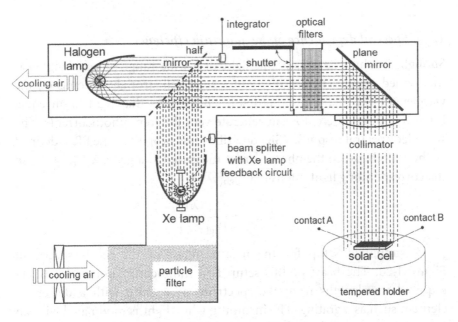

Figure 1.19.   Principle set-up of a sun simulator.

intensity in the AM1.5 spectrum. In the simplest case, water is used for filtering since a significant part of absorption in the earth's atmosphere is caused by water molecules. Furthermore, a shutter is implemented to enable measurements of $I–V$ characteristics in the dark. A beam splitter with a xenon lamp feedback circuit is needed to adjust the lamp current for keeping a constant and highly stabilized intensity of the xenon arc lamp. The integrator gives information about accumulated irradiation time. The solar cell is kept at a tempered holder at 25°C and contacted with a source measure unit. The contact resistance is usually tested with a second contact at the anode and cathode of the solar cell before the $I–V$ characteristics are measured.

Powerful light sources of the order of a kW are needed in a sun simulator for homogeneous illumination of solar cells with an area of $10 \times 10\,cm^2$ and larger. The light sources should be actively cooled with filtered air.

The life of lamps is limited. Therefore, a sun simulator needs regular service. For example, a halogen lamp has to be replaced after every 50 hours of operation. In addition, the sun simulator has to be calibrated regularly with calibrated solar cells that are sensitive in different spectral ranges.

1.5.6 *Spectral dependence of the quantum efficiency*

Sunlight is white, i.e. it contains light of a wide range of wavelengths as mentioned above. The sensitivity of a solar cell strongly depends on the wavelength (λ) of the exciting light. This means that light at an equal intensity but at different λ can generate very different photocurrents. The so-called spectral response (SR) describes this property. The SR is defined as the ratio between the photocurrent measured at a given λ ($I_{ph}(\lambda)$) and the corresponding light intensity ($P_{light}(\lambda)$).

$$SR(\lambda) = \frac{I_{ph}(\lambda)}{P_{light}(\lambda)} \tag{1.38}$$

A simplified setup for the measurement of the SR is shown in Figure 1.20. The heart of this setup is a monochromator which filters required wavelengths from the spectrum of a lamp with a dispersive element such as a grating. The incoming white light is modulated with an optical chopper and the photocurrent is measured with a lock-in amplifier.

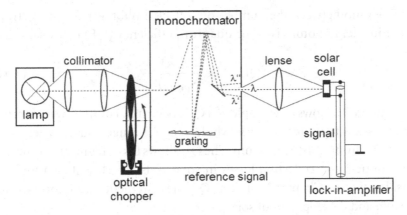

collimator

monochromator

lense   solar cell

λ''

λ'

λ

lamp

grating

signal

optical chopper

reference signal

lock-in-amplifier

Figure 1.20. Schematic of the simplest setup for spectral response measurements.

The lock-in amplifier registers only that part of the photocurrent which is caused by the modulated incoming light, i.e. noise signals and the sensitivity to bias light are strongly suppressed. The signal at the lock-in amplifier is measured as a function of λ.

The accurate absolute calibration of a monochromatic light source demands tremendous efforts. Therefore, it is much easier to compare a measured photocurrent spectrum of an unknown solar cell ($I_{ph,1}$) with that ($I_{ph,2}$) of a solar cell which spectral response spectrum is known (SR_2). The unknown spectral response (SR_1) is found from the following equation:

$$SR_1(\lambda) = \frac{I_{ph,1}(\lambda)}{I_{ph,2}(\lambda)} \cdot SR_2(\lambda) \tag{1.39}$$

A photocurrent is the photogenerated charge per time unit and the power is the energy per time unit. The photogenerated charge is the product of the number of collected photogenerated electrons ($N_e(\lambda)$) and the q and the energy of light is the product of the number of photons ($N_{ph}(\lambda)$) and the E_{ph}. Therefore, Equation (1.38) can be transformed to

$$SR(\lambda) = \frac{q \cdot \lambda}{h \cdot c} \frac{N_e(\lambda)}{N_{ph}(\lambda)} = SR_{max}(\lambda) \cdot \frac{N_e(\lambda)}{N_{ph}(\lambda)} \tag{1.40}$$

The maximum spectral response is proportional to the wavelength and amounts, for example, to 0.724 A/W at a λ of 900 nm.

The ratio between the number of collected photogenerated electrons and of incident photons is called quantum efficiency (QE) of a solar cell.

$$QE(\lambda) \equiv \frac{N_e(\lambda)}{N_{ph}(\lambda)} \qquad (1.41)$$

Figure 1.21 shows an example of a QE spectrum. For the given example, the QE is less than one over the whole spectral range and even zero at λ below 220 nm and above 1200 nm. The QE allows for obtaining information about optical and collection losses in solar cells, which will be a topic of consideration in Chapter 2. SR and QE measurements are important for research and development of solar cells.

The I_{SC} density of a solar cell can be calculated if the SR spectrum or the QE spectrum of a solar cell is known. For this purpose, the product of the SR spectrum and the intensity spectrum of the sun or the product of the QE spectrum and the photon flux spectrum ($\Phi_{sun}(\lambda)$) of the sun are integrated over the wavelength.

$$I_{SC} = q \cdot \int_0^\infty SR(\lambda) \cdot P_{sun}(\lambda) \cdot d\lambda \qquad (1.42')$$

$$I_{SC} = q \cdot \int_0^\infty QE(\lambda) \cdot \Phi_{sun}(\lambda) \cdot d\lambda \qquad (1.42'')$$

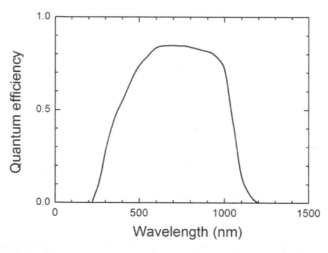

Figure 1.21. Example for a quantum efficiency spectrum of a solar cell.

The values of the I_{SC} density obtained from measurements with a sun simulator and obtained from the QE spectrum should be equal if the sun simulator and the reference solar cell for the analysis of the SR are well calibrated. Therefore, SR or QE measurements allow for an independent correct determination of I_{SC}.

1.6 Summary

The sun emits a huge amount of power or energy flux to earth characterized by the solar constant ($1356 \, W/m^2$) and by its spectral distribution with a maximum at photon energy around 1.4 eV. Total P_{sun} reaching the earth is about 1.3×10^8 GW. For purposes of comparison, a nuclear power plant has a power of about 1 GW and a person needs on average about 0.1 kW to sustain his or her biological life. The total power received on earth can be compared with the population of mankind (more than $7 \cdot 10^9$) and the energy demand per capita for comfortable life (2 kW per capita seems sufficient). The P_{sun} reaching the earth exceeds the energy demand of mankind by several thousand times. PV solar energy conversion is aimed at using a part of this energy for human needs in form of electricity. The solar energy conversion efficiency is the most important parameter of solar cells and PV power plants. A high η combined with a low degradation rate of η and a low energy payback time is decisive for sustainable solar energy conversion.

A solar cell has two basic functions: the generation of a photocurrent and the generation of a photovoltage for the production electric power. The maximum electric power of an illuminated solar cell is the product of the I_{SC}, the V_{OC} and the FF. Loads are used to extract the power from solar cells. From an illuminated solar cell, the maximum power is extracted with a load, the resistance of which has to be equal to the quotient of the potential and of the current in the mpp. The mpp of a solar cell changes with changing illumination. This has practical consequences such as mpp tracking. For worldwide comparison, solar cells and PV modules have to be characterized at STC (power at AM1.5 with $1000 \, W/m^2$ and temperature of the solar cell 25°C).

The behavior of solar cells can be analyzed with equivalent circuits. An ideal solar cell contains only a photocurrent generator and a diode. The diode is necessary for internal charge separation, i.e. for the generation of a photovoltage. Ideal solar cells can be completely described by two fundamental parameters, the I_{SC} and the I_0, i.e. all properties of materials and combinations of materials used in a solar cell are confined in I_{SC} and I_0 (Figure 1.22).

A major issue for PVs is to determine dependencies of I_{SC} and I_0 on limiting parameters related to materials and combinations of materials. The V_{OC} is a derived parameter and increases with increasing I_{SC} and/or decreasing I_0. The FF has to be obtained from the analysis of the I–V and power–voltage characteristics. The η of ideal solar cells increases with increasing I_{SC}, i.e. under concentrated sunlight (see also Figure 1.23).

Losses caused by resistive heating at R_s and R_p have to be considered for real solar cells under operation. A $R_{s,tol}$ of solar cells with high η should be equal or less than the quotient of V_{OC} and I_{SC} of the respective ideal solar cell divided by 100. A $R_{p,tol}$ of solar cells with high η should be equal or larger than the quotient of V_{OC} and I_{SC} of the respective ideal solar cell multiplied with 100. These $R_{s,tol}$ and $R_{p,tol}$ help to evaluate the potential of given materials and combinations of materials for reaching very high η at

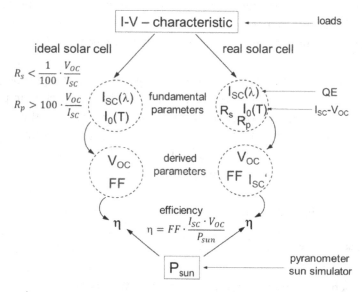

Figure 1.22. Summary of the basic characteristics of ideal and real solar cells.

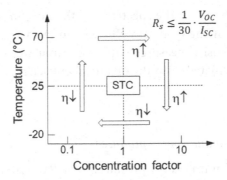

Figure 1.23. General dependence of the solar energy conversion efficiency on increasing and decreasing temperature and concentration factor of illumination.

certain operation conditions of solar cells and PV modules. For example, $R_{s,tol}$ are important economic factors for optimization of the amount of metals such as silver in bus bars on c-Si solar cells (see Chapter 7) and connecting cables in PV power plants.

At concentrated sunlight, the R_s limits FF and I_{SC}, whereas the R_p limits FF and V_{OC} at low light intensities. The η starts to reduce with increasing I_{SC} when R_s becomes larger than about a third of $R_{s,tol}$ (see also Figure 1.23). The minimum FF is 0.25. As examples, very low R_s can be realized in GaAs solar cells (see Chapter 8) and very high R_p are achieved in dye-sensitized solar cells (see Chapter 10).

The I_0 and the R_p of a solar cell can be easily obtained from I_{SC}–V_{OC} measurements. The E_A of I_0 can be derived from the dependence of the logarithm of I_0 on the reciprocal T. The thermal activation of the I_0 has the consequence that V_{OC} decreases with increasing T, i.e. η decreases with increasing T (see also Figure 1.23).

Load resistances are important for reliable measurements of I–V characteristics and therefore for the measurement of FF and R_s a solar cell. An I–V characteristic can be constructed from various working points measured precisely with different R_L. Electronic loads or constant current sources are widely used for I–V measurements. Fast I–V measurements are possible by measuring the current and the voltage during charging of a capacitor (flash operation) or by periodically applying variable R_L.

The P_{sun} can be precisely measured with a pyranometer. Standard test conditions are realized in laboratories with sun simulators. The I_{SC} densities can be calculated by using the spectrum of the photon flux from the sun and the spectrum of the quantum efficiency of a solar cell. For the same solar cell, the I_{SC} density measured with a sun simulator should be equal to the I_{SC} density calculated from the quantum efficiency.

1.7 Tasks

T1.1: Power installed and energy produced

Ascertain the maximum energy produced by a PV power plant of $1\,MW_p$ installed in the south of Spain compared with a PV power plant installed in Germany during the first year of operation.

T1.2: Degradation time of PV modules

Degradation rates of η of PV modules (k_{deg}) are between about 0.3–0.7%/a (c-Si) and about 0.8–1.5%/a (thin-film) (Jordan *et al.*, 2016). Plot the dependence of the degradation time of PV modules on k_{deg} after which η reduced to 90%, 80% and 50% of the initial value. Comment on the guarantee time for PV modules (usually 25 years) and on building integration of PV modules.

T1.3: Energy payback factor of PV modules

The energy payback time of PV modules (t_{EPB}) is often on the order of 1–2 years. Obtain an expression for the dependence of the energy payback factor of PV modules (EPBF) on k_{deg} and t_{EPB}. Plot EPBF for k_{deg} and t_{EPB} equal to 0.3 and 1.0%/a and to 0.5, 1 and 2 a, respectively, for a time range of up to 1000 years in logarithmic scales. Discuss the role of t_{EPB} and possible limits of EPBF.

T1.4: Photon energy in the maximum of blackbody radiation at T_s

Calculate the photon energy at which the blackbody radiation at 5800 K has a maximum.

T1.5: Wavelength of light in the maximum of blackbody radiation at T_s

Calculate the wavelength of light at which the blackbody radiation at 5800 K has a maximum.

T1.6: Maximum temperature of a solar cell under operation

Calculate the maximum temperature of solar cells with η of 10%, 20% and 40% under illumination at $1\,kW/m^2$ on earth when convection cooling is absent compared to solar cells on a satellite.

T1.7: Air mass

Ascertain the dependence of the air mass on the angle of incidence of sunlight and discuss the validity of the expression for large angles.

T1.8: Tolerable series and parallel resistances

Estimate the $R_{s,tol}$ and $R_{p,tol}$ for a solar cell with an I_0 of $10^{-19}\,A/cm^2$ operated at an I_{SC} density of $25\,mA/cm^2$ (AM1.5) and operated at concentrated sunlight with a concentration factor of 400.

T1.9: Current–voltage characteristics

Ascertain the V_{OC} and the FFs for solar cells with an I_0 of $10^{-18}\,A/cm^2$, I_{SC} densities of 10 or $100\,mA/cm^2$ and R_p of 1 of $100\,k\Omega cm^2$.

2

Photocurrent Generation and the
Origin of Photovoltage

The most important fundamental property of a photovoltaic (PV) absorber is its forbidden band gap (E_g). The E_g limits the maximum photocurrent and it is the prerequisite for realizing a photovoltage, i.e. a difference between potential energies at which electrons may be extracted from a solar cell and at which electrons may be transferred back after passing an external load. The ultimate efficiency is analyzed as a function of the E_g. Photocurrents are reduced by reflection losses and by transmission losses, which are determined by the reflectivity and by the absorption coefficient of a PV absorber. The minimization of optical losses by antireflection coatings and light trapping is discussed. The Boltzmann equations are given for calculating the Fermi-energies of free electrons and holes from their densities and from the effective densities of states at the valence and conduction band edges. Doping of semiconductors is explained. The condition for thermal equilibrium is described and the density of intrinsic charge carriers is introduced. The change of the Fermi-energy of minority charge carriers under illumination is a fundamental condition for a photovoltage.

2.1 Energy Gap of Photovoltaic Absorbers

2.1.1 Light absorption and band gap of photovoltaic absorbers

A photon disappears during an absorption event. The energy of a photon is transferred into the excitation energy of an electron, i.e. a photon excites an electron from the ground state (not excited) into a higher energetic state (excited).

Electrons in the excited state should be mobile for providing electric current in solar energy conversion by photovoltaics (PVs). The process of photoexcitation of electrons from the ground state into excited states in which electrons are mobile is called the photogeneration of free electrons.

During PV solar energy conversion, photogenerated electrons are extracted from the first contact of a solar cell, pass an external load and are transferred back into the solar cell at the second contact. For the generation of electric power it must be possible to extract photogenerated electrons at a potential energy which is higher than the potential energy at which the electrons are transferred back after passing the external load. Therefore, a PV absorber suitable for solar energy conversion should be able to increase the potential energy of photogenerated electrons.

The potential energy of photogenerated electrons can only be increased if the energy of excited electrons is separated from the energy of electrons in the ground state by an energy gap. For purposes of comparison, the potential energy of water is increased by evaporation and formation of clouds. Due to raining in the mountains, water can be collected at increased potential energy. The potential energy of falling water can be converted into electricity with a turbine. The energy gap of the falling water driving the turbine is proportional to the height from which the water falls. The analogy between the energy gaps of hydro-mechanical and solar energy conversion by PV is depicted in Figure 2.1.

The energy gap between electrons in the ground state and free electrons in the excited states gives the upper limit for the potential between the

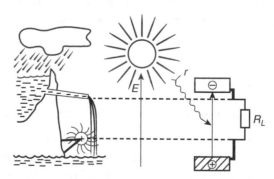

Figure 2.1. Sketch displaying the analogy between hydro-mechanical and PV solar energy conversion.

external leads of a solar cell, i.e. the upper limit of the open-circuit voltage (V_{OC}). The energy gap between electrons in the ground state and photogenerated free electrons in the excited state is called the forbidden band gap or band gap (E_g).

An PV absorber is a material in which the energy of absorbed photons is transferred into energy of photogenerated free electrons with increased potential energy. The most important fundamental property of any PV absorber is its E_g separating energetically photogenerated electrons from electrons in the ground state.

The energy band of all electrons in the ground state is called the valence band since it is formed by valence electrons participating in chemical bonds. The band in which excited mobile electrons can exist is called the conduction band. Electrons in the conduction band are free, i.e. they do not participate in chemical bonds but contribute to the electric conductivity.

In the dark most of the energetic states in the valence band are occupied and most of the energetic states in the conduction band are un-occupied. Under illumination, electrons (e^-) are excited from the valence band into the conduction band leaving a positive charge in the valence band. The missing electron in the valence band is called the hole (h^+). Holes are treated like mobile positive charge carriers since missing valence electrons are not fixed in chemical bonds.

Energy bands are depicted as a function of a coordinate in so-called band diagrams. Figure 2.2 shows the band diagram of an PV absorber at photoexcitation.

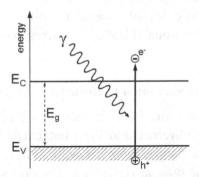

Figure 2.2.   Photogeneration of an electron–hole pair in a semiconductor. E_V, E_C and E_g denote the valence band edge, the conduction band edge and the band gap, respectively.

A material with a conduction band separated from a valence band by an E_g is also known as a semiconductor. The E_g of a semiconductor is defined as the difference between the energies at the conduction and valence band edges (E_C and E_V, respectively).

$$E_g = E_C - E_V \tag{2.1}$$

The potential energy depends on the sign of electric charge. Mobile electrons have the lowest energy at the E_C. Mobile holes carry a positive charge and therefore have the lowest energy at the highest energy in the valence band, i.e. at the E_V.

During an absorption event, a photon with an energy equal or above E_g creates an electron–hole pair by kicking out an electron from a chemical bond state in the valence band into the conduction band. Photons with energy below E_g are not absorbed by the semiconductor.

Semiconductors can be realized with materials based on inorganic or organic matter. The most prominent inorganic semiconductors are crystalline silicon (c-Si) and gallium arsenide (GaAs) with E_g at room temperature of 1.1 and 1.42 eV, respectively. The conduction and valence bands of inorganic semiconductors are semi-infinite in relation to photon energies in the sun spectrum. The value of E_g of most inorganic semiconductors usually decreases with increasing temperature. The negative temperature coefficient of E_g is of the order of half a meV per Kelvin for most crystalline semiconductors (see, for example, Chapter 8 for III–V semiconductors). The decrease of E_g with increasing temperature influences the performance of solar cells.

In organic semiconductors, the valence and conduction bands are formed by the highest occupied molecular orbital (HOMO) and lowest un-occupied molecular orbital (LUMO) bands, respectively (see Chapter 10 for details).

2.1.2 *Photon flux and maximum photocurrent of solar cells*

The part of the photon flux from the sun (Φ_{sun}) which is absorbed in the PV absorber is converted into the photocurrent of the solar cell. The Φ_{sun} provides a flow of photons with different energy (E_{ph}). In order to ascertain the Φ_{sun} at certain E_{ph}, the power of the sun at this E_{ph} has to be divided by the E_{ph}. The Φ_{sun} is given in the unit of

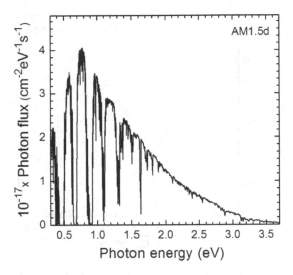

Figure 2.3. Photon flux spectrum of the sun at AM1.5d.

$1/(\text{cm}^2 \cdot \text{eV} \cdot \text{s})$. Figure 2.3 shows the spectrum of the Φ_{sun} at air mass (AM) 1.5d. The maximum Φ_{sun} of $4 \cdot 10^{17}$ $1/(\text{cm}^2 \cdot \text{eV} \cdot \text{s})$ is reached at a E_{ph} of about 0.8 eV.

Photogenerated electrons carry the photocurrent where one electron transports one elementary charge ($q = 1.6 \cdot 10^{-19}$ As). In order to ascertain the photocurrent, the Φ_{sun} has to be integrated over all photons participating in absorption events and multiplied by q. Photons with energy below the E_g do not participate in absorption events. Therefore, the fundamental limitation of the photocurrent is given by the E_g defining the lower integration boundary of the Φ_{sun}. The maximum photocurrent ($I_{\text{SC}}^{\text{max}}$) can be obtained if all photons with energy equal or above E_g are absorbed and if each absorbed photon generates one electron–hole pair, i.e. if the quantum efficiency is equal to 1 for photons with energy equal or larger than E_g.

$$I_{\text{SC}}^{\text{max}} = q \cdot \int_{E_g}^{\infty} \Phi_{\text{sun}}(E_{\text{ph}}) \cdot dE_{\text{ph}} \qquad (2.2)$$

The order of magnitude of $I_{\text{SC}}^{\text{max}}$ can be estimated if multiplying an average Φ_{sun} of $1 \cdot 10^{17}$ $1/(\text{cm}^2 \cdot \text{eV} \cdot \text{s})$ with an integration range of 2 eV and with q. Therefore $I_{\text{SC}}^{\text{max}}$ is of the order of tens of mA/cm^2.

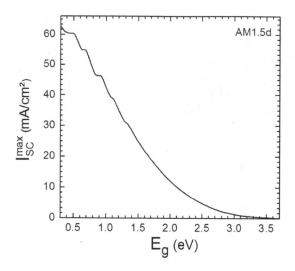

Figure 2.4.   Dependence of the I_{SC}^{max} at AM1.5d on the E_g of the PV absorber.

The I_{SC}^{max} at AM1.5d is plotted as a function of the E_g in Figure 2.4. The values of the I_{SC}^{max} have to be multiplied by a factor of 1.1 to determine I_{SC}^{max} at AM1.5G. The I_{SC}^{max} decreases with increasing E_g from about 60 mA/cm^2 at $E_g = 0.5$ eV to about 1 mA/cm^2 at $E_g = 3.1$ eV. Therefore it is not useful to apply semiconductors with E_g larger than 2.8–3.0 eV for solar cells.

It is interesting to compare the short-circuit current (I_{SC}) densities of record solar cells with I_{SC}^{max}. The I_{SC} densities are 41.8, 29.68 and 35.7 mA/cm^2 at AM1.5G for record c-Si, GaAs and Cu(In,Ga)Se$_2$ solar cells (Green *et al.*, 2016). The values of I_{SC}^{max} related to the corresponding E_g of 1.1, 1.42 and 1.2 eV amount to about 44, 31 and 39 mA/cm^2, respectively. Therefore, the I_{SC} of world record solar cells are almost very close to I_{SC}^{max}, which is an important fact for the theoretical analysis of limitations in different types of solar cells.

The I_{SC} of a given solar cell increases with increasing temperature due to the decrease of the E_g. The increase of the I_{SC} density can be between 1 and 2 mA/cm^2 for a quite strong increase of the temperature of the order of 50–70 K, depending on the absorber material (see task T2.2 at the end of this chapter).

An upper integration boundary ($E_g + \Delta E_g$) has to be considered for the calculation of I_{SC}^{max} for organic semiconductors due to the limited widths of the HOMO and LUMO bands (see Chapter 10). Therefore, the values of I_{SC}^{max} are lower for organic than for inorganic PV absorbers at the same E_g.

2.1.3 The ultimate efficiency of solar cells

The maximum I_{SC} density can be calculated as a function of E_g with reference to Equation (2.2). The maximum V_{OC} of a solar cell cannot exceed the value of the E_g divided by q. The upper limit of the fill factor (FF) of a solar cell is 1. The ultimate efficiency ($\eta_{ultimate}$) takes into account only the upper limits of I_{SC}, V_{OC} and FF, irrespective of whether these values are realistic for existing materials and combinations of materials suitable for PV absorbers. The $\eta_{ultimate}$ of a solar cell with one PV absorber depends only on the E_g of the absorber and can be calculated with the following equation:

$$\eta_{ultimate} = \frac{E_g \cdot \int_{E_g}^{\infty} \Phi_{sun}(E_{ph}) \cdot dE_{ph}}{P_{sun}} \tag{2.3}$$

The dependence of $\eta_{ultimate}$ on E_g is presented in Figure 2.5. The $\eta_{ultimate}$ is about 48% at E_g of 0.9 and 1.1 eV. The values of $\eta_{ultimate}$ are

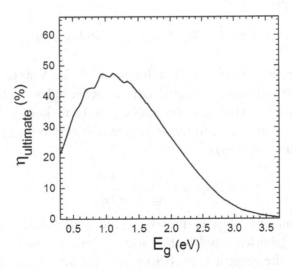

Figure 2.5. Dependence of the ultimate efficiency of a solar cell at AM1.5d on the E_g of the PV absorber.

larger than 40% for E_g between 0.7 and about 1.6 eV. The η_{ultimate} is lower than 10% for E_g larger than about 2.6 eV.

2.2 Photocurrent Limitation by Reflectivity and Transmission of Light

A part of the Φ_{sun} is lost in solar cells due to reflection and transmission of light (optical losses). The reflected and transmitted parts of the Φ_{sun} are denoted by $\Phi_{\text{sun}}^{\text{refl}}$ and $\Phi_{\text{sun}}^{\text{trans}}$, respectively.

$$\Phi_{\text{sun}}^{\text{abs}}(E_{\text{ph}}) = \Phi_{\text{sun}}(E_{\text{ph}}) - \Phi_{\text{sun}}^{\text{refl}}(E_{\text{ph}}) - \Phi_{\text{sun}}^{\text{trans}}(E_{\text{ph}}) \qquad (2.4)$$

Only the part of the Φ_{sun} which is absorbed in the PV absorber ($\Phi_{\text{sun}}^{\text{abs}}$) can be used for solar energy conversion by PV.

2.2.1 Reflectivity and antireflection coatings

The part of light being reflected is described by the reflection coefficient or reflectivity (R).

$$\Phi_{\text{sun}}^{\text{refl}}(E_{\text{ph}}) = R \cdot \Phi_{\text{sun}}(E_{\text{ph}}) \qquad (2.5)$$

The reflection losses are considered for the calculation of the I_{SC} density in the following equation:

$$I_{\text{SC}} = q \cdot \int_{E_g}^{\infty} \Phi_{\text{sun}}(E_{\text{ph}}) \cdot [1 - R(E_{\text{ph}})] \cdot dE_{\text{ph}} \qquad (2.6)$$

The reflection coefficient is a function of the wavelength (λ) of incoming light and can be calculated if the refractive indexes of the absorber (n_s) and of the surrounding ambience ($n_{\text{air}} \approx 1$) are known. For normal incidence of incoming light, the R is (see also textbooks in optics, for example (Grimsehl, 1988)):

$$R(\lambda) = \left(\frac{n_{\text{air}} - n_s(\lambda)}{n_{\text{air}} + n_s(\lambda)} \right)^2 \qquad (2.7)$$

For silicon and most of the crystalline semiconductors n_s ranges between 3.5 (shorter wavelengths) and 5 (longer wavelengths). As a consequence, the reflection coefficient of crystalline semiconductors is between 0.3 and 0.4.

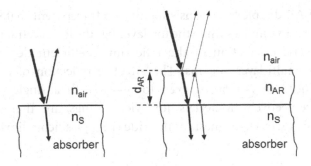

Figure 2.6. Principle of reflection and antireflection (AR) coating.

High reflection coefficients are reduced with transparent antireflection (AR) coatings (Figure 2.6). An AR coating is a surface layer with a thickness d_{AR} and a refractive index n_{AR} the value of which is between n_s and n_{air} (see, for example, Fa. Carl Zeiss, 1939). Reflection can be suppressed by destructive interference in the AR coating layer. The condition of destructive interference is:

$$\frac{\lambda}{4} = n_{AR} \cdot d_{AR} \qquad (2.8)$$

A minimum reflection coefficient can be obtained under condition (2.8)

$$R_{min} = \left(\frac{n_{AR}^2 - n_{air} \cdot n_s}{n_{AR}^2 + n_{air} \cdot n_s} \right)^2 \qquad (2.9)$$

The reflection coefficient can be reduced to 0 for the following condition:

$$R_{min} = 0 : \quad n_{AR} = \sqrt{n_{air} \cdot n_s} \qquad (2.10)$$

The refractive index of air is practically equal to 1. Many PV absorbers have a refractive index of the order of 3–5. Therefore the refractive index is about 1.7–2.2 for AR coatings.

The reflection coefficient can be minimized to a value close to 0 with regard to Equations (2.8)–(2.10) only at a certain wavelength. This wavelength is chosen in a range where the photocurrent generation has a maximum. The thickness of AR coatings is of the order of 100 nm. The reflection coefficient can be minimized at two different wavelengths by using an AR double layer consisting of two transparent materials with different refractive indices while the top layer has the lower refractive index.

The precise calculation of the reflection coefficient demands simulations for multi-layer systems. The effective reflection of a solar cell can be reduced to 7–10% or even to 3–4% with a single or double AR layer, respectively. Materials for AR coatings are titania (TiO_2), silicon nitride (Si_3N_4), magnesium fluoride (MgF_2), silicon dioxide (SiO_2) and others.

2.2.2 *Absorption coefficient and transmission losses*

The part of the Φ_{sun} which has not been reflected reduces depending on the length of the optical path of light which has been passed in the PV absorber. For normal incidence and planar geometry of the absorber the length of the optical path is equal to the thickness (x) passed by the light what will be considered in the following.

A photon flux (Φ_{ph}) decreases after passing a certain distance or optical path in a material (x) due to absorption (Figure 2.7). The decrease of the photon flux ($-d\Phi_{ph}(x)$) being absorbed in a very thin layer with the thickness dx around the distance x is proportional to the product of the dx and the photon flux at the distance x ($\Phi_{ph}(x)$) and can be expressed by

$$d\Phi_{ph}(x) = -\alpha \cdot \Phi_{ph}(x) \cdot dx \qquad (2.11)$$

The proportionality factor between the $d\Phi_{ph}(x)$ and the product of the $\Phi_{ph}(x)$ and dx is called the absorption coefficient (α). The absorption coefficient has the unit of cm^{-1}. The inverse absorption coefficient is called the absorption length (α^{-1}).

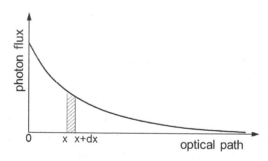

Figure 2.7.   Reduction of the photon flux with increasing optical path.

The solution of Equation (2.11) is the absorption or Lambert–Beer law.

$$\Phi_{ph}(x) = \Phi_{ph}(0) \cdot \exp(-\alpha \cdot x) \tag{2.12}$$

The photon flux at the surface ($\Phi_{ph}(0)$) is equal to the part of the photon flux that has not been reflected.

$$\Phi_{ph}(0) = \Phi_{sun} - \Phi_{sun}^{refl} \tag{2.13}$$

The product of the α and of the thickness of an absorber (d_{abs}) determines the part of the photon flux passing through the absorber and therefore the transmission losses. The part of the photon flux being absorbed in an absorber with a thickness d_{abs} is given by the following equation if it is assumed that the optical path is equal to d_{abs}:

$$\Phi_{sun}^{abs} = \Phi_{ph}(0) \cdot [1 - e^{-\alpha \cdot d_{abs}}] \tag{2.14}$$

The absorption coefficient is a function of the E_{ph} or λ of the incoming light and can vary over many orders of magnitude depending on the absorber material. The α is equal to 0 for transparent materials and can be as large as 10^6 cm^{-1} for strongly absorbing materials. Absorption spectra are shown as a function of E_{ph} for some conventional semiconductors in Figure 2.8.

Absorption occurs at photon energies above E_g. For c-Si the absorption length decreases from about 100 μm at 1.2 eV to about 10 μm at

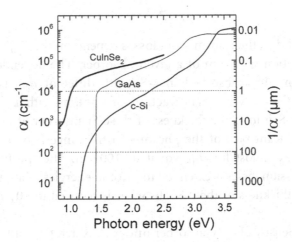

Figure 2.8. Absorption spectra and absorption length of c-Si, GaAs and CuInSe$_2$.

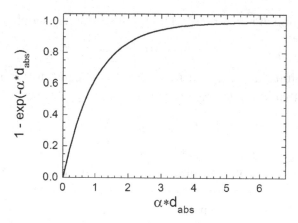

Figure 2.9. Amount of absorbed photons as a function of the product of the absorption coefficient and the absorber thickness.

1.5 eV and 1 μm at 2.2 eV. In contrast, the absorption length of CuInSe$_2$ is less than 1 μm over nearly the complete spectral range above the E_g of CuInSe$_2$.

The behavior of Equation (2.14) is expressed in Figure 2.9 as a function of $\alpha \cdot d_{abs}$. It can be seen that more than 95% of the incoming light is absorbed if the product of the α and the d_{abs} is larger than 3. This is an important design rule for simple solar cells. The d_{abs} should be at least three times larger than the α^{-1} of the absorber material.

$$d_{abs} \geq 3 \cdot \alpha^{-1} \tag{2.15}$$

At a fixed d_{abs}, the transmission losses depend on the E_{ph} with respect to the absorption spectrum of a given absorber. The dependence of the portion of the Φ_{sun} absorbed in c-Si with different d_{abs} is shown in Figure 2.10. Only a very small fraction of photons with E_{ph} very close to the E_g of c-Si is lost for a thickness of the c-Si absorber of 1 cm. Already a rather significant part of the photons in the range between the E_g of c-Si and 1.3 eV is lost for d_{abs} equal to 100 μm. The spectral range of large transmission losses is extended to photon energies of about 1.9 eV or 2.6 eV if the thickness of the c-Si absorber is reduced to 10 μm or 1 μm, respectively.

The I_{SC} density of a PV absorber under consideration of transmission losses (I_{SC}^{abs}) can be obtained by integrating the product of the Φ_{sun}, the

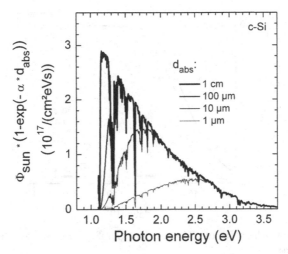

Figure 2.10. Photon flux absorbed in c-Si layers with thicknesses of 1 cm, 100, 10 and 1 μm (increasing thickness of solid lines). The reflectivity is set to 0.

right term of Equation (2.14) and multiplying with q. The normalized I_{SC} can be calculated after the following equation:

$$\frac{I_{SC}^{abs}}{I_{SC}^{max}} = \frac{\int_{E_g}^{\infty} \Phi_{sun}(E_{ph}) \cdot [1 - e^{-\alpha(E_{ph}) \cdot d_{abs}}] \cdot dE_{ph}}{\int_{E_g}^{\infty} \Phi_{sun}(E_{ph}) \cdot dE_{ph}} \tag{2.16}$$

The dependence of I_{SC}^{abs} normalized to I_{SC}^{max} on d_{abs} is plotted in Figure 2.11 for c-Si, GaAs and CuInSe$_2$ absorbers. For a c-Si absorber, a thickness of the absorber layer larger than 200 μm is required for obtaining at least 90% of I_{SC}^{max}. This is the reason why c-Si solar cells are based on wafers, i.e. on slices of c-Si crystals with a thickness of about 200 μm or larger (see Chapter 7). More than 95% of I_{SC}^{max} are reached for GaAs absorbers with a thickness of 2 μm (see Chapter 8), whereas only a 1 μm thick CuInSe$_2$ absorber is needed to obtain 95% of I_{SC}^{max}. An absorber with the thickness of the order of 1 μm cannot be mechanically self-supporting. PV absorbers like CuInSe$_2$ are well suited for thin-film solar cells which make use of a supporting foreign substrate (see Chapter 9).

2.2.3 *Increase of the optical path by light trapping*

Numerous measures employed in the architecture of solar cells aim to increase the optical path of light through the absorber layer. For example,

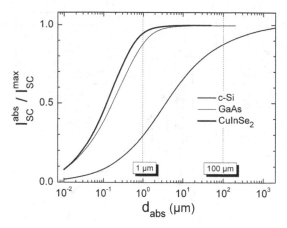

Figure 2.11. Normalized I_{SC} of c-Si, GaAs and CuInSe$_2$ absorbers as a function of the absorber thickness.

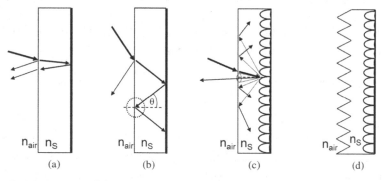

Figure 2.12. Increase of the optical path by a planar back reflector (a), total reflection (b), a scattering back reflector (c) and textured surface with scattering back reflector (d).

a mirror at the back side of an absorber increases the optical path by a factor of two (Figure 2.12(a)). Therefore the thickness of an absorber layer can be reduced by a factor of two without reducing the amount of absorbed photons, merely by applying a planar back reflector.

The optical path of light is further increased in case of total reflection (Figure 2.12(b)). The angle of incidence of the light should be larger than the critical angle for total reflection (θ_C) that is given by

$$\theta_C = \frac{n_{air}}{n_S} \qquad (2.17')$$

If the absorber is coated with an AR layer, then condition (2.17′) changes to

$$\theta_C = \frac{n_{AR}}{n_S} \tag{2.17″}$$

After total reflection at the interface between the absorber and the AR layer, photons can again pass through the absorber layer, can be reflected again at the mirror on the back side and can pass through the absorber layer a fourth time, can reach the interface between the absorber and AR layer where total reflection is possible again, and so on — photons are trapped in the absorber layer (light trapping).

The angle of incidence of sunlight can vary and therefore so can the condition for total reflection. It is useful to scatter the light at the back reflector. Ideally scattered light has a similar distribution for all angles (Lambertian back reflector) and therefore a large fraction of directions suitable for total reflection.

Light is scattered to all directions with the same probability for Lambertian light scattering. The area of a small scattering surface seen at a long distance reduces with increasing scattering angle (Figure 2.13(a)). The scattered photon flux of a small scattering surface is proportional to the area seen at a long distance from the scattering surface. For this reason the angular distribution of the flux of scattered photons is proportional to the cosine of the scattering angle (Figure 2.13(b)).

Only the light scattered into a cone with an opening angle equal to θ_C can escape the absorber, i.e. this fraction of light is lost for solar energy

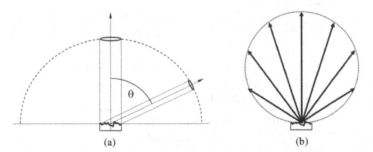

(a)   (b)

Figure 2.13. Lambertian light scattering: change of the area of a small scattering surface seen at a long distance with scattering angle θ (a) and corresponding distribution of the photon flux from a scattering surface presented as the length of arrows (b).

conversion by PV. For a Lambertian back reflector in a solar cell with an AR coating, the optical path can be increased by a factor of X_{path} (see also task T2.3 at the end of this chapter).

$$X_{path} = 2 \cdot \left(\frac{n_S}{n_{AR}}\right)^2 \tag{2.18}$$

Equation (2.18) describes the so-called Lambertian limit of light trapping which is the limit for light trapping in ray optics. The Lambertian limit of light trapping is of the order of 8–18 for most solar cells since the ratio between the refractive indices of the absorber and of the antireflection coating is of the order of 2–3.

The normalized I_{SC}^{abs} can be calculated by implementing the factor X_{path} into Equation (2.16).

$$\frac{I_{SC}^{abs}}{I_{SC}^{max}} = \frac{\int_{E_g}^{\infty} \Phi_{sun}(E_{ph}) \cdot [1 - e^{-\alpha(E_{ph}) \cdot X_{path} \cdot d_{abs}}] \cdot dE_{ph}}{\int_{E_g}^{\infty} \Phi_{sun}(E_{ph}) \cdot dE_{ph}} \tag{2.19}$$

Figure 2.14 shows the dependence of the normalized I_{SC} of c-Si, GaAs and CuInSe$_2$ absorbers on d_{abs} for an optical path increased by a factor of 10. Absorber thicknesses equal or larger than 50 μm, 200 and 100 nm are needed for obtaining I_{SC} densities of at least 95% of I_{SC}^{max}

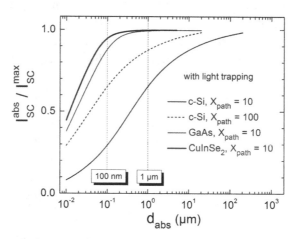

Figure 2.14. Normalized I_{SC} of c-Si, GaAs and CuInSe$_2$ absorbers as a function of the absorber thickness for an optical path increased by a factor of 10 (solid lines) and by a factor of 100 (dashed line, only for c-Si).

for c-Si, GaAs and CuInSe$_2$ absorbers, respectively. Intensive research is ongoing to develop methods to increase the optical path by values above the Lambertian limit. For example, c-Si absorbers could be reduced to 5 μm and even less if an increase of the optical path by a factor of 100 could be realized.

There are more geometries suitable for the increase of the optical path in solar cells, for example the use of textured absorber surfaces. The goal of the use of textured absorber surfaces is to increase the optical path of photons entering the absorber layer and to change the angle of incidence of incoming light in such a way that back reflected light undergoes total reflection.

Statistical ray optics (Yablonovich, 1982) are useful for considering the surface morphology. The optical path can also be increased by implementing structures into solar cells with smaller dimensions than the wavelength of light. By related measures, the optical path can be increased above the Lambertian limit of light trapping (see Chapter 10).

The reduction of d_{abs} by increasing the optical path of light through the absorber with light trapping is important for several reasons. First, the amount of absorber material can be reduced, which leads to a reduction of weight and costs. Second, the increase of the optical path of light opens opportunities to increase the energy conversion efficiency of solar cells under advanced conditions, as will be shown in Chapter 4. Third, the reduction of the d_{abs} to the range of 100 nm would allow the incorporation solar cells into very thin flexible structures and membranes suitable, for example, for lightweight applications or for applications in photocatalytic water-splitting systems. Fourth, light trapping is very useful for broadening the classes of suitable PV absorbers to materials with relatively short lifetimes (for definitions, see Chapter 3). In sum, therefore, the evident benefits of light trapping provide a clear impetus for the further development of solar cells.

2.3 Free Charge Carriers in Ideal Semiconductors

2.3.1 Densities of free charge carriers and Fermi-energy

For the analysis of solar cells it is necessary to describe the density of free charge carriers (see, for example Seeger, 2001). In an ideal semiconductor,

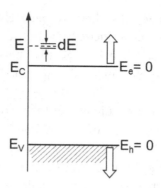

Figure 2.15. Energies of free electrons and holes in the band diagram of a semiconductor.

free electrons in the conduction band and free holes in the valence band are treated like atoms or molecules in an ideal gas while the free charge carriers are delocalized over the whole considered volume and follow the Pauli principle.

The density of free charge carriers in an interval of energy (dE, see Figure 2.15) around a considered energy (E) is the product of dE, of the density of states and of the occupation probability at E. The integral of this product over energy gives the density of the free carriers in the respective band.

The densities of free electrons and holes are denoted by n and p, respectively. The densities of states in the conduction and valence bands are denoted by $D_e(E)$ and $D_h(E)$, respectively. The occupation probability of an electron is denoted by f. The densities of free electrons and holes can be expressed by the following equations:

$$n = \int_{E_C}^{\infty} D_e(E) \cdot f(E) \cdot d(E) \tag{2.20'}$$

$$p = \int_{\infty}^{E_V} D_h(E) \cdot [1 - f(E)] \cdot d(E) \tag{2.20''}$$

The density of states at a certain energy is defined as the number of states in an energy interval around this energy ($dN_{e,h}(E)$, where the indices e and h denote the number of states in the conduction and valence bands, respectively) normalized to the considered volume (V) and to the energy interval

$$D_{e,h}(E) = \frac{1}{V} \cdot \frac{dN_{e,h}(E)}{dE} \tag{2.21}$$

The kinetic energies of free electrons and holes are 0 at the E_C and E_V, respectively. The energies of free electrons and holes in the conduction and valence bands are considered as kinetic energies ($E_{kin,e}$ and $E_{kin,h}$, respectively).

$$E_{kin,e} = E - E_C \tag{2.22'}$$

$$E_{kin,h} = E_V - E \tag{2.22''}$$

The $E_{kin,e}$ and $E_{kin,h}$ are given by the squared momentum ($k_{e,h}^2$) divided by the doubled effective mass of electrons or holes ($2 \cdot m_{e,h}^*$).

$$E_{kin,e,h}(k) = \frac{k_{e,h}^2}{2 \cdot m_{e,h}^*} \tag{2.23}$$

With regard to delocalization of free charge carriers and to the Heisenberg uncertainty principle, one state occupies the following volume in momentum space:

$$(\Delta k_{e,h})^3 = \frac{h^3}{V} \tag{2.24}$$

The number of states at a momentum equal to or lower than $k_{e,h}$ ($N(k_{e,h})$) is the volume of a sphere with radius $k_{e,h}$ divided by $(\Delta k_{e,h})^3$ and multiplied by two due to spin degeneracy (spin up and spin down).

$$N(k_{e,h}) = \frac{4\pi}{3} \cdot |k_{e,h}|^3 \cdot \frac{2}{(\Delta k_{e,h})^3} \tag{2.25}$$

Further, the momentum is substituted by the kinetic energy using Equation (2.23) so that the number of states becomes a function of energy and $\Delta k_{e,h}$ is substituted by transforming Equation (2.25). The kinetic energies are expressed by Equation (2.22). Following Equation (2.21), the number of states is divided by the volume and differentiated with respect to the energy. As a result, the densities of states in the conduction and valence bands are obtained:

$$D_e(E) = 4\pi \cdot \left(\frac{2 \cdot m_e^*}{h^2}\right)^{3/2} \cdot (E - E_C)^{1/2} \qquad (2.26')$$

$$D_h(E) = 4\pi \cdot \left(\frac{2 \cdot m_h^*}{h^2}\right)^{3/2} \cdot (E_V - E)^{1/2} \qquad (2.26'')$$

Equations (2.26') and (2.26'') were obtained from first principle and contain only the effective masses of electrons and holes in the conduction and valence bands, respectively, as material parameters.

The m_e^* and m_h^* do not necessarily correspond to the electron mass in free space (m_e) due to interactions of mobile charge carriers and atoms in the semiconductor. Effective masses can vary over a relatively wide range depending on the given semiconductor. For example, for c-Si m_e^* and m_h^* are $1.08 \cdot m_e$ and $0.5 \cdot m_e$, respectively (see, for example, Sze, 1981).

Equations (2.26') and (2.26'') can be written in the following manner:

$$D_e(E) = \frac{6.7 \cdot 10^{21}}{cm^3 \cdot eV} \cdot \left(\frac{m_e^*}{m_e}\right)^{3/2} \cdot (E - E_C)^{1/2} \qquad (2.27')$$

$$D_h(E) = \frac{6.7 \cdot 10^{21}}{cm^3 \cdot eV} \cdot \left(\frac{m_h^*}{m_e}\right)^{3/2} \cdot (E_V - E)^{1/2} \qquad (2.27'')$$

Equations (2.27') and (2.27'') describe very well the densities of free electron and hole states close to the edges of parabolic conduction and valence bands for which Equation (2.23) is valid.

The occupation probabilities of electrons and holes are given by the Fermi–Dirac statistics (Equation (2.18)), which follow from the Pauli principle. The decisive parameter in the Fermi–Dirac statistics is the Fermi-energy (E_F) at which the occupation probability is 0.5.

$$f(E) = \frac{1}{\exp\left(\frac{E - E_F}{k_B \cdot T}\right) + 1} \qquad (2.28)$$

Figure 2.16 shows the density of states, the electron occupation probability and the product of both of these for electrons and holes in energy diagrams. The energy scale is set to 0 at E_V. The electron occupation probability is close to 1 but not equal to 1 for the valence band and close to 0 but not equal to 0 for the conduction band. The shape of the density of free

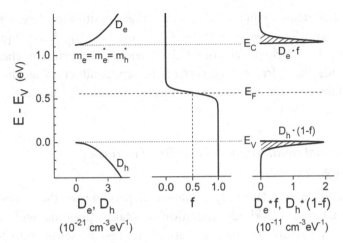

Figure 2.16. Densities of states of electrons and holes in the conduction and valence bands, respectively, occupation probability of electrons and product of both for free electrons in an energy diagram for the case that the Fermi-energy is in the middle of E_g.

electrons follows the exponential decrease of the occupation probability with increasing energy.

The effective densities of states at the conduction and valence band edges (N_C ad N_V, respectively) are obtained from the densities of states integrated from the E_C and E_V to the energies increased by $k_B \cdot T$, respectively.

$$N_{C,V} = 2 \cdot \left(2\pi \cdot k_B \cdot T \cdot \frac{m^*_{e,h}}{h^2} \right)^{3/2} \tag{2.29}$$

At room temperature, the effective densities of states at the conduction and valence bands edges are about $N_C = 3 \cdot 10^{19}\,\text{cm}^{-3}$ and $N_V = 1 \cdot 10^{19}\,\text{cm}^{-3}$ for c-Si (see, for example Sze, 1981).

Taking into account the definitions of N_C and N_V and the fact that the exponential in Equation (2.28) is much larger than 1, the densities of free electrons and holes are given by the so-called Boltzmann equations:

$$n = N_C \cdot \exp \left(-\frac{E_C - E_{Fn}}{k_B \cdot T} \right) \tag{2.30'}$$

$$p = N_V \cdot \exp \left(-\frac{E_{Fp} - E_V}{k_B \cdot T} \right) \tag{2.30''}$$

The Boltzmann equations connect the densities of free electrons and holes with their Fermi-energies (E_{Fn} and E_{Fp}) which are equivalent to the activity or chemical potential of chemical species in chemistry. More details about free charge carriers in semiconductors can be found in textbooks about semiconductors and semiconductor devices (see, for example, Sze, 1981).

2.3.2 *Thermal equilibrium and density of intrinsic charge carriers*

Any material emits blackbody radiation depending on the temperature of the material. Blackbody radiation contains photons with energy above E_g. Therefore, e^- and h^+ are generated in semiconductors by absorption of photons from blackbody radiation with E_{ph} above the E_g. This process is called thermal generation of free charge carriers. The thermal generation rate of free charge carriers is constant for a fixed absolute temperature (T). Obviously, the n and p cannot increase to infinite values. The process limiting the n and p is called recombination, meaning the annihilation of e^- and h^+ (see Chapter 3). Radiative recombination is defined as the annihilation of electron–hole pairs by emitting photons. The radiative recombination rate is proportional to the product of the n and p since both carriers are involved in the radiative recombination process (see Chapter 3).

In thermal equilibrium, the thermal generation rate is equal to the radiative recombination rate. The densities of free electrons or holes in thermal equilibrium are denoted by n_0 or p_0, respectively. It follows from the definitions of the thermal equilibrium and of radiative recombination that the product of n_0 and p_0 is constant. The product of n_0 and p_0 can be calculated by using Equations (2.30′) and (2.30″).

The Fermi-energies of free electrons and holes can change. Therefore, the difference of the Fermi-energies of free electrons and holes must be 0 in thermal equilibrium since the product of n_0 and p_0 is constant. As a consequence, there is one common Fermi-energy of free electrons and holes in thermal equilibrium (E_{F0}).

$$E_{F0} = E_{Fn} = E_{Fp} : \quad n_0 \cdot p_0 = N_C \cdot N_V \cdot \exp\left(-\frac{E_g}{k_B \cdot T}\right) \qquad (2.31)$$

The densities of free electrons and holes are equal in a pure or so-called intrinsic semiconductor in thermal equilibrium. The density of intrinsic charge carriers (n_i) is defined as:

$$n_i \equiv \sqrt{N_C \cdot N_V} \cdot \exp\left(-\frac{E_g}{2 \cdot k_B \cdot T}\right) \qquad (2.32)$$

The n_i is the minimum density of free charge carriers that can be reached in a semiconductor. The n_i increases exponentially with decreasing E_g and with increasing T. Figure 2.17 shows the T dependence of n_i for GaAs, c-Si and Ge (germanium) under consideration of the T dependence of their band gaps.

At room temperature, the n_i of pure c-Si, GaAs and Ge are about 10^{10}, 10^6 and 10^{13} cm^{-3}, respectively. For comparison, the density of electrons in metals is of the order of 10^{22} cm^{-3}. Therefore, the conductivity of pure c-Si, GaAs and Ge is much lower than the conductivity of metals but much higher than the conductivity of insulators, which is the reason why pure c-Si, GaAs and Ge belong to the class of semiconductors.

Semiconductors with very large E_g such as diamond (5 eV) or quartz (9 eV) are insulators at room temperature due to the very low n_i. Insulators become semiconducting at high T due to the exponential increase of n_i with increasing T. *Vice versa*, all semiconductors are insulators at low T.

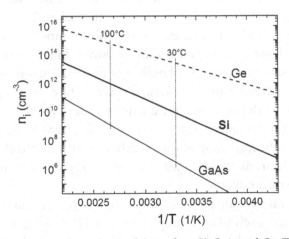

Figure 2.17. Temperature dependence of the n_i for c-Si, GaAs and Ge. The temperature dependence of E_g is considered.

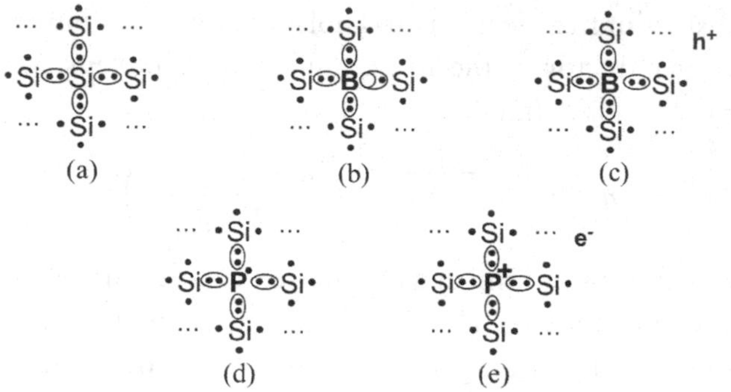

Figure 2.18. Bond configuration of a Si atom in a silicon crystal (a) replaced by a boron (b, c) or phosphorus (d, e) atoms before (b, d) and after ionization (c, e).

2.3.3 *Doping, majority and minority equilibrium charge carriers*

Doping of materials is the purposeful replacement of host atoms by impurity atoms. The goal of doping semiconductors is a controlled change in conductivity. The density of free charge carriers in thermal equilibrium can be changed in semiconductors over many orders of magnitude by doping, which makes doping of semiconductors very important for electronic and PV applications.

For doping of semiconductor crystals, host atoms are replaced by impurity atoms with a higher or with a lower valence than the valence of the host atoms (see also the example in Figure 2.18). Doping of semiconductors leads to extra electrons, which cannot be incorporated into chemical bonds if the valence of the dopant is higher than the valence of the host atom, for example, phosphorus in c-Si. The un-bonded extra electrons can be free, i.e. they can occupy states in the conduction band. On the other hand, if the valence of the dopant is lower than the valence of the host atom — for example, boron in c-Si — doping leads to electrons which are missing for the formation of chemical bonds with shared electron pairs. The missing electrons can become free holes occupying states in the valence band of the semiconductor.

Dopants causing an increase of the n_0 or p_0 are called donors or acceptors, respectively. For example, Si and Ge belong to the fourth group of the periodic table of elements. Atoms of elements of the third or fifth group of the periodic table of elements form acceptors

(B, Al, Ga, In) or donors (N, P, As, Sb) in Si or Ge crystals, respectively. Semiconductors doped with donors or acceptors are called n-type or p-type semiconductors, respectively.

Donor (D) or acceptor (A) atoms can be electrically neutral or ionized depending on whether extra or missing electrons fill states in the conduction or valence bands or not, respectively.

$$D \leftrightarrow D^+ + e^- \tag{2.33'}$$

$$A \leftrightarrow A^- + h^+ \tag{2.33''}$$

Free electrons or holes experience Coulomb attractive force with ionized donors or acceptors, similarly to an electron experiencing the Coulomb attractive force with the proton in a hydrogen atom (see also Figure 2.19).

The electron in a hydrogen atom can become free if it becomes excited to an energy equal or above the ionization energy of the hydrogen atom ($E_{\text{ion}}^H = 13.56\,\text{eV}$). With regard to the Bohr model of atoms (see, for example, Vogel, 1995), the ionization energy is proportional to the quotient of the electron mass and the squared dielectric constant. Coulomb potentials are strongly screened in inorganic semiconductors due to the quite large relative dielectric constant (ε_r) which is, for example, 11.8 for c-Si. The value of ε_r is equal to 1 in a hydrogen atom since the electron is surrounded only by vacuum. Therefore, the ionization energies of electrons or holes in doped inorganic semiconductors are more than 100 times less than the ionization energy of the hydrogen atom. The ionization energies of donors and acceptors ($E_{\text{ion}}^{D,A}$) can be estimated by taking into account

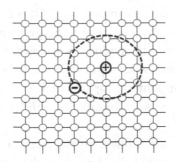

Figure 2.19. Hydrogen model of an ionized donor in a crystal.

$m^*_{e,h}$, ε_r and E^H_{ion} (hydrogen model of donors and acceptors)

$$E^{D,A}_{ion} \approx E^H_{ion} \cdot \frac{m^*_{e,h}}{m_e} \cdot \frac{1}{\varepsilon_r^2} \qquad (2.34)$$

The ionization energies obtained from Equation (2.34) are of the same order as the thermal energy at room temperature. Therefore, the thermal energy is sufficient to ionize practically all donors or acceptors in doped semiconductors at room temperature. As a consequence, the n_0 or p_0 are practically equal to the density of donor (N_D) or acceptor (N_A) atoms at room temperature. Related donors or acceptors are also called shallow donors or shallow acceptors. The so-called doping regime is relevant for solar cells.

$$N_D = n_0 = N_D^+ \qquad (2.35')$$

$$N_A = p_0 = N_A^- \qquad (2.35'')$$

At very low temperature the thermal energy is much lower than the ionization energy of the donors and acceptors. In this case, the electrons and holes cannot escape from the region of the Coulomb potential of the ionized donors and acceptors, i.e. the electrons and holes are no longer free. This so-called freeze-out regime is not relevant for solar cells.

The n_i increases exponentially with increasing T as mentioned before. At high T the n_i exceeds the density of donors or acceptors, i.e. all semiconductors become intrinsic at high T. This so-called intrinsic regime can become relevant for PV absorbers with low E_g at relatively high T.

The Fermi-energy shifts towards the E_C for n-type and towards the E_V for p-type semiconductors (Figure 2.20).

A semiconductor is called compensated if it is doped with donors and acceptors at the same time. The Fermi-energy cannot shift widely over the E_g for compensated semiconductors despite possible high densities of free electrons and holes. Compensation can become important for semiconductors with amphoteric dopants such as Si which can replace a Ga or As atoms in GaAs or for semiconductors with strong self-compensation such as $CuInSe_2$.

Condition (2.31) is also fulfilled for doped semiconductors in thermal equilibrium. This means that the density of the free charge carriers not belonging to the dopants can be calculated with respect to Equation (2.31).

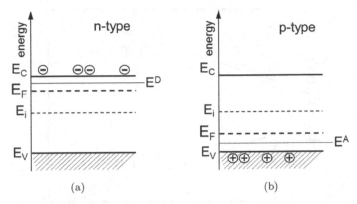

Figure 2.20. Energy diagrams for *n*-type (a) and *p*-type (b) doped semiconductors. E_i denotes the so-called intrinsic Fermi-energy of the un-doped semiconductor. E^D and E^A are the energies of the unoccupied donor and occupied acceptor states. The energy differences E_C–E^D and E^A–E_V correspond to the ionization energies of the donor and acceptor, respectively.

The density of these charge carriers is very low, and consequently they are called minority charge carriers. The free charge carriers originating from the ionized dopants are called majority charge carriers.

$$n_i^2 = n_0 \cdot p_0 \tag{2.36}$$

Free electrons are the majority charge carriers in *n*-type semiconductors, and free holes are the majority charge carriers in *p*-type semiconductors. Conversely, free holes and electrons are minority charge carriers in *n*-type and *p*-type semiconductors, respectively. For example, the absorber of c-Si solar cells is usually *p*-type doped with a density of majority carriers of $p_0 = 10^{16}\,\text{cm}^{-3}$. The density of minority charge carriers is $n_0 \approx 10^4\,\text{cm}^{-3}$ since $n_i \approx 10^{10}\,\text{cm}^{-3}$ for c-Si at room temperature. The density of minority charge carriers depends strongly on T due to the T dependence of n_i.

Figure 2.21 shows the dependence of the Fermi-energy in thermal equilibrium (E_{F0}) on the densities of majority charge carriers for *n*-type and *p*-type doped c-Si at 300 K. In the doping range, the E_{F0} can be calculated following:

$$E_{F0} = E_C - k_B \cdot T \cdot \ln\left(\frac{N_C}{n_0}\right) = E_V + k_B \cdot T \cdot \ln\left(\frac{N_V}{p_0}\right) \tag{2.37}$$

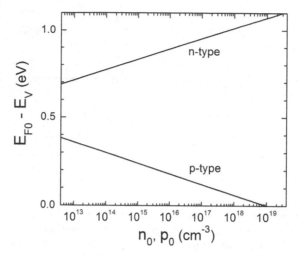

Figure 2.21. Dependence of the Fermi-energy on the densities of majority charge carriers for *n*-type and *p*-type doped c-Si at 300 K in thermal equilibrium.

The E_{F0} shifts towards the middle of the E_g with increasing T for a given density of dopants. Any doped semiconductor becomes intrinsic at high temperatures depending on E_g and doping. This limits the operation range of semiconductor devices.

The E_{F0} shifts into the valence or conduction bands when the density of dopants becomes larger than N_V or N_C, respectively. Such highly doped semiconductors are said to be degenerated. Very high densities of majority charge carriers are sometimes denoted by $n^{+(++)}$ or $p^{+(++)}$.

The density of free charge carriers in semiconductors can be varied by doping over a range from n_i to the solubility limit of a given dopant in a given semiconductor. The solubility limit is of the order of 10^{20} cm^{-3} for many dopants. If taking into account minority and majority charge carriers, the n_0 or p_0 can be varied in c-Si, for example, over a range of about 20 (!) orders of magnitude.

2.3.4 Fermi-level splitting for photogenerated charge carriers

Non-equilibrium free electrons and holes with the densities Δn and Δp ($\Delta n = \Delta p$) are generated under illumination. Their total densities are

$$n = n_0 + \Delta n \tag{2.38'}$$

$$p = p_0 + \Delta p \tag{2.38''}$$

The Fermi-energies of free electrons and holes become different under illumination and can be calculated if Δn or Δp is known

$$E_{Fn} = E_C - k_B \cdot T \cdot \ln \left(\frac{N_C}{n_0 + \Delta n} \right) \qquad (2.39')$$

$$E_{Fp} = E_V + k_B \cdot T \cdot \ln \left(\frac{N_V}{p_0 + \Delta p} \right) \qquad (2.39'')$$

The potential energy at which charge carriers can be extracted from a semiconductor is related to their Fermi-energy. As already mentioned, the Fermi-energy of free charge carriers is equivalent to their chemical potential. The difference of chemical potentials between two different materials is used in batteries, for example. In PV absorbers, two different chemical potentials arise in the same material under illumination.

In solar cells, E_{Fn} and E_{Fp} are separately contacted for the same PV absorber (Figure 2.22) with, for example, a *pn*-homo-junction (see Chapter 4). Extracted free electrons flow from the PV absorber contacted at E_{Fn} via an external load back to the PV absorber contacted at E_{Fp}, where the extracted electrons recombine with free holes.

The assumption of separate contacts for E_{Fn} and E_{Fp} implies two fundamentally different functionalities: (i) separation of free charge carriers in space, for example, with a *pn*-junction (see Chapter 4) and (ii) transfer of free charge carriers to external leads via ohmic contacts (see Chapter 5).

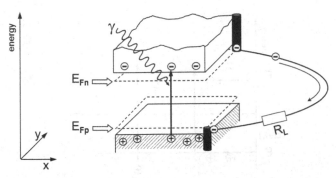

Figure 2.22.   Energetic scheme of a PV absorber under illumination with separate contacts at E_{Fn} and E_{Fp} and with an external load.

The change of the Fermi-energy in equilibrium and under illumination is very different for majority and minority charge carriers. The following equations can be obtained by taking into account Equations (2.37) and (2.39)

$$E_{Fn} - E_{F0} = k_B \cdot T \cdot \ln \left(\frac{n_0 + \Delta n}{n_0} \right) \qquad (2.40')$$

$$E_{F0} - E_{Fp} = k_B \cdot T \cdot \ln \left(\frac{p_0 + \Delta p}{p_0} \right) \qquad (2.40'')$$

For p-type (n-type) semiconductors, p_0 (n_0) is usually much larger than $\Delta p(\Delta n)$ so that the change of E_{Fp} (E_{Fn}) in comparison to E_{F0} is low and can be neglected (see also Figure 2.23). Therefore, the illumination-induced change of the Fermi-energy of the majority charge carriers does not significantly contribute to the photovoltage of a solar cell. Conversely, the density of minority charge carriers changes over many orders of magnitude under illumination, and so the Fermi-energy of minority charge carriers increases strongly. For example, Δn is of the order of 10^{15} cm^{-3} for c-Si solar cells illuminated at AM1.5 which is 11 orders of magnitude larger than n_0 since $p_0 = 10^{16}$ cm^{-3}. A difference of 11 orders of magnitude in the density of free electrons corresponds to a change of E_{Fn} by 0.66 eV which is almost close to $q \cdot V_{OC}$ of the world record c-Si solar cell. Therefore, the illumination-induced change of the Fermi-energy of minority charge carriers is a fundamental reason for the photovoltage of a solar cell under operation.

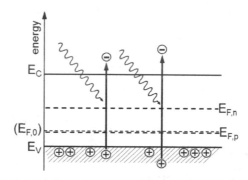

Figure 2.23. Energy diagram for p-type semiconductors under illumination.

The difference between the Fermi-energies of free electrons and holes (also known as Fermi-level splitting) in a PV absorber is of pivotal importance for solar energy conversion by PV since it defines the maximum potential energy difference at which a solar cell can be operated, i.e. V_{OC} multiplied by q.

$$E_{Fn} - E_{Fp} = k_B \cdot T \cdot \ln\left(\frac{(n_0 + \Delta n) \cdot (p_0 + \Delta p)}{n_i^2}\right) \qquad (2.41)$$

The Fermi-level splitting increases with increasing Δn for a given semiconductor at fixed n_i, i.e. at fixed T. This is the physical reason for the increase of V_{OC} with increasing light intensity (see Chapter 1).

Equation (2.41) implements a T dependence of the Fermi-level splitting. Figure 2.24 shows the T dependencies of the Fermi-level splitting under identical p_0 and Δn in Ge, Si and GaAs. The largest Fermi-level splitting is obtained for the largest value of E_g, i.e. the lowest n_i. The Fermi-level splitting decreases with increasing T and disappears for Ge at T above 150°C.

The decrease of the Fermi-level splitting with increasing temperature is the reason for the decrease of V_{OC} and therefore for the decrease of the energy conversion efficiency of solar cells with increasing temperature.

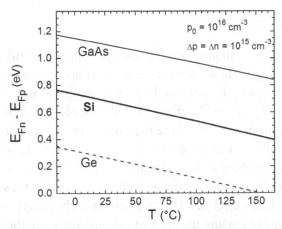

Figure 2.24. Temperature dependencies of the Fermi-level splitting for $p_0 = 10^{16}$ cm^{-3} and $\Delta n = 10^{15}$ cm^{-3} in PV absorbers with $E_g = 0.78$, 1.1 and 1.42 eV corresponding to the E_g of Ge, c-Si and GaAs, respectively. The T dependence of E_g is taken into account.

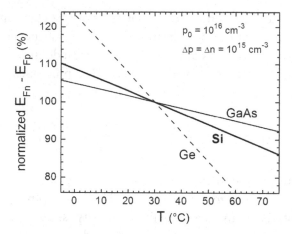

Figure 2.25. Temperature dependencies of the normalized Fermi-level splitting at 30°C for $p_0 = 10^{16}$ cm^{-3} and $\Delta n = 10^{15}$ cm^{-3} in PV absorbers with $E_g = 0.78, 1.1$ and 1.42 eV for Ge, c-Si and GaAs, respectively.

The relative decrease of the Fermi-level splitting is higher for lower E_g. Therefore the relative decrease of the η with increasing T is less for solar cells based on PV absorbers with higher E_g. Figure 2.25 gives an impression about the relative T dependence of the Fermi-level splitting for GaAs, c-Si and Ge in a temperature range relevant for most terrestrial applications of solar cells.

2.4 Summary

Solar cells have to be made from PV absorbers or semiconductors with a forbidden band gap. The E_g of a semiconductor is decisive for solar energy conversion by PV. On the one hand, the V_{OC} of a solar cell cannot exceed the value of the E_g divided by the q. On the other hand, photons with E_{ph} below the E_g cannot not contribute to photocurrent generation.

During an absorption event, a photon is converted into an electron–hole pair where the hole and the electron are mobile charge carriers in the valence and conduction bands, respectively. In an illuminated solar cell with maximum efficiency, electrons are extracted at the chemical potential of free electrons and, after passing through an external load, are then transferred back into the semiconductor at the chemical potential of the free holes where the electrons then recombine with free holes.

The η_{ultimate} takes into account the upper limit of V_{OC} and a hypothetical FF of 1, which is, in theory, impossible (see Chapter 1). However, the analysis of the η_{ultimate} gives the range of E_g suitable for very high η. The η_{ultimate} is above 40% for semiconductors with E_g between 0.7 and 1.6 eV. There are numerous inorganic and organic semiconductors with E_g in this range. The η_{ultimate} is below 10% for PV absorbers with E_g above 2.5 eV, i.e. so-called wide-gap semiconductors are not useful as PV absorbers.

According to the η_{ultimate}, realistic maximum η of the order of 25–30% can be expected if assuming a realistic FF of the order of 85% (see Chapter 1) and a realistic V_{OC} of the order of 70–80% of E_g divided by q.

The number of photons that are absorbed in the semiconductor is limited by the E_g, reflection and transmission. The reflection coefficient (R) and the absorption coefficient (α) are material parameters depending on the wavelength of incident light or the E_{ph}. The R and the α of semiconductors are typically of the order of 0.3–0.4 and 10^2–10^5 cm^{-1}, respectively.

Losses in photogeneration caused by E_g cannot be minimized. Optical losses caused by reflection and transmission can be minimized by implementing antireflection coatings and by increasing the optical path of light in the PV absorber, respectively. The optical path of light can be increased by increasing the thickness of the PV absorber and/or by implementing measures for light trapping such as Lambertian light scattering combined with total reflection. The optical path can be increased by about 8–18 times for this so-called Lambertian limit of light trapping, depending on the ratio of the refractive index of the PV absorber and of the antireflection coating.

The densities of states in the valence and conduction bands have been derived by taking into account only delocalization of free charge carriers and the Heisenberg uncertainty principle. The density of states increases by the square root of the difference between the energy of free charge carriers and the energy of the corresponding band edge. The density of states and the effective density of states (N_C, N_V) can be calculated if the E_g and the effective masses of free electrons and holes of the semiconductor are known. The effective mass of the electron, for example, is about $0.067 \cdot m_e$ for GaAs and $1.08 \cdot m_e$ for c-Si. The value of N_C, for example, is about $5 \cdot 10^{17}$ cm^{-3} for GaAs and $3 \cdot 10^{19}$ cm^{-3} for c-Si at room temperature.

In order to calculate the densities of free electrons and holes, the densities of states have to be multiplied with the corresponding occupation probability which is given for Fermions (following the Pauli principle meaning that two particles cannot have the same quantum numbers) by the Fermi distribution function. As a result, the Boltzmann equations are obtained for calculating the densities of free electrons and holes from their chemical potentials (which are called Fermi-energies) and from the N_C and N_V, respectively.

The n_i (intrinsic density of free charge carriers) of an ideal semiconductor follows from the absorption of blackbody radiation and radiative recombination by the semiconductor. The value of n_i^2 is proportional to the product of N_C and N_V and of the exponential of the — E_g divided by $k_B \cdot T$. Typical values of n_i^2 are about 10^{20} cm^{-3} for c-Si and 10^{12} cm^{-3} for GaAs at room temperature. The exponential increase of n_i^2 with increasing T is a major limiting factor of the η at increased T.

Semiconductors can be p-type or n-type doped by replacing atoms in the lattice with atoms of a lower or higher valance, respectively. There is one common Fermi-energy for a semiconductor in thermal equilibrium. The product of free electrons and free holes is equal to n_i^2 in thermal equilibrium. Under illumination, the Fermi-energy splits into two different values (E_{Fn} and E_{Fp}). The difference between E_{Fn} and E_{Fp} is called Fermi-level splitting and is dominated by the change of the Fermi energy of minority charge carriers.

The Fermi-level splitting decreases with increasing T where the relative change is larger for semiconductors with lower E_g. Therefore, the V_{OC} and the η decrease at a greater pace with increasing temperature for solar cells made from semiconductors with lower E_g than for solar cells made from semiconductors with higher E_g.

2.5 Tasks

T2.1: I_{SC} density

Ascertain the I_{SC} current density for a PV absorber with E_g of 1.6 eV and a width of the absorption band of 0.3 eV.

T2.2: Temperature-dependent I_{SC} density

Estimate the increase of the I_{SC} density for a c-Si solar cell, the T of which is increased from 25°C ($E_g = 1.12$ eV) to 65°C ($E_g = 1.115$ eV).

T2.3: Lambertian limit of light trapping

Get an expression for the Lambertian limit of light trapping by taking into account a Lambertian back reflector and total internal reflection at the interface between the absorber and the antireflection layer.

T2.4: Temperature-dependent effective density of states

Solar cells are usually operated in a temperature range between 0 and 70°C. Calculate the effective density of states at the E_C of c-Si at both temperatures. The effective mass of free electrons is $1.08 \cdot m_e$.

T2.5: Temperature-dependent density of minority charge carriers

Solar cells are usually operated in a temperature range between 0 and 70°C. Calculate the density of minority charge carriers in p-type doped c-Si at both temperatures ($p_0 = 3 \cdot 10^{16}$ cm^{-3}). The effective mass of free holes is $0.55 \cdot m_e$. The E_g of c-Si is 1.131 eV at 0°C and 1.113 eV at 70°C.

T2.6: Temperature-dependent Fermi-energy

Solar cells are usually operated in a temperature range between 0 and 70°C. Calculate the Fermi-energy in a p-type doped c-Si absorber ($p_0 = 3 \cdot 10^{16}$ cm^{-3}) in thermal equilibrium at both these temperatures.

T2.7: Fermi-level splitting

Calculate the Fermi-level splitting at 25°C in p-type doped c-Si with $p_0 = 3 \cdot 10^{16}$ cm^{-3} under illumination with $\Delta n = 10^{14}$, 10^{15} and 10^{16} cm^{-3}.

T2.8: Temperature-dependent Fermi-level splitting

Compare the temperature-dependent decrease of the Fermi-level splitting in a c-Si absorber ($p_0 = 3 \cdot 10^{16}$ cm^{-3}, $\Delta n = 10^{15}$ cm^{-3}) with the temperature-dependent increase of I_{SC}^{max} if the temperature is increased from 25°C to 65°C.

3

Influence of Recombination on the Minimum Lifetime

Recombination is the annihilation of free electrons and holes. Recombination limits the density of photogenerated free charge carriers and therefore the photocurrent and the photovoltage of a solar cell. The recombination rate is described by the reciprocal lifetime of free charge carriers. The minimum lifetime condition follows from the requirement of light absorption in relation to charge transport. The bulk and the interfaces of photovoltaic (PV) absorbers have to be conditioned in such a way that the lifetime of photogenerated free charge carriers is equal or longer than the minimum lifetime. The upper limits of doping of PV absorbers follow from the radiative and/or Auger-recombination lifetimes. Shockley–Read–Hall (SRH) and surface recombination are caused by defect states in the band gap in the bulk or at the surface of a PV absorber, respectively. The acceptable maximum densities of bulk and surface defects follow from the minimum lifetime condition. Strategies of the purposeful elimination of defect states in the forbidden band gap are illuminated for passivation of defects in the bulk and at the surface of a PV absorber.

3.1 Minimum Lifetime

3.1.1 Decay of photocurrent, recombination rate and lifetime

A photocurrent does not disappear immediately after switching off illumination or, in other words, a photocurrent still lasts for a certain time after switching off the light (see, for example, Ryvkin, 1964). To get an impression of this time one has to look at the time-dependent density of photogenerated free charge carriers.

The time-dependent change of the densities of free electrons (n) and holes (p) is expressed by the continuity equations. In the simplest case, n and p depend only on generation and recombination (generation and recombination rates: G and R, respectively). This case is related to the spatially homogeneously distributed densities of free electrons and holes when gradients of densities and electric fields are absent, i.e. diffusion and drift of free charge carriers can be neglected.

$$\frac{\partial n}{\partial t} = G - R \qquad (3.1')$$

$$\frac{\partial p}{\partial t} = G - R \qquad (3.1'')$$

The densities of free electrons and holes are the sums of the densities of free electrons or holes in thermal equilibrium (n_0 and p_0) and of the density of photogenerated free electrons (Δn) or density of photogenerated free holes (Δp). Similarly, the generation rate is the sum of the thermal generation rate (G_0) and of the photogeneration rate (ΔG).

A photocurrent is proportional to the density of photogenerated free charge carriers. Under stationary illumination, i.e. at constant ΔG, a stationary photocurrent is reached, i.e. Δn is constant (Figure 3.1).

The decay of photocurrent transients depends on the recombination rate of photogenerated charge carriers (ΔR), which is in turn

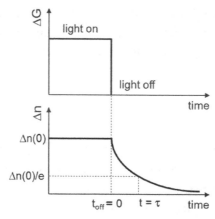

Figure 3.1. Photogeneration rate (top) and density of photogenerated free electrons (bottom) before and after switching off illumination at $t_{off} = 0$.

proportional to Δn. The proportionality factor between ΔR and Δn has the unit of inverse time. The reciprocal proportionality factor between ΔR and Δn is called the lifetime of photogenerated free charge carriers (τ). Therefore,

$$\Delta R \equiv \frac{\Delta n}{\tau} \qquad (3.2)$$

The continuity equation (3.1′) of photogenerated free electrons can be written after switching off illumination ($\Delta G = 0$) and taking into account definition (3.2) as

$$\frac{\partial(\Delta n)}{\partial t} = -\frac{\Delta n}{\tau} \qquad (3.3)$$

The differential equation (3.3) has the following solution:

$$\Delta n(t) = \Delta n(0) \cdot \exp\left(-\frac{t}{\tau}\right) \qquad (3.4)$$

Equation (3.4) describes an exponential decay with the exponential decay time constant τ, i.e. the photocurrent or the Δn decreases for a constant τ to the value 2.718 times less than the stationary value ($\Delta n(0)$) after passing a time interval of decay equal to τ. It is interesting to note that the τ is not necessarily constant but can depend on Δn and even on time, depending on the recombination and transport processes involved.

3.1.2 *Diffusion length and lifetime*

Photogenerated free charge carriers are generated in the photovoltaic (PV) absorber and collected at the corresponding charge-selective electron or hole contacts. In order to be collected, photogenerated free charge carriers have to pass from the place of photogeneration to the place of collection. The passage of photogenerated free charge carriers to the corresponding charge-selective electron or hole contacts takes some time, during which the photogenerated free charge carriers should not recombine. Only those photogenerated free charge carriers can be collected that can pass to the corresponding charge-selective electron or hole contacts within a time shorter than their lifetime.

The passage of photogenerated free charge carriers to the corresponding charge-selective electron or hole contacts can be driven by diffusion or by drift. Concentration gradients and electric fields are the origin for diffusion and drift of photogenerated free charge carriers, respectively. Diffusion drives the passage of most photogenerated free charge carriers to the corresponding charge-selective electron or hole contacts for many types of PV absorbers, for example, for crystalline silicon (c-Si, see Chapter 7). Diffusion will be considered in the following. In a diffusion process the squared average displacement of a diffusing species is proportional to the considered time interval.

Photogenerated free charge carriers can move until they recombine, i.e. they can diffuse only within their lifetime. The average distance over which free charge carriers can diffuse within their lifetime is known as the diffusion length (L). The squared L is proportional to the τ. The proportionality factor between the squared diffusion length and the lifetime is called the diffusion coefficient (D). The diffusion length can be calculated if D and τ are known, and *vice versa* the τ can be calculated for given D and L.

$$L = \sqrt{D \cdot \tau} \tag{3.5}$$

Incidentally, the D of crystalline semiconductors is practically constant for densities of free charge carriers below about $10^{17}\,\text{cm}^{-3}$ and decreases for larger densities of free charge carriers.

3.1.3 *Minimum lifetime condition for PV absorbers*

Photogenerated free charge carriers can be collected at their corresponding charge-selective contact if they are photogenerated closer to the contact than L. It was shown in Chapter 2 (see Equation (2.15)) that more than 95% of incoming light is absorbed if the thickness of the PV absorber (d_{abs}) is equal to or larger than three times the absorption length (α^{-1}). However, absorption is only useful if the photogenerated charge carriers can be collected at their corresponding charge-selective contacts, i.e. if not only d_{abs} but also L is equal to or larger than $3 \cdot \alpha^{-1}$. Therefore the following condition for the minimum lifetime of a PV absorber (τ_{min}) can be written.

$$\tau_{\text{min}} = \frac{9 \cdot \alpha^{-2}}{D} \tag{3.6}$$

Equation (3.6) is the minimum lifetime condition and connects the optical properties with the electronic properties of a PV absorber. It is, after Equation (2.15), the second important demand for PV absorbers. The minimum lifetime condition has to be fulfilled for reaching reliable solar energy conversion efficiencies (η), and therefore defines the demands to those electronic properties, which limit the lifetime of photogenerated free charge carriers for a PV absorber with a given α^{-1} and D.

With regard to Figure 2.8, the α^{-1} of c-Si and of GaAs (gallium arsenide) absorbers is of the order of 100 and 1 μm, respectively. The D is about 15 and 60 cm^2/s for c-Si and GaAs, respectively. Therefore, the following τ_{min} is obtained for PV absorbers based on c-Si or GaAs:

$$\tau_{min}(\text{c-Si}) \approx 100\,\mu s \qquad (3.7')$$

$$\tau_{min}(\text{GaAs}) \approx 10\,\text{ns} \qquad (3.7'')$$

These conditions are rather harsh with respect to doping and the purity of the materials, as will be shown in the following. It is worth noting that, since the electronic properties of PV absorbers sensitively depend on their production technology, the minimum lifetime condition also defines the demands for suitable technologies and therefore the limits for basic cost factors of PV absorbers.

The minimum lifetime condition can be used for the estimation of the density of photogenerated free charge carriers from the maximum short circuit current (I_{SC}^{max}) density given by Equation (2.2). A current is defined as the charge (ΔQ) passing through a contact within a certain time interval (Δt). If the Δt is limited by the τ_{min}, the charge that can pass through the contact is roughly the areal density of charge carriers photogenerated within a layer of thickness of the L, i.e. Δn multiplied by the L.

$$I = \frac{\Delta Q}{\Delta t} \approx q \cdot \frac{\Delta n \cdot L}{\tau_{min}} \qquad (3.8)$$

With respect to the minimum lifetime condition, the following equation can be applied to estimate Δn:

$$\Delta n \approx \frac{\tau_{min}}{q \cdot 3 \cdot \alpha^{-1}} \cdot I_{SC}^{max} \qquad (3.9)$$

For a PV absorber based on c-Si, I_{SC}^{max} is about $44 \, mA/cm^2$ (air mass (AM) 1.5). Therefore Δn is about $1 \cdot 10^{15} \, cm^{-3}$ for solar cells based on c-Si and operated under illumination at AM1.5.

The L and τ_{min} can be reduced if increasing the optical path of light in the PV absorber by light trapping (see, for example, Equation (2.18) in Chapter 2). For example, the increase of the optical path of light by a factor of ten would allow a reduction of τ_{min}, referring to Equation (3.6), by a factor of 100 (!). The reduction of τ_{min} due to application of light trapping opens opportunities for technologies with softer requirements, for example, for maximum defect densities.

3.2 Radiative Recombination

Radiative recombination (see, for example, van Roesbroeck and Shockley, 1954) is the annihilation of an electron–hole pair under emission of a photon (Figure 3.2), i.e. a free electron (e^-) falls from the conduction band into the valence band by filling the place of a missing electron (hole, h^+) and by transferring the excess energy to the emission of a photon (γ', Equation (3.10)). Radiative recombination cannot be avoided in PV absorbers.

$$e^- + h^+ \rightarrow \gamma' \qquad (3.10)$$

The radiative recombination rate (R_{rad}) is proportional to the densities of free electrons and holes since both of them are involved in the radiative

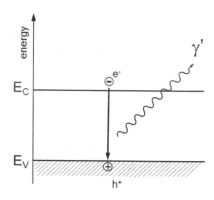

Figure 3.2.   Energy scheme for radiative recombination.

recombination process

$$R_{\text{rad}} = B \cdot n \cdot p \tag{3.11'}$$

The proportionality factor (B) is the radiative recombination rate constant and amounts to about $3 \cdot 10^{-15}$ cm^3/s for c-Si (Varshni, 1967a) and about $2 \cdot 10^{-10}$ cm^3/s for GaAs (Strauss *et al.*, 1993). The large difference between the radiative recombination rate constants of c-Si and GaAs is caused by the large difference in the absorption coefficients of both semiconductors.

Since the densities of photogenerated electrons and holes are equal, Equation (3.11') can be written in the following form by using Equation (2.38):

$$R_{\text{rad}} = B \cdot n_0 \cdot p_0 + B \cdot \Delta n \cdot (n_0 + p_0 + \Delta n) \tag{3.11''}$$

With reference to definition (3.2) and Equation (3.11''), the radiative recombination lifetime of photogenerated free charge carriers (τ_{rad}) can be written as

$$\tau_{\text{rad}} = \frac{1}{B \cdot (n_0 + p_0 + \Delta n)} \tag{3.12'}$$

The density of majority charge carriers is much higher than the density of minority charge carriers. At AM1.5 the density of photogenerated electrons and holes is usually much higher than the densities of minority or intrinsic free charge carriers but much smaller than the density of majority charge carriers. High injection is reached when the density of photogenerated charge carrier becomes higher than the density of majority charge carriers. Equation (3.12') can be simplified for n-type and p-type semiconductors and for high injection as

$$\tau_{\text{rad}} = \begin{cases} (B \cdot p_0)^{-1} & (p\text{-type}) \\ (B \cdot n_0)^{-1} & (n\text{-type}) \\ (B \cdot \Delta n)^{-1} & (\Delta n \gg n_0, p_0) \end{cases} \tag{3.12''}$$

Equation (3.12'') shows that τ_{rad} depends on the density of majority charge carriers in PV absorbers. Figure 3.3 shows the dependence of τ_{rad} on the density of majority charge carriers in c-Si and GaAs absorbers.

The τ_{min} is reached for c-Si and GaAs at densities of majority charge carriers of about $3 \cdot 10^{18}$ cm^{-3} and $5 \cdot 10^{17}$ cm^{-3}, respectively. This means

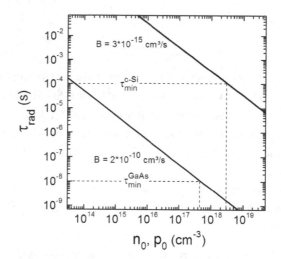

Figure 3.3. Dependence of the radiative recombination lifetime on the density of majority charge carriers in c-Si and GaAs.

that, under the assumption that radiative recombination would be the only limiting recombination process, PV absorbers based on c-Si or GaAs can be doped with majority charge carriers up to these densities without losses in the η. Doping at higher densities of majority charge carriers is not useful as a result of collection losses.

3.3 Auger Recombination

Auger recombination (Auger, 1932) is the annihilation of an electron–hole pair under excitation of a third free charge carrier up to a higher kinetic energy. In a PV absorber, the kinetic energy is dissipated by a cascade of phonons (denoted by Γ) emission, i.e. heating of the PV absorber (Figure 3.4). The energy of the recombining electron can be transferred to another free electron or to a second free hole (Equation (3.13)). Auger recombination cannot be avoided in PV absorbers.

$$2 \cdot e^- + h^+ \rightarrow e^- + i \cdot \Gamma \tag{3.13'}$$

$$e^- + 2 \cdot h^+ \rightarrow h^+ + j \cdot \Gamma \tag{3.13''}$$

The Auger recombination rate (R_{Auger}) is proportional to the product of the squared density of free electrons and the density of free holes or

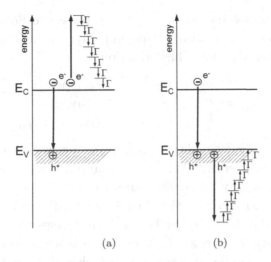

(a)   (b)

Figure 3.4.   Energy schemes for Auger recombination with $n \gg p$ (a) and $n \ll p$ (b).

to the product of the density of free electrons and the squared density of free holes since a third free charge carrier is involved in the Auger recombination process.

$$R_{\text{Auger}} = C_{A,e} \cdot n^2 \cdot p \qquad (3.14')$$

$$R_{\text{Auger}} = C_{A,h} \cdot n \cdot p^2 \qquad (3.14'')$$

The proportionality factors ($C_{A,e}$ and $C_{A,h}$) are the Auger recombination rate constants. The Auger recombination rate constant does not strongly depend on the band structure of a given PV absorber since the third free charge carrier is excited deep into the conduction or valence bands where the density of states is very large. Therefore, there is practically no need to distinguish between C_A for the process when the energy is transferred to a second free electron or to a second free hole ($C_A = C_{A,e} \approx C_{A,h}$). C_A is similar for most semiconductors and amounts to about $2 \cdot 10^{-30}$ cm^6/s for c-Si (Yablonovitch and Gmitter, 1986; Wang and Macdonald, 2012) and $7 \cdot 10^{-30}$ cm^6/s (Strauss *et al.*, 1993) or $1.3 \cdot 10^{-30}$ cm^6/s (Picozzi *et al.*, 2002) for GaAs.

As already mentioned, the densities of photogenerated free electrons and holes are equal and are much higher than the densities of minority free charge carriers. Therefore, the following equations can be written for the

Auger recombination lifetimes (τ_{Auger}) with respect to Equations (2.38), (3.2) and (3.14) in the cases of p-type and n-type semiconductors under illumination at AM1.5 and of high injection.

$$\tau_{Auger} = \begin{cases} (C_A \cdot (p_0)^2)^{-1} & (p\text{-type}) \\ (C_A \cdot (n_0)^2)^{-1} & (n\text{-type}) \\ (C_A \cdot (\Delta n)^2)^{-1} & (\Delta n \gg n_0, p_0) \end{cases} \qquad (3.15)$$

Equation (3.15) shows that τ_{Auger} depends on the squared density of majority charge carriers in PV absorbers. Figure 3.5 shows the dependence of τ_{Auger} on the density of majority charge carriers.

The τ_{min} is reached for c-Si and GaAs at densities of majority charge carriers of about $6 \cdot 10^{16}$ cm^{-3} and $4 \cdot 10^{18}$ cm^{-3}, respectively. The density of majority charge carriers at which the τ_{min} is reached for c-Si is less for Auger recombination than for radiative recombination. On the other side the density of majority charge carriers at which the τ_{min} is reached for GaAs is larger for Auger recombination than for radiative recombination. Therefore, under the assumption that radiative and Auger recombination are the only limiting recombination process, PV absorbers based on c-Si or GaAs can be doped with majority charge carriers up to about $5 \cdot 10^{16}$ cm^{-3} for c-Si (limitation by Auger recombination) and $5 \cdot 10^{17}$ cm^{-3} for GaAs (limited by radiative recombination).

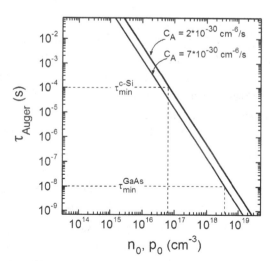

Figure 3.5. Dependence of the Auger recombination lifetime on the density of majority charge carriers.

3.4 Shockley–Read–Hall Recombination

3.4.1 *Elementary processes at a trap state*

An ideal PV absorber does not have electronic states in the forbidden band gap. However, real PV absorbers have localized defect states in the forbidden band gap due to imperfections, arising, for example, in a crystalline lattice during the growth process. Defect states in the forbidden band gap of a semiconductor are also known as trap states and cause the so-called Shockley–Read–Hall (SRH) recombination (Hall, 1952; Shockley and Read, 1952).

A trap or defect state with an energy level in the forbidden band gap (E_t) interacts with both the valence and conduction bands (see Figure 3.6). These interactions lead to four elementary processes at a trap state. A free electron can be captured from the conduction band at the unoccupied defect state. This process is called electron trapping. The trapped electron is no longer mobile, but localized at the trap state.

An electron can be emitted from an occupied trap state into the conduction band, a process which is known as electron detrapping or electron emission. Similarly, a free hole can be trapped at an occupied trap state (hole trapping). The trapped hole is also localized. A trapped hole can be emitted from an unoccupied trap state into the valence band (detrapping or emission of a hole). A defect state is characterized by the rates of all four processes: capture rate of an electron (R_e), capture rate of a hole (R_h), emission rate of an electron (G_e) and emission rate of

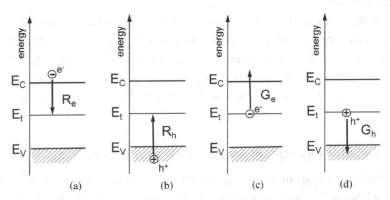

Figure 3.6. Elementary processes at a trap state: electron trapping (a), hole trapping (b), electron emission (c) and hole emission (d).

a hole (G_h). An electron–hole pair recombines at a trap state if both the free electron and the free hole are trapped at the trap state at the same time.

The density of occupied traps (n_t) can be calculated by using the Boltzmann statistics. Traps are defects localized in space. The density of defects in the bulk is denoted by N_t.

$$n_t = N_t \cdot \exp\left(-\frac{E_t - E_F}{k_B \cdot T}\right) \tag{3.16}$$

For SRH recombination, the rate Equations (3.1) have to be extended by the rate equation of the density of occupied trap states and G, Ge, Gh, R_e and R_h have to be taken into account.

$$\frac{\partial n}{\partial t} = G - R_e + G_e \tag{3.17'}$$

$$\frac{\partial p}{\partial t} = G - R_h + G_h \tag{3.17''}$$

$$\frac{\partial n_t}{\partial t} = R_e - R_h - G_e + G_h \tag{3.17'''}$$

The equations under (3.17) are not independent of each other, as they are coupled by the five rates. In a solar cell, trapped electrons and trapped holes can be emitted by excitation with photons originating from thermal emission or from sunlight, similar to the generation of free electrons and holes due to excitation of electrons from the valence into the conduction band.

A trap state can be charged differently depending on the nature of the trap and on occupation. An acceptor-like trap is negatively charged when it is occupied and neutral when it is unoccupied. *Vice versa* a donor-like trap state is neutral when it is occupied and positively charged when it is unoccupied. This has to be considered for charge neutrality of a PV absorber.

3.4.2 *Emission and capture rates of electrons and holes*

The electron emission rate is proportional to the density of occupied traps whereas the hole emission rate is proportional to the density of unoccupied traps, i.e. the difference between the density of traps and the density of occupied traps. The proportionality factors are the electron and hole emission rate constants, β_e and β_h, respectively.

$$G_e = \beta_e \cdot n_t \tag{3.18'}$$

$$G_h = \beta_h \cdot (N_t - n_t) \tag{3.18''}$$

The capture rate of a free electrons is proportional to the density of free electrons, to the density of unoccupied traps and to the thermal velocity (v_{th}). The capture rate of a free holes is proportional to the density of free holes, to the density of occupied trap states and to v_{th}. The thermal velocity (about 10^7 cm/s at room temperature) has influence on the probability of a free charge carrier to meet a localized trap. The proportionality factors have the unit of the area and are the electron and hole capture cross sections, σ_e and σ_h, respectively. The capture cross sections of traps can be understood as the extension of a trap perpendicular to the direction of motion of a free charge carrier. Capture cross sections are of the order of 10^{-15} cm^2 for traps with states close to the middle of the forbidden band gap.

$$R_e = \sigma_e \cdot v_{th} \cdot (N_t - n_t) \cdot n \tag{3.19'}$$

$$R_h = \sigma_h \cdot v_{th} \cdot n_t \cdot p \tag{3.19''}$$

The emission rates are not independent parameters. They can be obtained if assuming steady state, i.e. when the densities of free holes, free electrons and occupied traps are constant, and in the absence of photogeneration, i.e. Δn is 0. In this case the electron or hole emission rates are equal to the electron or hole capture rates, respectively. The following equations are obtained for β_e and β_h by considering Equations (3.18) and (3.19), and by taking into account the Boltzmann statistics for free electrons and holes and for occupied traps:

$$\beta_e = \sigma_e \cdot v_{th} \cdot N_C \cdot \exp\left(-\frac{E_C - E_t}{k_B \cdot T}\right) \tag{3.20'}$$

$$\beta_h = \sigma_h \cdot v_{th} \cdot N_V \cdot \exp\left(-\frac{E_t - E_V}{k_B \cdot T}\right) \tag{3.20''}$$

Equations (3.20') and (3.20'') show that the emission rates depend exponentially on the energy position of the trap state and on temperature.

The electron or hole emission rates are large for traps with states close to the conduction or valence band edges, respectively.

3.4.3 *Minimum Shockley–Read–Hall recombination lifetime*

The maximum capture rates are reached for electrons or holes when all trap states are unoccupied or occupied, respectively. The minimum SRH lifetimes are obtained for the conditions of the maximum capture rates with respect to Equations (3.2) and (3.19):

$$\tau_{e,\mathrm{SRH}}^{\min} = \frac{1}{\sigma_e \cdot v_{\mathrm{th}} \cdot N_t} \tag{3.21'}$$

$$\tau_{h,\mathrm{SRH}}^{\min} = \frac{1}{\sigma_h \cdot v_{\mathrm{th}} \cdot N_t} \tag{3.21''}$$

The recombination lifetime of free electrons and holes is the sum of the SRH electron and hole lifetimes since both an electron and a hole have to be captured during recombination of an electron–hole pair.

Figure 3.7 shows the dependence of the minimum SRH lifetimes as a function of the density of trap states. The minimum lifetimes are reached

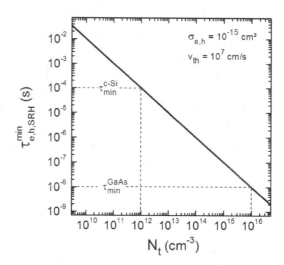

Figure 3.7. Dependence of the minimum SRH lifetime on the density of traps in the bulk under the assumption of equal capture cross sections for electrons and holes with $\sigma_e = \sigma_h = 10^{-15}\ \mathrm{cm}^2$.

for c-Si and GaAs at densities of traps with states in the forbidden gap of about 10^{12} cm^{-3} and 10^{16} cm^{-3}, respectively.

The condition for a minimum density of traps with states in the band gap of a PV absorber, which is established in Figure 3.7, is very harsh for c-Si. It means that only one defect per 10^{11} host Si atoms is acceptable for high-efficiency c-Si solar cells. Extremely pure silicon crystals can be produced by using distillation of precursor molecules and float zone crystal growth (see Chapter 7). For this reason, for example, a highly enriched ^{28}Si crystal has been proposed as a new standard to determine the Avogadro constant and the kilogram definition (Becker *et al.*, 2007).

3.4.4 *Shockley–Read–Hall recombination rate*

The SRH recombination rate can be calculated for a single state in accordance to its energetic position in the forbidden band gap in the case of steady state. In steady state two equations with two unknown variables n_t and G are obtained from Equations (3.17)

$$G = R_e - G_e = R_h - G_h \qquad (3.22)$$

The density of occupied trap is obtained if the equation containing the capture and emission rates of electrons and holes is solved in accordance with Equations (3.18)–(3.20) and (3.22):

$$n_t = \frac{\sigma_e \cdot N_t \cdot n + \sigma_h \cdot N_t \cdot N_V \cdot \exp\left(-\frac{E_t - E_V}{k_B \cdot T}\right)}{\sigma_e \cdot \left[n + N_C \cdot \exp\left(-\frac{E_C - E_t}{k_B \cdot T}\right)\right] + \sigma_h \cdot \left[p + N_V \cdot \exp\left(-\frac{E_t - E_V}{k_B \cdot T}\right)\right]}$$

$$(3.23)$$

The SRH recombination rate (R_{SRH}) is equal to the generation rate in steady state so that R_{SRH} can be calculated considering Equations (3.21), (3.22) and the Boltzmann statistics for intrinsic free charge carriers (2.28):

$$R_{SRH} = \frac{n \cdot p - n_i^2}{\tau_{h,SRH}^{min} \cdot \left[n + N_C \cdot \exp\left(-\frac{E_C - E_t}{k_B \cdot T}\right)\right]} \qquad (3.24)$$
$$+ \tau_{e,SRH}^{min} \cdot \left[p + N_V \cdot \exp\left(-\frac{E_t - E_V}{k_B \cdot T}\right)\right]$$

Equation (3.24) (Sze, 1981) is rather complex. The energy of the trap state and the densities of majority and photogenerated free charge carriers are the most important parameters in Equation (3.24).

In the following, identical capture cross sections ($\sigma_e = \sigma_h = 10^{-15}$ cm^2) are assumed in order to facilitate a simplified analysis. As an example, a p-type c-Si absorber is chosen and the ratio between the density of photogenerated free charge carriers and the density of majority charge carriers is varied. The SRH recombination lifetime is calculated with reference to Equation (3.24) and the definition of the recombination rate. The dependence of the SRH recombination lifetime is plotted in Figure 3.8(a) as a function of E_t for intrinsic c-Si at three different densities of photogenerated free charge carriers.

For intrinsic c-Si, the minimum SRH recombination lifetime is the doubled value of $\tau_{e,h,\text{SRH}}^{\min}$. The lowest SRH recombination lifetime is reached for trap states with energies in the middle of the forbidden band gap independent of the density of majority charge carriers and/or of the density of photogenerated free charge carriers. Trap states around the middle of the band gap are formed in c-Si, for example, by copper, gold or iron impurities ($E_t - E_V = 0.52, 0.54$ and 0.55 eV, respectively, Sze and Irvine, 1968). Incidentally, copper, gold and iron also form other states

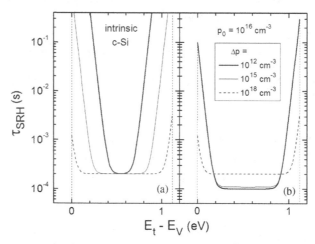

Figure 3.8. SRH recombination lifetime as a function of the energy of the trap state in the forbidden band gap for intrinsic (a) and for p-type c-Si with $p_0 = 10^{16}$ cm^{-3} (b) at densities of photogenerated free charge carriers of 10^{12}, 10^{15} and 10^{18} cm^{-3} (thick solid line, thin solid line and dashed line, respectively).

in addition to the states around the middle of the band gap. Impurities forming trap states around the middle of the band gap of a PV absorber are also called "lifetime killers" since they limit — and do so very efficiently — the lifetime of photogenerated free charge carriers.

The SRH recombination lifetime increases when E_t shift towards the valence or conduction band edges, i.e. decreasing $E_t - E_V$ or with decreasing $E_C - E_t$, respectively. This means that defects with trap states closer to the conduction or valence band edges are less recombination active. However, the SRH recombination lifetime strongly decreases with increasing density of majority charge carriers for defects with trap states close to the conduction or valence band edges (Figure 3.8(b)). This is also the case for increasing density of photogenerated free charge carriers. Therefore, any trap state in the forbidden band gap of a PV absorber becomes a "lifetime killer" at high densities of majority charge carriers and/or of photogenerated free charge carriers.

3.4.5 *Strategies for minimizing Shockley–Read–Hall recombination*

The SRH recombination rate has to be minimized in PV absorbers for reaching high solar energy conversion efficiencies. The SRH recombination rate of photogenerated free charge carriers is proportional to their density. The density of photogenerated charge carriers should be as large as possible in a PV absorber and cannot serve therefore for minimizing SRH recombination.

The capture cross section of a trap state is given by the nature of a given defect and cannot be changed. Therefore, the most efficient way to reduce the SRH recombination rate is the reduction of the density of intrinsic defects and of defects introduced by unwanted impurities. Intrinsic defects can be, for example, host atoms in a crystal, which are displaced from a lattice position to an interstitial position, or atoms which are part of extended structural defects such as dislocations or grain boundaries in crystalline or polycrystalline PV absorbers, respectively.

The generation of intrinsic defects is thermally activated and requires high activation energies of the order of the band gap in semiconductor crystals. For example, massive displacement of silicon atoms in c-Si occurs only at temperatures very close to the melting point of c-Si. The density of intrinsic defects in a c-Si lattice at room temperature can be estimated if is assumed that there is an activation energy of about 1.1 eV and a density of host atoms of about $5 \cdot 10^{22}$ cm^{-3} (with respect to the density of c-Si equal to 2.33 g/cm^3 and a molar mass of Si atoms of 28 g/mol). The expected density of intrinsic point defects in a c-Si lattice is about 19 orders of magnitude less than the density of host atoms. This is extremely small in comparison to the density of free charge carriers.

So-called defect passivation technologies have been developed for numerous cases when the incorporation of unwanted impurities (important, for example, in c-Si) or the generation of intrinsic defects (important, for example, in amorphous silicon) cannot be suppressed sufficiently during the production of the PV absorber. Defect passivation means the transformation of a recombination active defect into an electronically passive defect, i.e. a defect that does not have electronic states in the forbidden band gap. Passivation strategies are explained in Figure 3.9 for bulk defects in PV absorbers.

The first strategy of defect passivation aims to modify the chemical nature of bulk defects. A chemical reaction between a defect and a passivating atom leads to the removal of trap states from the band gap. Simultaneously, electronic bonding and anti-bonding states are formed, the energy of which is inside the valence and conduction bands of the PV absorber or very close to them, respectively (Figure 3.9(a)). Defects can be passivated, for example, with hydrogen (hydrogen passivation), which is decisive especially for passivation of amorphous silicon absorbers (see Chapter 9).

Second, the local density of defects can be reduced by so-called gettering (Figure 3.9(b)). A getterer is additionally introduced into a PV absorber. Getterers serve as sinks for diffusing impurity atoms which form defects with electronic states in the band gap. Related sinks can be formed by atoms, complexes of atoms, precipitates or structural defects. The application of getterers allows the SRH recombination lifetime and therefore the diffusion length of photogenerated free charge carriers to increase in a certain region of a PV absorber. Gettering based on SiO$_2$ precipitates is applied, for example, in silicon crystals produced by Czochralski growth (see Chapter 7).

Recombination active defect states are usually formed along extended structural defects such as dislocations. Third, extended structural defects

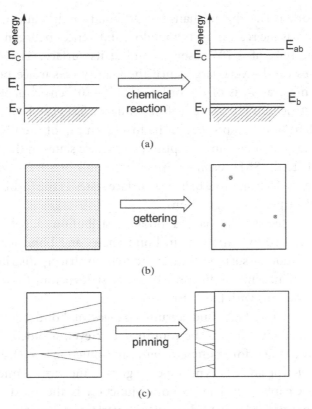

Figure 3.9. Passivation of defects in PV absorbers: (a) removal of trap states from the band gap by chemical reaction, E_b and E_{ab} denote the energies of the bonding and anti-bonding states, respectively, (b) decrease of the local density of defects by gettering and (c) pinning of dislocations in a layer outside the PV absorber.

such as dislocations can be stopped by pinning them, for example, in defect-rich layers outside the active PV absorber layer (Figure 3.9(c)). The pinning of dislocations is a very efficient method to reduce defects in PV absorbers based especially on III-V semiconductors (see Chapter 8).

3.5 Surface Recombination

3.5.1 *Surface recombination velocity*

Surfaces and interfaces are part of solar cells. Well-defined interfaces are required, for example, for charge separation or extraction of charge carriers to external leads. The bond configuration changes abruptly at the surface of a PV absorber. The abrupt change of the bond configuration at surfaces and interfaces is the reason for tremendous differences between electronic states in the bulk of a PV absorber and at its surface. The density of surface states can be very large since the density of surface bonds with changed configuration is of the order of $10^{14}-10^{15}$ cm^{-2} depending on the surface structure. Therefore surfaces and interfaces can be regions with very high densities of trap states in the forbidden gap of the PV absorber. Free electrons and holes can be captured at surface states in the forbidden gap similarly to the SRH recombination in the bulk of a PV absorber. The annihilation of electron–hole pairs via surface states in the forbidden band gap is called surface recombination.

Surface states usually have a broad distribution due to disorder. Disorder is caused by variations in bond angles and bond lengths. The broad distribution of surface states is different to sharply distributed trap states in crystalline semiconductors where trap states belong to well-defined bond configurations with fixed structure.

Disorder at interfaces and abrupt changes of bond configurations between semiconductors can be minimized or even avoided. Defects at interfaces are avoided, for example, at *pn*-homo-junctions which are formed between an *n*-type and a *p*-type doped region of the same semiconductor crystal (see Chapter 4). The *pn*-homo-junction is the most perfect of interfaces in solar cells. The density of interface states is negligible at hetero-junctions between two different semiconductors with the same lattice constants grown by epitaxy on each other (decisive for solar cells with III–V semiconductors, see Chapter 8).

In analogy to SRH, surface states are characterized by the energy and capture and emission rates of electrons and holes (Figure 3.10). Surface states with energy around the middle of the band gap are highly recombination active since the capture rates for both free electrons and free holes are large. Surface states can be donor- or acceptor-like depending whether the occupied state is neutral and the unoccupied state is positively charged (donor-like) or whether the unoccupied state is neutral and the occupied state is negatively charged (acceptor-like).

Photogenerated free electrons and holes have to reach the surface where they can recombine. This means that the surface recombination rate decreases with increasing distance between the surface and the

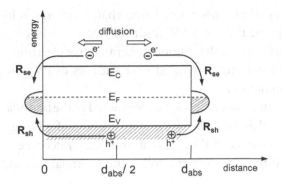

Figure 3.10. Schematic of surface recombination for a symmetric PV absorber.

place where the photogenerated free charge carriers are photogenerated. Therefore, the capture rates of free electrons (R_{se}) and holes (R_{sh}) at surface states are proportional to the half of the reciprocal thickness of a homogeneous PV absorber with two equivalent surfaces.

The capture rates of free electrons or holes are proportional to the densities of free electrons (n_s) or holes (p_s) at the surface, respectively, and to the product of the density of surface states (N_{st}) and thermal velocity (v_{th}). The proportionality factors are, in analogy to SRH recombination, the surface capture cross sections for electrons (σ_{se}) and holes (σ_{sh}):

$$R_{se} = \sigma_{se} \cdot v_{th} \cdot N_{st} \cdot \frac{n_s}{d/2} \tag{3.25$'$}$$

$$R_{sh} = \sigma_{sh} \cdot v_{th} \cdot N_{st} \cdot \frac{p_s}{d/2} \tag{3.25$''$}$$

The product of the surface capture cross section, v_{th} and N_{st} has the unit has the unit of velocity. The electron and hole surface recombination velocities (s_n and s_p, respectively), which are actually capture velocities, are defined as:

$$s_n = \sigma_{se} \cdot v_{th} \cdot N_{st} \tag{3.26$'$}$$

$$s_p = \sigma_{sh} \cdot v_{th} \cdot N_{st} \tag{3.26$''$}$$

The capture rates of free electrons and holes at surface states are traditionally written with the surface recombination velocities as the proportionality factors (see, for details, Eades and Swanson, 1985).

The density of photogenerated free charge carriers is lower near the surface than in the bulk of a PV absorber due to surface recombination. The surface serves as a sink for photogenerated free electrons and holes, i.e. the Fermi-level splitting is reduced near surfaces of PV absorbers due to surface recombination.

Surface recombination has to be considered in the boundary conditions for suitable models describing the distribution of photogenerated free charge carriers in space. If the access of photogenerated free charge carriers is limited by diffusion, then the following boundary conditions can be written:

$$-D_n \cdot \left.\frac{\partial n}{\partial x}\right|_{x=0,d_{abs}} = s_n \cdot n|_{x=0,d_{abs}} \tag{3.27$'$}$$

$$-D_p \cdot \left.\frac{\partial p}{\partial x}\right|_{x=0,d_{abs}} = s_p \cdot p|_{x=0,d_{abs}} \tag{3.27$''$}$$

3.5.2 Surface recombination lifetime

The surface recombination lifetimes of photogenerated free electrons and holes are obtained by using the definition of the lifetime of photogenerated free charge carriers and Equations (3.26$'$) and (3.26$''$):

$$\tau_{se} = \frac{d_{abs}}{2 \cdot s_n} \tag{3.28$'$}$$

$$\tau_{sh} = \frac{d_{abs}}{2 \cdot s_p} \tag{3.28$''$}$$

The surface recombination lifetimes described by Equations (3.28$'$) and (3.28$''$) have the sense of the minimum surface recombination lifetime in analogy to the minimum SRH recombination lifetime. The surface recombination lifetimes are plotted as a function of N_{st} in Figure 3.11 for identical surface capture cross sections for free electrons and holes and for two thicknesses of the absorber.

The minimum lifetime condition is reached for c-Si absorbers with a thickness of $100\,\mu$m for densities of surface defects between 10^9 and 10^{10} cm^{-2}, which corresponds to surface recombination velocities between 10 and 100 cm/s. A density of surface defects below 10^{10} cm^{-2} is very low and demands sophisticated technologies for surface and interface

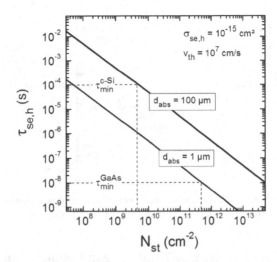

Figure 3.11. Surface recombination lifetimes as a function of the density of surface defects for $\sigma_{s,e} = \sigma_{s,h} = 10^{-15}$ cm^2 and for two thickness (1 and 100 μm) of the PV absorber.

treatment. For a GaAs absorber with the thickness of 1 μm, the minimum lifetime condition is reached for N_{st} below 10^{12} cm^{-2}.

Light trapping can be applied to reduce the thickness of PV absorbers as mentioned earlier. However, surface recombination becomes more crucial with a reduction of d_{abs} and the density of surface defects must be reduced in accordance with the reduction of d_{abs} (see Chapter 4).

3.5.3 Strategies for minimizing surface recombination

The purposeful reduction of the surface recombination rate is called surface passivation. Strategies of surface passivation are directed to the reduction of the density of free charge carriers at the surface as well as of the density of surface defects with states in the forbidden band gap. Surface passivation plays a very important role for solar cells with high solar energy conversion efficiency.

A very effective way of surface passivation is the reduction of the density of surface defects in the forbidden band gap of the PV absorber (see Figure 3.12).

The density of defects can be strongly reduced if surface atoms participate in chemical bonds with electronic states in the valence or conduction band or at least very close to the valence or conduction band edges of the PV absorber. For example, surface atoms can be partially

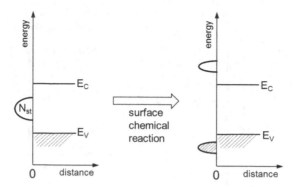

Figure 3.12.   Surface passivation by surface chemical reactions removing surface states from the region around the middle of the band gap of the PV absorber.

oxidized (see, for example, Si/SiO_2 interface in Chapter 7) or unsaturated bonds at interfaces can be passivated with hydrogen.

The total surface recombination rate is proportional to the product of the densities of free electrons and holes at the surface. Therefore the surface recombination rate can be strongly reduced if one of the densities of free electrons or holes is very low at the surface. The reduction of n_S and/or p_S is possible by incorporating barriers repelling free charge carriers from the surface. Barriers can be realized, for example, with doping profiles at the side of the PV absorber (back surface field, see Chapter 7) or with fixed electrostatic charge in an additional layer close to the absorber surface (see, for example, Chapter 7).

Concepts of surface passivation for solar cells are based on the incorporation of so-called passivation layers (Figure 3.13). The aims of a passivation layer are (i) the reduction of the density of recombination-active surface states at the interface between the passivation layer and the PV absorber and (ii) the formation of barriers for photogenerated charge carriers between the PV absorber and the external surface of the passivation layer. The first aim is achieved by choosing the right chemical bond configuration at the interface between the passivation layer and the PV absorber. The second aim is achieved by realizing so-called type I hetero-junctions (see Chapter 8 for details). The passivation layer should be much thicker than the tunneling length (for details on tunneling see Chapter 5).

The barriers for free electrons and holes at interfaces between a passivation layer and PV absorber are given by the differences between

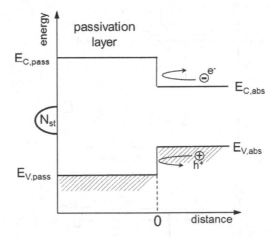

Figure 3.13. Surface passivation with a surface passivation layer avoiding defect states at the interface between the passivation layer and the absorber and avoiding the penetration of free electrons and holes from the PV absorber to the external surface of the passivation layer.

the valence and conduction band edges of the passivation layer ($E_{V,\text{pass}}$ and $E_{C,\text{pass}}$, respectively) and of the PV absorber ($E_{V,\text{abs}}$ and $E_{C,\text{abs}}$, respectively). The following condition should be fulfilled for effective surface passivation:

$$E_{C,\text{pass}} - E_{C,\text{abs}} \gg k_B \cdot T \tag{3.29$'$}$$

$$E_{V,\text{abs}} - E_{V,\text{pass}} \gg k_B \cdot T \tag{3.29$''$}$$

Historically, the most prominent passivation layer is SiO_2 grown on c-Si (see Chapter 7). Practically perfect surface passivation can be reached by epitaxial growth of a passivation layer having the same lattice constant as the crystal of the PV absorber (for example, passivation of a GaAs surface with an $Al_xGa_{1-x}As$ layer, see Chapter 8).

The Fermi-energy of majority charge carriers is contacted to the external leads of a solar cell with ohmic contacts and the surface recombination velocity must be very high at ohmic contacts (see Chapter 5). Therefore, the surface recombination rate can only be reduced by reducing the density of minority charge carriers at the ohmic contact. Barriers for minority charge carriers at ohmic contacts can be differently implemented (Figure 3.14, electrons are the majority charge carriers). The passivation of ohmic contacts is crucial for reaching very high η.

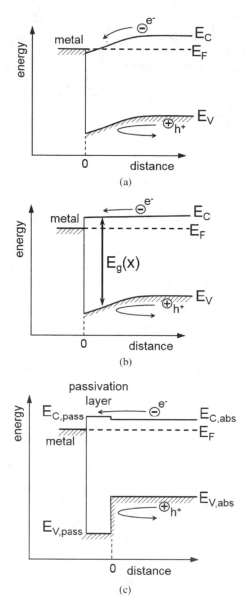

Figure 3.14. Surface passivation of an ohmic contact with a back surface field (a), a graded band gap (b) and a surface passivation layer resulting in a barrier for minority but not for majority charge carriers (c).

At the ohmic contact, the Fermi-energy of majority charge carriers is above the edge of the corresponding band. In the PV absorber, the Fermi-energy is within the band gap. The difference between the band edge of minority charge carriers and the Fermi-energy increases towards the ohmic contact and acts therefore as a barriers for minority charge carriers. This so-called back surface field is widely applied in silicon solar cells (see Chapter 7).

Certain classes of PV absorbers such as chalcogenides (see Chapter 9) cannot be doped by purpose. In quaternary chalcopyrites, for example, a barrier can be introduced for minority charge carriers by increasing the band gap (graded band gap) towards to ohmic contact. The grading of a band gap is realized by a systematic variation of the stoichiometry of the PV absorber.

Hetero-junctions (see Chapter 8) between the PV absorber and a very thin passivation layer are also used for the passivation of ohmic contacts. At the hetero-junction the difference between the band edges of the majority charge carriers should be low enough in order to avoid the formation of a barrier for majority charge carriers but large enough in order to avoid losses in the chemical potential of majority charge carriers. In contrast, the difference between the band edges of the minority charge carriers should be as large as possible to form a barrier for minority charge carriers.

3.6 Summary

The density of charge carriers that are photogenerated in a PV absorber is limited by recombination processes. The recombination rate increases with decreasing lifetime of minority charge carriers in a PV absorber. The minimum lifetime condition has been introduced as a criterion with which to analyze materials requirements for PV absorbers.

The minimum lifetime follows from the demands of strong optical absorption and of sufficient collection of photogenerated free charge carriers. Strong absorption means that d_{abs} should be equal or larger than $3 \cdot \alpha^{-1}$. The thickness of the PV absorber can be reduced by a factor X_{path} giving the ratio by which the optical path is increased under light trapping (see Chapter 2). Sufficient collection of photogenerated free charge carriers means that photogeneration takes place at a distance equal to or shorter than the diffusion length (L) of minority charge carriers. Under these conditions, L is given by the diffusion coefficient (D) and the minimum lifetime (τ_{min}) and the minimum lifetime condition can be written as:

$$\frac{3 \cdot \alpha^{-1}}{X_{path}} = \sqrt{D \cdot \tau_{min}} \tag{3.30}$$

The density of majority charge carriers should be as high as possible since the density of minority charge carriers in thermal equilibrium should be as low as possible for reaching a high Fermi-level splitting in an illuminated PV absorber (see Chapter 2). However, the maximum useful density of majority charge carriers is limited by radiative and/or Auger recombination, which cannot be avoided in PV absorbers.

SRH and surface recombination are proportional to the densities of bulk and surface defects with states in the forbidden band gap of the PV absorber. Related defects have to be minimized by technological measures. Passivation strategies are directed to the reduction of SRH and surface recombination rates by the elimination of corresponding defects. For this purpose bond configurations with defect states in the forbidden band gap can be transformed by chemical reactions into bond configurations without states in the band gap. Further, regions with very low densities of defects can be obtained in PV absorbers by gettering of atoms forming defect states or by pinning of extended defects outside the photoactive part of the absorber.

Figure 3.15 summarizes the lifetimes for the various recombination mechanisms as a function of the density of dopants for c-Si and GaAs absorbers.

The minimum lifetimes are about 10 ns for GaAs) and 100 μs for c-Si absorbers. For a GaAs absorber the density of majority charge carriers is limited by radiative recombination to about $5 \cdot 10^{17}$ cm^{-3}. For a c-Si absorber the density of majority charge carriers is limited by Auger-recombination to about 2–$5 \cdot 10^{16}$ cm^{-3}. For c-Si absorbers the densities of bulk and surface defects with states in the forbidden band gap should be equal or less than 10^{11} cm^{-3} and 10^{10} cm^{-2}, respectively.

The total lifetime follows from the lifetime of all recombination processes in a PV absorber. The reciprocal total lifetime (τ_{total}) is the sum

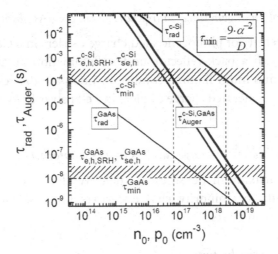

Figure 3.15. Summarized lifetimes for the various recombination mechanisms as a function of the density of dopants.

of the reciprocal lifetimes of all recombination processes since the rates of the different recombination processes have to be added.

$$\frac{1}{\tau_{total}} = \frac{1}{\tau_{rad}} + \frac{1}{\tau_{Auger}} + \frac{1}{\tau_{SRH}} + \frac{2 \cdot s_{n,p}}{d} \qquad (3.31)$$

The total lifetime of photogenerated free charge carriers is limited by the recombination process with the shortest lifetime. The total lifetime of a given PV absorber has to be longer than the minimum lifetime obtained from the minimum lifetime condition.

Strategies and technological measures to minimize SRH and surface recombination are under permanent development in order to reach new record solar energy conversion efficiencies and in order to further reduce the energy required for production of solar cells without losses in the solar energy conversion efficiency.

3.7 Tasks

T3.1: Minimum lifetime and light trapping

Ascertain the minimum lifetimes for c-Si absorbers with a single pass of light, a planar back reflector and a Lambertian back reflector for light trapping.

T3.2: Density of photogenerated charge carriers in a GaAs absorber

Estimate the density of photogenerated charge carriers in a GaAs absorber without and with a back reflector under illumination at AM1.5G with regard to the minimum lifetime condition. Compare the result with the density of photogenerated charge carriers in a c-Si absorber without a back reflector.

T3.3: Density of majority charge carriers in a c-Si absorber

Ascertain the maximum densities of majority charge carriers and the diffusion lengths of minority charge carriers for c-Si absorbers with lifetimes of 7 ms and 7 μs.

T3.4: Intrinsic carrier lifetime

The intrinsic carrier lifetime is related to the hypothetic lifetime, which one would obtain for an absolutely defect-free undoped semiconductor. Calculate the intrinsic carrier lifetimes for c-Si and GaAs.

T3.5: Maximum limit of lifetime

Is it realistic to produce a silicon crystal with a lifetime of minority charge carriers of 157 s? If so, why?

T3.6: Lifetime at the solubility limit of doping

The solubility limit is of the order of 10^{20} cm^{-3} for many dopants. Calculate the lifetime of c-Si and GaAs absorbers at the solubility limit. Get the density of defects in the bulk which may be accepted at the solubility limit with respect to the lifetime, and estimate the related diffusion length.

T3.7: Total lifetime

Calculate the total lifetime of photogenerated electrons in a *p*-type doped c-Si absorber ($p_0 = 3 \cdot 10^{16}$ cm^{-3}, thickness 100 μm) with a density of bulk defects of 10^{12} cm^{-3} ($\sigma_e = 10^{-15}$ cm^2) and a surface recombination velocity of 10 cm/s at both sides of the absorber.

4

Charge Separation Across *pn*-Junctions

The charge-selective contact is the heart of each solar cell since charge separation at the charge-selective contact provides the driving force for directed motion of photogenerated charge carriers. At a charge-selective contact photogenerated charge carriers are separated towards the side with lower potential energy. Junctions between *n*- and *p*-type doped semiconductors (*pn*-junctions) are the most prominent charge-selective contacts since they can be realized as nearly ideal charge-selective contacts. Electrostatic properties such as the width of the space charge regions and the diffusion potential of a *pn*-junction are introduced. Charge-selective contacts are characterized by the diode saturation current density (I_0), which gives the fundamental limitation of the open circuit voltage of an illuminated solar cell. The dependence of I_0 on the density of intrinsic charge carriers, on the density of donors at the *n*-type doped and on the density of acceptors at the *p*-type doped sides of the *pn*-junction as well as the dependence of I_0 on the diffusion coefficients and diffusion lengths of minority charge carriers is obtained. The role of Shockley–Read–Hall recombination at a *pn*-junction is illuminated and considered in the two-diode model with recombination in the neutral regions of the absorber and with recombination in the region of the charge-selective contact. The influence on I_0 of surface recombination concomitant in addition to the thickness of the *n*-type or *p*-type doped semiconductor layers is analyzed. Principles to achieve very high solar energy conversion efficiencies of solar cells are elucidated.

4.1 Concept of the Ideal Charge-Selective Contact

The splitting of the Fermi-energies of electrons and holes in an illuminated photovoltaic absorber is the first prerequisite for getting a photovoltage in a solar cell (see Chapter 2). The second prerequisite is the ability to extract photogenerated electrons from an illuminated photovoltaic (PV) absorber at a high chemical potential (Fermi-energy of electrons, E_{Fn}) and to transfer the electrons back to the PV absorber at a reduced chemical potential (Fermi-energy of holes, E_{Fp}) after passing an external load. The electrons transferred back to the PV absorber recombine with holes.

The extraction of electrons at E_{Fn} is realized by electron separation with charge-selective electron contacts, where holes are extracted by separation with charge-selective hole contacts at E_{Fp}.

Charge separation in a solar cell is a directed transfer of photogenerated charge carriers from a PV absorber across an interface into electronic states of a contacting material (see Figure 4.1). The driving force for a directed charge transfer is given by the decrease of the potential energy of transferred mobile charge carriers in their final state, i.e. transferred photogenerated charge carriers have a lower potential energy after the charge transfer than before.

Transferred photogenerated electrons or holes should remain mobile in the contacting material for getting them out to external leads. Mobile

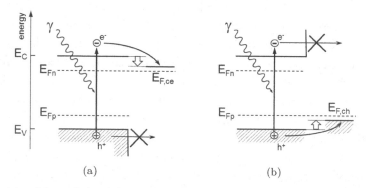

(a)   (b)

Figure 4.1. Schematic band diagrams of ideal charge-selective electron (a) and hole (b) contacts acting as sinks for photogenerated electrons and holes, respectively. The Fermi-energies E_{Fn} and $E_{F,ce}$ (a) and E_{Fp} and $E_{F,ch}$ (b) are aligned at charge-selective electron and hole contacts. The transfer of photogenerated holes or electrons is blocked from entering contacting material of charge-selective electron or hole contacts, respectively.

electrons and holes in the contacting material are characterized by their Fermi-energies ($E_{F,ce}$ and $E_{F,ch}$). The nature of electronic states in the contacting material is not important. Contacting materials can be inorganic or organic semiconductors, metals and electrolytes.

Charge-selective electron or hole contacts act as sinks for photo-generated electrons or holes, respectively. In contrast, the transfer of photogenerated holes or electrons should be blocked from the PV absorber into the contacting materials of charge-selective electron or hole contacts, respectively. The transfer of photogenerated holes or electrons is blocked by barriers.

In ideal charge-selective contacts the (i) potential energy of transferred photogenerated electrons (holes) is reduced, (ii) transfer of photogenerated holes (electrons) is blocked and (iii) $E_{F,ce}$ ($E_{F,ch}$) is aligned with E_{Fn} (E_{Fp}).

Electronic defect states in the forbidden band gap must be avoided at the contact for realizing charge-selective contacts close to ideal. In practicality, this is only possible with *pn*-junctions made in one semiconductor crystal with an extremely low density of defects such as crystalline silicon (c-Si) or gallium arsenide (GaAs) (*pn*-homojunction or *pn*-junction). Charge-selective contacts can also be realized with hetero-junctions between two different semiconductors (hetero-junction, for example, the very well passivated c-Si/a-Si:H hetero-junction) or between semiconductors and metal or electrolyte contacts (Schottky contacts). The most important charge-selective contact is the *pn*-junction.

4.2 The Ideal Semi-infinite *pn*-Junction in Thermal Equilibrium

4.2.1 *Formation of a pn-junction*

A *pn*-junction is formed at the contact between an *n*-type and *p*-type doped semiconductors. If the *n*-type and *p*-type doped semiconductors have the same forbidden band gap, the *pn*-junction is defined as a homo-junction. A homo-junction is considered as metallurgic sharp if inter-diffusion of dopants across the junction can be neglected. In the following, an ideally flat and metallurgically sharp *pn*-junction will be analyzed following Sze (1981).

The density of free electrons is much higher at the *n*-type doped side than at the *p*-type doped side, and similarly the density of free holes is much higher at the *p*-type doped side than at the *n*-type doped side of the *pn*-junction. Therefore gradients of the densities of free electrons and holes exist at a *pn*-junction.

The gradients of the densities of free electrons and holes cause diffusion of free electrons from the *n*-type to the *p*-type doped and of free holes from the *p*-type to the *n*-type doped sides of the *pn*-junction. There are permanent diffusion currents of free electrons and holes across the *pn*-junction (see Figure 4.2). Free electrons and holes diffusing to the *p*-type and *n*-type doped sides of the *pn*-junction, respectively, recombine with the majority charge carriers.

The disappearance of majority charge carriers from the *pn*-junction is called depletion. The ionized donors and acceptors at the *n*-type and *p*-type doped sides of the *pn*-junction are left in the so-called depletion regions. The charge of ionized donors or acceptors is not compensated by free electrons or holes in the depletion regions at the *n*-type and *p*-type doped sides of the *pn*-junction, respectively. Therefore the diffusion of free electrons and holes to opposite sides of the *pn*-junction leads to electrostatic charging of the *pn*-junction. The depletion regions at the *n*-type and *p*-type sides of the *pn*-junction are also known as space charge regions for this reason.

As a result of electrostatic charging, electric fields are formed at a *pn*-junction. Free electrons or holes thermally generated in the space charge regions are accelerated in the electric field towards the *n*-type and *p*-type doped sides of the *pn*-junction, respectively. These drift currents

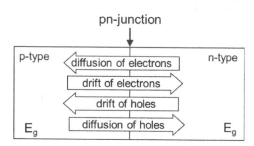

Figure 4.2. Diffusion and drift of electrons and holes across a *pn*-junction in thermal equilibrium.

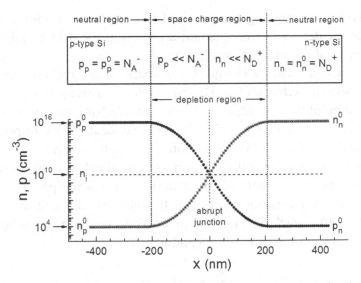

Figure 4.3. Densities of free electrons and holes across a *pn*-junction of c-Si at room temperature for densities of majority carriers of 10^{16} cm^{-3} and extensions of the respective neutral, depletion and space charge regions. The symbols p_p^0, n_p^0, n_n^0 and p_n^0 denote the densities of free holes and electrons at the *p*-type doped side and of the free electrons and holes at the *n*-type doped sides, respectively, in thermal equilibrium.

have the opposite sign of the diffusion currents. The drift currents compensate the diffusion currents so that the net current is 0 in thermal equilibrium.

Figure 4.3 illustrates the densities of free electrons and holes across a *pn*-junction in the case of c-Si for equal densities of donors and acceptors at the *n*-type and *p*-type sides, respectively. The densities of mobile charge carriers and ionized dopants are equal in the neutral regions. Majority charge carriers become minority charge carriers at the *pn*-junction. The densities of free electrons and holes are equal to the density of intrinsic free charge carriers at the *pn*-junction.

4.2.2 *Space charge regions of a pn-junction*

The densities of free charge carriers decrease strongly between the neutral and space charge regions within a very short distance, and so the densities of free charge carriers are much lower than the densities of ionized dopants practically over the entirety of the space charge regions. Therefore the space charge regions can be approximated by layers with certain thicknesses (widths of the space charge regions) and with constant densities of charge equal to the densities of donors and acceptors at the n-type and p-type doped sides of the pn-junction, respectively.

The uncompensated charge in the space charge regions causes an electric field and a change of the electric potential across a pn-junction. The distributions of the electric field (ε) and of the electric potential (φ) can be calculated by integrating the Poisson equation once and twice, respectively. The one-dimensional Poisson equation connects the second derivative of the electric potential over the position (x) with the density of charge (ρ) and the dielectric constant of the material which is the product of the relative dielectric constant (ε_r) and of the dielectric constant in vacuum ($\varepsilon_0 = 8.8 \cdot 10^{-14}$ As/Vcm). The relative dielectric constant ranges between 10 and 12 for most inorganic semiconductors.

$$\frac{\partial^2 \varphi(x)}{\partial x^2} = -\frac{\rho(x)}{\varepsilon_r \cdot \varepsilon_0} \tag{4.1}$$

The density of charge depends upon the position of the charge carriers in relation to the pn-junction. The widths of the space charge regions at the p-type and n-type doped sides of the pn-junction are denoted by x_p and x_n, respectively. The density of charge is equal to 0 in the neutral regions, i.e. when x is less than $-x_p$ or larger than x_n. The density of charge is equal to the negative product of the elementary charge (q) and the density of acceptors in the space charge region at the p-type doped side of the pn-junction, i.e. for x ranging between $-x_p$ and 0. In the space charge region at the n-type doped side of the pn-junction, i.e. for x ranging between 0 and x_n, the density of charge is equal to the product of q and the density of donors (see also Figure 4.4(a)).

$$x < -x_p : \rho = 0 \tag{4.2'}$$

$$x \in (-x_p, 0) : \rho = -q \cdot N_A \tag{4.2''}$$

$$x \in (0, x_n) : \rho = q \cdot N_D \tag{4.2'''}$$

$$x > x_n : \rho = 0 \tag{4.2''''}$$

The electric field is equal to 0 in the neutral regions, and the density of charge is constant in the space charge regions. Therefore, integration of the Poisson equation leads to a linear dependence of the electric field on the

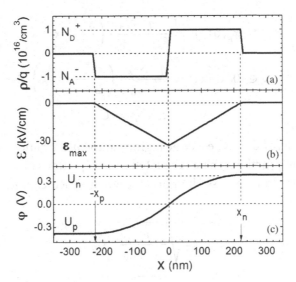

Figure 4.4. Distributions of the density of charge (a), electric field (b) and electric potential (c) across a *pn*-junction in the case of c-Si with $N_D = N_A = 10^{16}$ cm^{-3}.

position in the space charge regions.

$$x < -x_p : \varepsilon = 0 \qquad (4.3')$$

$$x \in (-x_p, 0) : \varepsilon = -\frac{q \cdot N_A}{\varepsilon_r \cdot \varepsilon_0} \cdot (x + x_p) \qquad (4.3'')$$

$$x \in (0, x_n) : \varepsilon = \frac{q \cdot N_D}{\varepsilon_r \cdot \varepsilon_0} \cdot (x - x_n) \qquad (4.3''')$$

$$x > x_n : \varepsilon = 0 \qquad (4.3'''')$$

The electric potential is a smooth function. It is reasonable to set the electric potential to zero at the *pn*-junction. The integration constants are set to $-U_p$ and U_n in the neutral regions at the *p*-type and *n*-type doped sides of the *pn*-junction, respectively. The values of U_p and U_n correspond to the potential drops across the space charge regions at the *p*-type and *n*-type doped sides of the *pn*-junction, respectively. The second integration of the Poisson equation leads to the distribution of the electric potential

$$x < -x_p : \varphi = -U_p \qquad (4.4')$$

$$x \in (-x_p, 0) : \varphi = -\frac{q \cdot N_A}{\varepsilon_r \cdot \varepsilon_0} \cdot \left(\frac{x^2}{2} + x_p \cdot x \right) \qquad (4.4'')$$

$$x \in (0, x_n) : \varphi = \frac{q \cdot N_D}{\varepsilon_r \cdot \varepsilon_0} \cdot \left(\frac{x^2}{2} - x_n \cdot x \right) \qquad (4.4''')$$

$$x > -x_n : \varphi = U_n \qquad (4.4'''')$$

The widths of the space charge regions at the p-type and n-type doped sides of the pn-junction can be calculated from Equations (4.4'') and (4.4''') if the voltage drops across the corresponding space charge regions are known.

$$x_p = \sqrt{\frac{2 \cdot \varepsilon_r \cdot \varepsilon_0}{q \cdot N_A} \cdot U_p} \qquad (4.5')$$

$$x_n = \sqrt{\frac{2 \cdot \varepsilon_r \cdot \varepsilon_0}{q \cdot N_D} \cdot U_n} \qquad (4.5'')$$

Charge neutrality demands that the product of the density of acceptors and the width of the space charge region at the p-type doped side is equal to the product of the density of donors and the width of the space charge region at the n-type doped side of the pn-junction.

$$N_A \cdot x_p = N_D \cdot x_n \qquad (4.6)$$

The widths of the space charge regions and the potential drops across the space charge regions are not independent from each other (see task T4.1 at the end of this chapter).

Figures 4.4(b) and 4.4(c) give an example for the distributions of the electric field and of the electric potential across a pn-junction in the case of c-Si for equal densities of donors and acceptors (10^{16} cm^{-3}). The space charge regions are extended over about 200 nm for the given example. The electric field is negative since it is directed from the negative to the positive charge. The maximum electric filed is reached directly at the pn-junction. The maximum electric field is identical at the p-type and n-type doped sides of the pn-junction (see task T2). For the given example, the maximum electric field is about -34 kV/cm. This is a typical value for electric fields across charge-selective contacts in solar cells. Across the space charge regions the electric potential increases continuously from $-U_p$ in the neutral region of the p-type doped side to U_n in the neutral region at the n-type doped side of the pn-junction.

The width of the space charge region decreases with increasing density of donors or acceptors with regard to Equations (4.5′) and (4.5″). Depending on the densities of dopants at both sides of the pn-junction, the potential drop across a space charge can vary significantly.

Figure 4.5 shows the dependence of the width of the space charge region as a function of the density of donors or acceptors for a semiconductor with a relative dielectric constant of 12 and for different values of U_p or U_n.

The densities of donors or acceptors amount to the order of $5 \cdot 10^{16}$ cm^{-3} or $5 \cdot 10^{17}$ cm^{-3} in solar cells based on c-Si or GaAs absorbers, respectively (see Chapter 3). Therefore the width of the space charge region is of the order of 100–200 nm or 30–50 nm for pn-junctions in solar cells based on c-Si or GaAs absorbers, respectively. These widths are much less than the corresponding absorption lengths of c-Si or GaAs (see Chapter 2). It can be concluded that more than 90% or 80% of the incoming photons are not absorbed within the space charge regions in solar cells based on c-Si or GaAs absorbers.

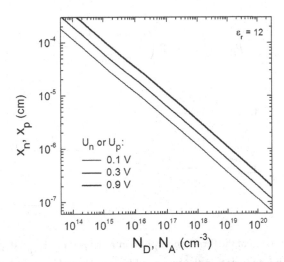

Figure 4.5. Dependence of the width of the space charge region on the density of donors and density of acceptors for a semiconductor with a relative dielectric constant equal to 12 and for different values of the potential drops across the space charge region.

The width of the space charge region decreases to values of the order of only 2–4 nm for highly doped semiconductors. Space charge regions with widths this small are not suitable charge-selective contacts, but they are very important for the formation of ohmic contacts (see Chapter 5).

4.2.3 Diffusion potential of a pn-junction

The Fermi-energy has to be constant across a *pn*-junction in thermal equilibrium since there is no net current flow.

$$E^0_{F(p)} = E^0_{F(n)} = E_{F0} \tag{4.7}$$

The Fermi-energy is closer to the valence band edge in the neutral region at the *p*-type doped side ($E^0_{F(p)}$) and closer to the conduction band edge in the neutral region at the *n*-type doped side ($E^0_{F(n)}$) of the *pn*-junction (see Figure 4.6).

The difference between $E^0_{F(p)}$ and the valence band edge in the neutral region at the *p*-type doped side and the difference between the conduction band edge and $E^0_{F(n)}$ at the *n*-type doped side of the *pn*-junction are calculated by the Boltzmann equations (2.30′) and (2.30″). The Boltzmann equations for the densities of majority charge carriers are in the neutral regions (p^0_p and n^0_n) are:

$$p^0_p = N_V \cdot \exp\left(-\frac{E_{F0} - E_{V(p)}}{k_B \cdot T}\right) \tag{4.8'}$$

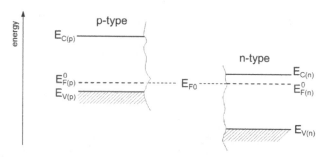

Figure 4.6. Alignment of the Fermi-energy of contacted *p*-type ($E^0_{F(p)}$) doped and *n*-type ($E^0_{F(n)}$) doped semiconductors in thermal equilibrium. The valence and conduction band edges at the *p*-type and *n*-type doped sides are denoted by $E_{V(p)}$, $E_{C(p)}$, $E_{V(n)}$ and $E_{C(n)}$, respectively.

$$n_n^0 = N_C \cdot \exp\left(-\frac{E_{C(n)} - E_{F0}}{k_B \cdot T}\right) \tag{4.8''}$$

The respective densities of the minority charge carriers in the neutral regions (n_p^0 and p_n^0) can be written as:

$$p_n^0 = N_V \cdot \exp\left(-\frac{E_{F0} - E_{V(n)}}{k_B \cdot T}\right) \tag{4.9'}$$

$$n_p^0 = N_C \cdot \exp\left(-\frac{E_{C(p)} - E_{F0}}{k_B \cdot T}\right) \tag{4.9''}$$

The energy of an electron in the conduction band in the neutral region at the p-type doped side is higher than the energy of an electron in the neutral region at the n-type doped side of the pn-junction due to the fact that the Fermi-energy is constant in thermal equilibrium. The kinetic energy of free electrons is much smaller than the forbidden band gap of the semiconductor. The kinetic energy of free electrons is equal in the neutral regions at the p-type and n-type doped sides of the pn-junction. Therefore the difference between the energies of the conduction band edge in the neutral regions at the p-type and n-type doped sides of the pn-junction corresponds to the difference of the potential energy of free electrons at both sides.

The potential energy of an electron decreases with increasing electrostatic potential due to the definition of the direction of the electric field. The potential energy of a free electron in the conduction band therefore follows exactly the negative electrostatic potential across a pn-junction (see Figure 4.7).

The difference between the conduction band edges in the neutral regions at the p-type ($E_{C(p)}$) and n-type ($E_{C(n)}$) doped sides of the pn-junction is caused by the diffusion of majority charge carriers during the formation of the pn-junction. The diffusion potential (U_D) of a pn-junction is defined as the difference of $E_{C(n)}$ and $E_{C(p)}$ divided by q. The value of U_D is equal to the negative sum of the potentials drops across the space charge regions of a pn-junction.

$$U_D \equiv \frac{E_{C(p)} - E_{C(n)}}{q} = -(U_p + U_n) \tag{4.10}$$

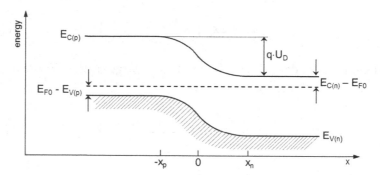

Figure 4.7. Schematic band diagram of a *pn*-junction in thermal equilibrium. U_D denotes the diffusion potential of a *pn*-junction.

The diffusion potential of a *pn*-junction can be obtained if the band gap of the semiconductor (E_g) and the densities of the acceptors and donors at the *p*-type and *n*-type sides of the *pn*-junction are known. According to Figure 4.7:

$$U_D = \frac{1}{q} \cdot [E_g - (E_{F0} - E_{V(p)}) - (E_{C(n)} - E_{F0})] \tag{4.11}$$

Equation (4.11) can be transformed by using the Boltzmann equations (2.30′) and (2.30″) and the definition of the density of intrinsic charge carriers (Equation (2.32)) and by considering complete ionization of donors and acceptors, i.e. p_p^0 and n_n^0 are equal to N_A and N_D, respectively.

$$U_D = \frac{k_B \cdot T}{q} \cdot \ln \frac{N_A \cdot N_D}{n_i^2} \tag{4.12}$$

The open-circuit voltage (V_{OC}) of a solar cell cannot exceed the U_D, i.e. U_D is the upper limit for V_{OC}. Equation (4.12) shows that the higher N_A and/or N_D the larger U_D. However, radiative or Auger recombination limit the useful doping range of a PV absorber (see Chapter 3). Furthermore, U_D is larger for *pn*-junctions based on semiconductors with lower n_i, i.e. with larger values of the E_g. However, as already mentioned, the value of the E_g limits the maximum short circuit current density (see Chapter 2).

The temperature dependence of V_{OC} of a solar cell with a *pn*-junction is dominated by the temperature dependence of U_D. Figure 4.8 shows the

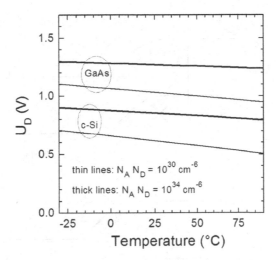

Figure 4.8. Temperature dependence of the U_D for ideal c-Si and GaAs solar cells with $N_A \cdot N_D$ equal to 10^{30} and 10^{34} cm^{-3}.

temperature dependence of U_D for ideal c-Si and GaAs solar cells with two different products of N_A and N_D. The U_D of a pn-junction decreases with increasing temperature. For a given semiconductor, the value of U_D is less reduced with increasing temperature if the product of N_A and N_D is increased.

4.3 Collection of Photogenerated Charge Carriers at a *pn*-Junction

4.3.1 *pn-Junction as an ideal charge-selective electron and hole contact*

Photogeneration takes place at the n-type doped and p-type doped sides of a solar cell with a pn-junction as the charge-selective contact. Electrons are minority charge carriers at the p-type doped side. The potential energy of electrons photogenerated at the p-type doped side (Figure 4.9(a)) is higher than the potential energy of electrons at the n-type doped side where the electrons are majority charge carriers. As a consequence, electrons photogenerated at the p-type doped side is transferred to the n-type doped side where the Fermi-energy of the charge-selective electron contact corresponds to the Fermi-energy of electrons at the n-type doped side of

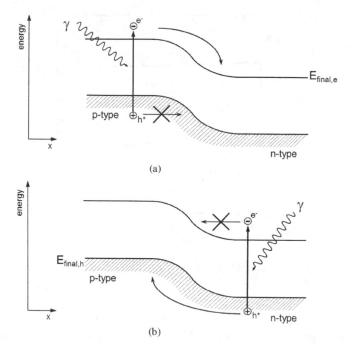

Figure 4.9. Separation of photogenerated electrons and holes at a *pn*-junction.

the *pn*-junction. The potential energy of holes photogenerated at the *p*-type doped side is lower than the potential energy of holes at the *n*-type doped side. This potential barrier rejects the transfer of holes photogenerated at the *p*-type doped side to the *n*-type doped side. Therefore the *n*-type doped side of a *pn*-junction is a charge-selective contact for electrons photogenerated at the *p*-type doped side.

The potential energy of holes photogenerated at the *n*-type doped side of the *pn*-junction is higher than the potential energy of holes at the *p*-type doped side. Therefore holes photogenerated at the *n*-type doped side can be easily transferred to the *p*-type doped side of the *pn*-junction (Figure 4.9(b)). In contrast, the potential energy of electrons photogenerated at the *n*-type doped side of the *pn*-junction is lower than the potential energy of electrons at the *p*-type doped side, i.e. the transfer of electrons photogenerated at the *n*-type doped side to the *p*-type doped side is blocked. Consequently, the *p*-type doped side is a charge-selective contact for holes photogenerated at the *n*-type doped side of the *pn*-junction.

The majority charge carriers electrically contact the photogenerated minority charge carriers at the opposite side of the *pn*-junction. Therefore, the majority charge carriers probe the densities of the minority charge carriers at the opposite side of the *pn*-junction, which is equivalent to the contact of the chemical potential of the minority charge carriers, i.e. their Fermi-levels. Ideal *pn*-junctions are ideal charge-selective contacts since the potential energy of the blocking barrier is much larger than the thermal energy of photogenerated charge carriers.

4.3.2 *Charge carriers photogenerated in a space charge region*

A proportion of photons from sunlight is absorbed in space charge regions of charge-selective contacts. Mobile charge carriers photogenerated in a space charge region are driven by the electric field. The maximum transit time of a mobile charge carrier photogenerated in the electric field of a *pn*-junction can be estimated by taking into account the potential drop across a space charge region, the extension of a space charge region and the drift mobility of photogenerated charge carriers. For example, the electron drift mobility (μ_e) is of the order of 1000 cm^2/Vs (Jacoboni *et al.*, 1977).

The electron drift mobility is defined as the proportionality factor between the electron drift velocity (v_e) and the electric field.

$$v_e \equiv \mu_e \cdot \varepsilon \tag{4.13}$$

In the simplest case, v_e can be estimated as the width of the space charge region divided by the transit time (t_{tr}). The value of $t_{tr,e}$ can be estimated for electrons photogenerated in the space charge region at the *p*-type doped side of the *pn*-junction by using the following equation:

$$t_{tr,e} \approx \frac{x_p^2}{\mu_e \cdot U_p} \tag{4.14}$$

The resulting value of $t_{tr,e}$ as well as of $t_{tr,h}$ is of the order of one ps for c-Si solar cells and even less for GaAs solar cells. The lifetimes of photogenerated free charge carriers are much longer than 1 ps in c-Si or GaAs solar cells. Therefore, all charge carriers photogenerated in the space charge region of a *pn*-junction are collected.

4.3.3 *Charge carriers photogenerated in a neutral region*

The space charge region of a *pn*-junction is usually much thinner than the absorption length for most photons in the spectrum of sunlight. Therefore, the vast majority of excess charge carriers is photogenerated in the neutral regions of a solar cell. For example, less than 10% or 20% of the maximum photocurrent is photogenerated in c-Si or GaAs solar cells within a space charge region with a thickness of 100 or 30 nm, respectively. Therefore, the collection of photogenerated charge carriers is mainly limited by diffusion of minority charge carriers.

Minority charge carriers photogenerated in the neutral regions diffuse towards the *pn*-junction and disappear when reaching the space charge region. The relation between the absorption length (α^{-1}) and the diffusion lengths of minority charge carriers ($L_{n,p}$) is crucial for the collection of photogenerated free charge carriers and has been considered for the definition of the minimum lifetime condition (see Chapter 3). However, the minimum lifetime condition does not give information about losses of carrier collection.

The dependence of the photocurrent on the ratio between α^{-1} and $L_{n,p}$ will be derived for dominant light absorption in the neutral region at the *p*-type doped side of a *pn*-junction (see also Gärtner, 1959). The thickness of a PV absorber is given by the thicknesses of the *p*-type (H_p) and *n*-type (H_n) doped regions. It is assumed in the following that H_n as well as x_n (or transparent *n*-type doped side) and x_p are much thinner than α^{-1} and that H_p is much thicker than α^{-1} (see Figure 4.10).

The density of photogenerated electrons is denoted by Δn. The stationary diffusion equation of electrons photogenerated in the neutral region of the *p*-type doped side of the *pn*-junction and the boundary conditions can be written in the following way:

$$0 = D_n \cdot \frac{\partial^2 (\Delta n)}{\partial x^2} + G(x) - R_n \tag{4.15}$$

$$\Delta n|_{x=x_p} = 0 \tag{4.16'}$$

$$\Delta n|_{x \to \infty} = 0 \tag{4.16''}$$

The photogeneration rate $G(x)$ at the depth x is proportional to the photon flux from the sun (Φ_{sun}) and the part of photons that are not

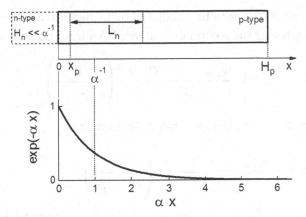

Figure 4.10.   Diffusion length and absorption length at a *pn*-junction with a transparent *n*-type doped side.

reflected, and to the amount of transmitted photons

$$G(x) = \Phi_{sun} \cdot (1 - R) \cdot \alpha \cdot \exp(-\alpha \cdot x) \qquad (4.17)$$

The recombination rate (R_n) is given by definition (3.2). The product of the diffusion coefficient of electrons (D_n) and the electron lifetime can be replaced by the diffusion length of electrons (L_n). Equation (4.15) can be rewritten as

$$0 = \frac{\partial^2 (\Delta n)}{\partial x^2} + \frac{\Phi_{sun} \cdot (1 - R) \cdot \alpha \cdot \exp(-\alpha \cdot x)}{D_n} - \frac{\Delta n}{L_n^2} \qquad (4.18)$$

The approach for the solving this differential equation is

$$\Delta n = A1 \cdot \exp(-\alpha \cdot x) + A2 \cdot \exp\left(-\frac{x}{L_n}\right) + A3 \cdot \exp\left(\frac{x}{L_n}\right) \qquad (4.19)$$

The constant A3 is equal to 0 with respect to boundary condition (4.16″). The constant A2 is obtained from the boundary condition (4.16′):

$$A2 = -A1 \cdot \exp(-\alpha \cdot x_p) \cdot \exp\left(\frac{x_p}{L_n}\right) \qquad (4.20)$$

The constant A1 has to be calculated by taking into account Equation (4.20) and by taking the second derivative of Δn from Equation (4.18).

$$A1 = \frac{\Phi_{sun} \cdot (1 - R) \cdot \alpha}{D_n} \cdot \left(\frac{1}{\alpha^2 - \frac{1}{L_n^2}} \right) \tag{4.21}$$

The density of the photogenerated electrons is

$$\Delta n = \frac{\Phi_{sun} \cdot (1 - R) \cdot \alpha}{D_n} \cdot \left(\frac{1}{\alpha^2 - \frac{1}{L_n^2}} \right)$$
$$\cdot \left[\exp(-\alpha \cdot x) - \exp\left(-\alpha \cdot x_p + \frac{x_p - x}{L_n} \right) \right] \tag{4.22}$$

The diffusion current densities of photo-generated electrons (I_n) and holes (I_p) are defined as:

$$I_n = -q \cdot D_n \cdot \frac{\partial(\Delta n)}{\partial x} \tag{4.23}$$

$$I_p = -q \cdot D_p \cdot \frac{\partial(\Delta p)}{\partial x} \tag{4.23'}$$

The I_n is calculated following Equation (4.23), by taking into account Equation (4.22) and setting x equal to x_p:

$$I_n = q \cdot \Phi_{sun} \cdot (1 - R) \cdot \exp(-\alpha \cdot x_p) \cdot \left(\frac{1}{\frac{\alpha^{-1}}{L_n} + 1} \right) \tag{4.24}$$

Incidentally, solution (4.24) is the part of the diffusion current density related to the neutral region in depletion layer photoeffects (Gärtner, 1959). The last term in Equation (4.24) describes the dependence of I_n on the ratio between the absorption length (α^{-1}) and the diffusion length of electrons photogenerated in the neutral region at the p-type doped side of the pn-junction (L_n). The last term in Equation (4.24) is very important since it reflects the collection losses of photogenerated charge carriers for any type of solar cell. It shows, for example, that only half of the possible photocurrent is reached at the minimum carrier lifetime condition.

The dependence of the last term in Equation (4.24) is plotted as a function of α^{-1}/L_n in Figure 4.11. It can be seen that the L_n should exceed the α^{-1} by a factor of 10 or 100 in order to achieve photocurrents above 90%

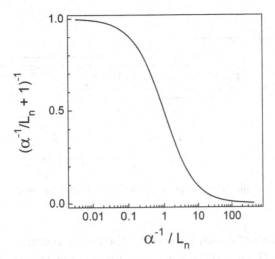

Figure 4.11. Plot of the reciprocal sum of 1 and the ratio between the absorption length (α^{-1}) and the diffusion length of photogenerated minority charge carriers (L_n) as a function of α^{-1}/L_n.

or 98% of the maximum photocurrent, respectively. Therefore, c-Si wafers with a diffusion length of even more than 1 mm are needed for the realization of c-Si solar cells with very high solar energy conversion efficiency.

4.4 Diode Saturation Current Density of a *pn*-Junction

4.4.1 *Fermi-level splitting and density of minority charge carriers*

Fermi-level splitting occurs under illumination at both the *n*-type doped and *p*-type doped sides of a *pn*-junction. For the minority charge carriers the Fermi-energies change strongly, in contrast to the very small change of the Fermi-energies of the majority charge carriers. The Fermi-energies of the minority electrons and holes at the *p*-type doped and *n*-type doped sides of the *pn*-junction are denoted by $E_{Fn(p)}$ and $E_{Fp(n)}$, respectively. The Fermi-energies of the majority electrons and holes at the *n*-type doped and *p*-type doped sides of the *pn*-junction are denoted by $E_{Fn(n)}$ and $E_{Fp(p)}$, respectively. At the *pn*-junction, the Fermi-energies of the majority charge carriers contact the Fermi-energies of the minority charge carriers

$$E_{Fn(n)}q = E_{Fn(p)} \tag{4.25'}$$

$$E_{Fp(p)} = E_{Fp(n)} \tag{4.25''}$$

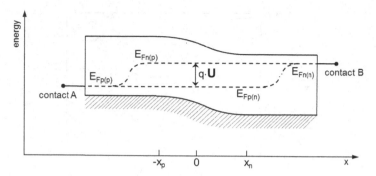

Figure 4.12. Schematic band diagram of an illuminated *pn*-junction with external contacts.

External leads of a solar cell contact the Fermi-energies of the majority charge carriers (Figure 4.12). For ideal ohmic contacts (see Chapter 5) the Fermi-energies in the external leads are identical to the Fermi-energies of the majority charge carriers. The difference between the Fermi-energies in two external leads is measured as a potential energy, i.e. as the voltage drop between two wires (U) multiplied by q. Therefore the photovoltage measured between the external leads of an illuminated solar cell with an ideal *pn*-junction can be written as

$$q \cdot U \equiv E_{\text{Fn}(n)} - E_{\text{Fp}(p)} \qquad (4.26)$$

The Fermi-energies of the minority charge carriers can be expressed by the Fermi-energies of the majority charge carriers and U by using Equations (2.25) and (2.26):

$$E_{\text{Fn}(p)} = E_{\text{Fp}(p)} + q \cdot U \qquad (4.27')$$

$$E_{\text{Fp}(n)} = E_{\text{Fn}(n)} - q \cdot U \qquad (4.27'')$$

The densities of the minority charge carriers in the illuminated *pn*-junction are, according to Equations (4.9) and (4.27):

$$n_p = n_p^0 \cdot \exp\left(\frac{q \cdot U}{k_{\text{B}} \cdot T}\right) \qquad (4.28')$$

$$p_n = p_n^0 \cdot \exp\left(\frac{q \cdot U}{k_{\text{B}} \cdot T}\right) \qquad (4.28'')$$

Therefore the densities of the photogenerated minority charge carriers are:

$$\Delta n_p = n_p^0 \cdot \left[\exp\left(\frac{q \cdot U}{k_B \cdot T} \right) - 1 \right] \qquad (4.29')$$

$$\Delta p_n = p_n^0 \cdot \left[\exp\left(\frac{q \cdot U}{k_B \cdot T} \right) - 1 \right] \qquad (4.29'')$$

4.4.2 Diode saturation current density of a semi-infinite pn-junction

An illuminated solar cell is operated under steady state condition. The diffusion coefficients and the recombination rates of the minority charge carriers are denoted by D_n, D_p, R_p and R_n, respectively. Here the steady state diffusion equations are considered for minority charge carriers injected at an external forward potential:

$$0 = D_p \cdot \frac{\partial^2 (\Delta p_n)}{\partial x^2} - R_p \qquad (4.30')$$

$$0 = D_n \cdot \frac{\partial^2 (\Delta n_p)}{\partial x^2} - R_n \qquad (4.30'')$$

The diffusion equations (4.30') and (4.30'') can be transformed by taking into account the definitions of the recombination rate (3.2) and of the diffusion length (3.5).

$$\frac{\Delta p_n}{L_p^2} = \frac{\partial^2 (\Delta p_n)}{\partial x^2} \qquad (4.31')$$

$$i\frac{\Delta n_p}{L_n^2} = \frac{\partial^2 (\Delta n_p)}{\partial x^2} \qquad (4.31'')$$

The density of diffusing minority charge carriers becomes 0 behind the pn-junction since the pn-junction acts as an ideal sink. The diffusion equations (4.31') and (4.31'') can be solved with the following approach:

$$\Delta p_n = B_p \cdot \exp\left(-\frac{x}{L_p} \right) \qquad (4.32')$$

$$\Delta n_p = B_n \cdot \exp\left(-\frac{x}{L_n} \right) \qquad (4.32'')$$

The constants B_p and B_n in Equations (4.32') and (4.32'') are obtained from the densities of minority charge carriers at the *pn*-junction at $x = 0$, and Δp_n and Δn_p are given by Equations (4.29') and (4.29''), respectively.

The diffusion current density of the electrons (I_n) is given by Equation (4.23). For getting the diffusion current density of holes (I_p), D_n and Δn_p should be replaced by D_p and Δp_n in Equation (4.23), respectively.

The I_n and I_p are obtained from Equations (4.23) and (4.23') by introducing the first derivative of Equations (4.32') and (4.32'') and using Equations (4.29') and (4.29''). Furthermore, it is considered that the diffusion current is independent of x and that the net current density (I) is the sum of I_p and I_n. In addition, the density of the minority charge carriers can be expressed by the density of the majority charge carriers in in thermal equilibrium (Equation (2.31)). The densities of the majority charge carriers are equal to the densities of donors and acceptors if assuming complete ionization. The following final expression is obtained for the current density across the semi-infinite *pn*-junction under forward bias:

$$I = I_0 \cdot \left[\exp\left(\frac{q \cdot U}{k_B \cdot T} \right) - 1 \right] \tag{4.33}$$

Equation (4.33) is the diode equation of the ideal *pn*-junction with I_0 given by

$$I_0 = I_{0,n} + I_{0,p} = q \cdot n_i^2 \cdot \frac{D_n}{L_n \cdot N_A} + q \cdot n_i^2 \cdot \frac{D_p}{L_p \cdot N_D} \tag{4.34}$$

The I_0 limits V_{OC} (see Chapter 1) and should be therefore as low as possible. Equation (4.34) is the simplest model for a I_0 of a solar cell and already depends on eight parameters: temperature and E_g, densities of donors and acceptors at the *n*-type doped and *p*-type doped sides of the *pn*-junction and the diffusion coefficients and diffusion lengths of photogenerated minority charge carriers at the *n*-type doped and *p*-type doped sides of the *pn*-junction. The parameters are not independent from each other. For example, an increase of N_A or N_D leads to a decrease of L_n or L_p, respectively. The interdependence of other parameters makes the minimization of I_0 yet more complex.

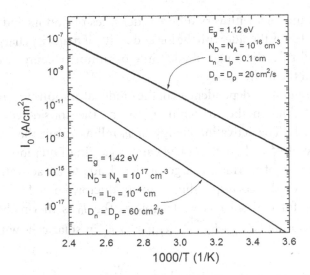

Figure 4.13. Arrhenius plots of the I_0 of two solar cells with fixed band gaps of 1.12 and 1.42 eV. Densities of donors and acceptors of 10^{17} cm^{-3} or 10^{16} cm^{-3}, diffusion lengths of photogenerated charge carriers of 1 μm or 1 mm and diffusion coefficients of 60 cm^2/s or 20 cm^2/s were assumed for the solar cell with the band gap of 1.42 eV or 1.12 eV, respectively.

Figure 4.13 gives an impression about possible orders of magnitude of I_0 for GaAs and c-Si solar cells. The I_0 of a semi-infinite pn-junction is plotted following Equation (4.34) as a function of the reciprocal temperature for two approximate sets of parameters equal for both electrons and holes.

The diffusion lengths of minority charge carriers are kept at only 1 μm for the considered GaAs solar cell. The value of I_0 at room temperature is about $3 \cdot 10^{-18}$ A/cm^2 for the considered GaAs solar cell, which is larger by more than one order of magnitude in comparison to the I_0 estimated for a record GaAs solar cell (see Equation (1.22'')).

Relatively long diffusion lengths of minority charge carriers of 1 mm are chosen for the considered c-Si solar cell while the densities of donors and acceptors are kept at only 10^{16} cm^{-3}. The chosen parameters result in an I_0 at room temperature of about $3 \cdot 10^{-13}$ A/cm^2, which is about 3 times larger than the estimated value of I_0 of the world record c-Si solar cell (see Equation (1.22')). The value of I_0 would be three times less if the densities of donors and acceptors were increased by 3 times.

In *pn*-junctions with very different densities of donors and acceptors, the I_0 is limited by the side with the lower density of majority charge carriers. As a general rule, the longer the lifetime of minority charge carriers, the lower I_0 and the higher V_{OC}.

The temperature dependencies of the diode saturation current densities are straight lines in the Arrhenius plots for the considered GaAs and c-Si solar cells. The activation energy of I_0 follows from the temperature dependence of n_i^2 in Equation (4.34) and regarding Equation (2.32). The squared density of intrinsic charge carriers leads to an activation energy equal to E_g of the corresponding semiconductor, i.e. 1.42 eV for GaAs and 1.12 eV for c-Si solar cells. Therefore, the I_0 is thermally activated with an activation energy $E_A^{\text{ideal-}pn}$ equal to E_g in solar cells with an ideal *pn*-junction.

$$E_A^{\text{ideal-}pn} = E_g \qquad (4.35)$$

It is worth noting that the conclusion expressed in (4.35) follows from the condition that the thermal generation rate is only balanced by radiative recombination in the thermal equilibrium (see Chapter 2).

4.4.3 *Shockley–Read–Hall recombination in a pn-junction*

PV absorbers are not perfect, they can have defect states around the middle of the forbidden band gap for which Shockley–Read–Hall (SRH) recombination rates are large (see Chapter 3). The SRH recombination rate (R_{SRH}) is given by the rather complex Equation (3.24). For the analysis of the influence of SRH recombination on the I_0 of a *pn*-junction some simplifying assumptions are introduced following Goetzberger *et al.* (1997) (see Figure 4.14).

First, it is assumed that the energy of the defect states is equal to the Fermi-energy of the intrinsic semiconductor.

$$E_t = E_i \qquad (4.36)$$

Second, similar effective densities of states at the valence and conduction band edges are assumed.

$$N_C = N_V \qquad (4.37)$$

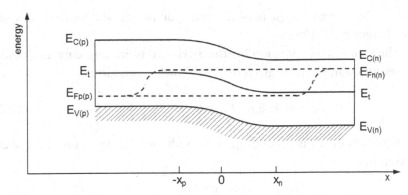

Figure 4.14. Schematic energy band diagram of a *pn*-junction with defect states at the middle of the forbidden band gap in non-equilibrium.

Third, it is assumed that the minimum SRH lifetimes are equal for electrons and holes.

$$\tau_{e,SRH}^{min} = \tau_{h,SRH}^{min} = \tau_0 \tag{4.38}$$

The following expression is obtained for R_{SRH} if taking into account Equation (3.24), the simplifications (4.36)–(4.38), the Boltzmann equation (2.30) and the definition of the density of intrinsic charge carriers (2.32).

$$R_{SRH} = \frac{(n_0 + \Delta n) \cdot (p_0 + \Delta n) - n_i^2}{\tau_0 \cdot (n_0 + p_0 + 2 \cdot \Delta n + 2 \cdot n_i)} \tag{4.39}$$

The maximum SRH recombination rate at the *pn*-junction is obtained from Equation (4.39) by transforming it with the binomial equation:

$$R_{SRH,pn}^{max} = \frac{n - n_i}{2 \cdot \tau_0} \tag{4.40}$$

The density of photogenerated electrons at the *pn*-junction corresponds to the half of the Fermi-level splitting under illumination. Therefore Equation (4.40) can be transformed to:

$$R_{SRH,pn}^{max} = \frac{n_i}{2 \cdot \tau_0} \cdot \left[\exp\left(\frac{q \cdot U}{2 \cdot k_B \cdot T} \right) - 1 \right] \tag{4.41}$$

SRH leads to a current density which is the product of the R_{SRH}, q and the thickness of the region where the SRH recombination rate is large

(δ_{SRH}). The value of δ_{SRH} is of the order of 10 nm (for more details see Goetzberger *et al.*, 1997).

The simplified expression for the SRH recombination current density of a *pn*-junction (I_{SRH}) is given by the following equation:

$$I_{SRH} = I_{0,SRH} \cdot \left[\exp \left(\frac{q \cdot U}{2 \cdot k_B \cdot T} \right) - 1 \right] \qquad (4.42)$$

Equation (4.42) is a diode equation with the ideality factor of 2 and an I_0 given by

$$I_{0,SRH} = \frac{q \cdot \delta_{SRH}}{2 \cdot \tau_0} \cdot n_i \qquad (4.43)$$

The diode saturation current density for SRH recombination is of the order of 10^{-12}–10^{-8} A/cm^2 for c-Si solar cells with respective SRH lifetimes in the c-Si absorber (τ_0 between 1 μs and 10 ms). The value of $I_{0,SRH}$ can even be much larger if the charge-selective contact is formed, for example, at a hetero-junction with a very high density of defects at the interface. Further $I_{SRH,0}$ is proportional to n_i and therefore has an activation energy of:

$$E_A^{SRH,pn} = \frac{E_g}{2} \qquad (4.44)$$

Incidentally, defect states causing SRH recombination are important for any charge-selective contact independent on whether they are formed by *pn*-junctions or hetero-junctions. Equations (4.42) and (4.43) describe a rather simplified case. SRH recombination can be distinguished experimentally by analyzing the ideality factor and the thermal activation energy of the I_0. In reality the ideality factors of c-Si solar cells, for example, are usually between 1.1 and 1.3.

4.4.4　*The two-diode model of an illuminated pn-junction*

A part of photogenerated charge carriers recombines in the *n*-type and *p*-type doped neutral regions while the other part of photogenerated charge carriers recombines via deep defect states at the *pn*-junction of a solar cell. Therefore both recombination processes are considered in the equivalent circuit of a solar cell by two diodes connected in parallel (Figure 4.15). The diode saturation current densities of the two-diode model of a solar

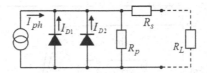

Figure 4.15. Equivalent circuit of the two-diode model of a solar cell.

cell are limited by so-called bulk recombination in the neutral regions (I_0) given by Equation (4.34) for a semi-infinite *pn*-junction and by SRH recombination at the charge-selective contact ($I_{0,\text{SRH}}$) given in a simplified form by Equation (4.43) for a *pn*-junction.

$$-I = I_0 \cdot \left[\exp\left(\frac{q \cdot (U - I \cdot R_s)}{k_B \cdot T} \right) - 1 \right]$$
$$+ I_{0,\text{SRH}} \cdot \left[\exp\left(\frac{q \cdot (U - I \cdot R_s)}{2 \cdot k_B \cdot T} \right) - 1 \right] + \frac{U - I \cdot R_s}{R_p} - I_{\text{SC}}$$

$$(4.45)$$

Figure 4.16 shows examples for the influence of SRH recombination on current–voltage characteristics for a solar cell with I_0 equal to 10^{-10} A/cm^2 (a) and 10^{-12} A/cm^2 (b). The current–voltage characteristics remain practically unchanged for very low values of $I_{0,\text{SRH}}$.

The fill factor (FF) and V_{OC} of a solar cell decrease with increasing $I_{0,\text{SRH}}$. The decrease of FF and V_{OC} sets in at lower values of $I_{0,\text{SRH}}$ for solar cells with lower I_0. This is shown in more detail in Figure 4.17 where the dependencies of V_{OC} and FF on $I_{0,\text{SRH}}$ are given for solar cells with I_0 equal to 10^{-9} and 10^{-13} A/cm^2. The V_{OC} of the solar cell with I_0 equal to 10^{-13} A/cm^2 starts to decrease at values of $I_{0,\text{SRH}}$ of about 10^{-8} A/cm^2 while the decrease of FF sets in already at values below 10^{-9} A/cm^2. The value of V_{OC} decreases from 0.65 V at $I_{0,\text{SRH}}$ equal to 10^{-7} A/cm^2 to 0.54 V at $I_{0,\text{SRH}}$ equal to 10^{-6} A/cm^2. At the same time the FF decreases only from about 75% to 70% in the same range of $I_{0,\text{SRH}}$. Therefore, low values of V_{OC} are an indication for strong recombination at the charge-selective contact in solar cells with relatively high FF.

Solar cells with large I_0 are less sensitive to recombination at the charge-selective contact. For a solar cell with I_0 equal to 10^{-9} A/cm^2 the value of

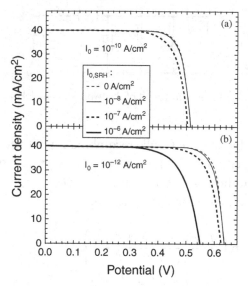

Figure 4.16. Current–voltage characteristics for solar cells with two diodes ($I_0 = 10^{-10}$ and 10^{-12} A/cm^2 (a) and (b), respectively, and $I_{SRH,0} = 0$, 10^{-8}, 10^{-7} and 10^{-6} A/cm^2, thin dotted, thin solid, thick dotted and thick solid lines, respectively), very low R_s and very large R_p.

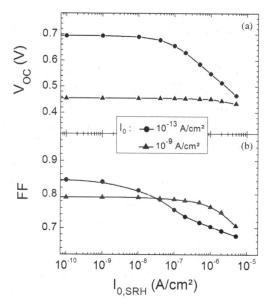

Figure 4.17. Dependence of V_{OC} (a) and FF (b) on the I_0 of SRH recombination for solar cells with I_0 equal to 10^{-9} (triangles) and 10^{-13} (circles) A/cm^2, very low R_s and very large R_p.

V_{OC} starts to reduce remarkably for $I_{0,SRH}$ larger than 10^{-6} A/cm^2 while FF already reduces from 78% at $I_{0,SRH}$ equal to 10^{-7} A/cm^2 to 76% at $I_{0,SRH}$ equal to 10^{-6} A/cm^2.

4.4.5 *Consideration of surface recombination*

Solar cells have a finite thickness. Photogenerated charge carriers can reach the surfaces opposite to the *pn*-junction at the *p*-type and *n*-type doped sides of a solar cell. These surfaces are often known as front surface and back surface, depending on the direction of incoming sunlight or external surfaces of the absorber.

Photogenerated charge carriers can recombine at the front and back surfaces of a solar cell. The thickness of the *n*-type and *p*-type doped sides of the solar cell (H_n and H_p, respectively) and the corresponding surface recombination velocities of the minority charge carriers (s_p and s_n at the external surface of the *n*-type and *p*-type doped regions of the solar cell, respectively) have to be considered for the calculation of the I_0 of a solar cell in addition to the E_g, temperature, densities of donors and acceptors, diffusion constants and diffusion lengths of minority charge carriers (Figure 4.18).

The stationary diffusion equations (Equations (4.31)) have to be solved for minority charge carriers for the following boundary conditions:

$$\Delta n_p|_{x=x_p} = 0 \tag{4.46'}$$

$$\Delta p_n|_{x=x_n} = 0 \tag{4.46''}$$

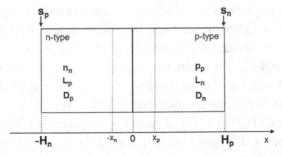

Figure 4.18. Scheme of a *pn*-junction with limited thickness of the *p*-type and *n*-type doped regions and given surface recombination velocities of the minority charge carriers.

and

$$\Delta n_p|_{x=H_p} \cdot s_n = -D_n \cdot \frac{\partial(\Delta n_p)}{\partial x} \tag{4.47'}$$

$$\Delta p_n|_{x=H_n} \cdot s_p = -D_p \cdot \frac{\partial(\Delta p_n)}{\partial x} \tag{4.47''}$$

The boundary conditions (4.47′) and (4.47″) take into account the limitation of surface recombination by the access of minority charge carriers to the surface (see also Chapter 3). The densities of minority charge carriers can be obtained by taking into account negative and positive arguments in the exponentials of the approach for solving the diffusion equations. The boundary conditions bring in the functions of sinh and cosh. The diffusion currents of the minority charge carriers can be calculated following Equations (4.23) and (4.23′) for photogenerated electrons and holes, respectively. The solutions are given for the current densities for minority electrons and holes (Goetzberger *et al.*, 1997):

$$I_n = I_{0,n} \cdot G_{fn} \cdot \left[\exp\left(\frac{q \cdot U}{k_B \cdot T} \right) - 1 \right] \tag{4.48'}$$

$$I_p = I_{0,p} \cdot G_{fp} \cdot \left[\exp\left(\frac{q \cdot U}{k_B \cdot T} \right) - 1 \right] \tag{4.48''}$$

The diode saturation current densities are a product of the diode saturation current densities of the ideal *pn*-junction (Equation (4.34)) and of the so-called geometry factors (G_{fn} and G_{fp}). Equations (4.48′) and (4.48″) show that the I_0 of a solar cell can increase or decrease depending on the behavior of the geometry factors.

The geometry factors are a function of the ratios of the thickness of the *p*-type or *n*-type doped regions and the diffusion length of the respective minority charge carriers (H_p/L_n and H_n/L_p for G_{fp} and G_{fn}, respectively). Further, G_{fp} and G_{fn} are a function of the ratios of the quotients of the diffusion coefficient and diffusion length of the corresponding minority charge carriers and of the corresponding surface recombination velocities ((D_p/L_p)/s_p and (D_n/L_n)/s_n for G_{fp} and G_{fn}, respectively). The quotients of the corresponding diffusion constant and diffusion length are denoted by $s_{\infty p}$ and $s_{\infty n}$ and can be treated as recombination velocities in the *p*-type and *n*-type doped regions. The ratios $H_{p(n)}/L_{n(p)}$ and $s_{n(p)}/s_{\infty n(p)}$ have an influence on the diode saturation current densities and therefore play an important role for the value of V_{OC}.

The values of the geometry factors can be calculated by using the following equations (Goetzberger *et al.*, 1997):

$$G_{fn} = \frac{\cosh\left(\frac{H_p}{L_n}\right) + \frac{S_{\infty n}}{S_n} \cdot \sinh\left(\frac{H_p}{L_n}\right)}{\frac{S_{\infty n}}{S_n} \cdot \cosh\left(\frac{H_p}{L_n}\right) + \sinh\left(\frac{H_p}{L_n}\right)} \qquad (4.49')$$

$$G_{fp} = \frac{\cosh\left(\frac{H_n}{L_p}\right) + \frac{S_{\infty p}}{S_p} \cdot \sinh\left(\frac{H_n}{L_p}\right)}{\frac{S_{\infty p}}{S_p} \cdot \cosh\left(\frac{H_n}{L_p}\right) + \sinh\left(\frac{H_n}{L_p}\right)} \qquad (4.49'')$$

The geometry factors G_{fn} and G_{fp} are plotted in Figure 4.19 as a function of the ratios H_p/L_n or H_n/L_p, respectively. The geometry factors are independent of surface recombination, i.e. equal to 1, if the n-type or p-type doped regions are thicker than the diffusion length of the corresponding minority charge carriers or if the ratios $s_p/s_{\infty p}$ or $s_n/s_{\infty n}$ are equal to 1.

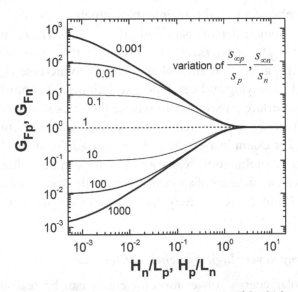

Figure 4.19. Dependence of the geometry factors on the ratio of the thickness of the doped region and the diffusion length of the corresponding minority charge carriers for different ratios of the surface recombination velocity and the bulk recombination velocity of the corresponding minority charge carriers.

There are two very different regimes if H_n or H_p are thinner than L_p or L_n, respectively. The geometry factors are larger than 1 for large surface recombination velocities, i.e. when $s_{\infty p}/s_p$ or $s_{\infty n}/s_n < 1$. This is the case, for example, for surface recombination velocities with values above 1000 cm/s or 10 cm/s for $D_{n(p)}$ of about 10 cm^2/s and $L_{n(p)}$ of about 100 μm or 1 cm, respectively. Incidentally, surface recombination velocities of even more than 10^6 cm/s are required for ohmic contacts (see Chapter 5). The geometry factors can increase by more than two orders of magnitude if, for example, H_p/L_n or H_n/L_p are about 0.01 and $s_{\infty p}/s_p$ or $s_{\infty n}/s_n$ are about 0.001. An increase of the I_0 by a factor of 100, for example, leads to a decrease of V_{OC} by 0.12 V. Therefore there is no sense, with respect to a maximum of V_{OC}, in reducing H_p/L_n or H_n/L_p to values below 1 if surface recombination cannot be controlled in a way that $s_{\infty p}/s_p$ or $s_{\infty n}/s_n$ are larger than 1, i.e. a strong increase of the diffusion length to values much larger than α^{-1} can improve carrier collection but reduce V_{OC}. In addition, advanced light trapping is only useful if it is combined with advanced surface passivation.

The ratios $s_{\infty p}/s_p$ and $s_{\infty n}/s_n$ are larger than 1 in the regime of low surface recombination. In this regime the geometry factors can be reduced by more than one order of magnitude if H_p/L_n or H_n/L_p are less than 0.1–0.01 and $s_{\infty p}/s_p$ or $s_{\infty n}/s_n$ are larger than 10–100. The reduction of the geometry factor gives an additional opportunity to increase V_{OC}. This can be achieved by applying and developing techniques for sophisticated light trapping and surface passivation to reduce the thickness of the absorber and therefore $H_{p(n)}$ and to reduce the surface recombination velocity, respectively. For example, electron diffusion lengths of the order of 1 mm and a surface recombination velocity of 10 cm/s can be realized in p-type doped c-Si with a thickness of about 100 μm. In this case H_p/L_n and $s_{\infty n}/s_n$ are about 0.1 and 20, respectively, leading to a decrease of G_{fn} by about 10 times.

4.4.6 *The way to very high efficiencies of solar cells*

Very high solar energy conversion efficiencies can be reached for solar cells if all losses including resistive, optical and recombination losses are reduced to a minimum. Resistive and optical losses can already be reduced to a very few percent of the overall solar energy conversion efficiency.

Recombination losses are important for carrier collection, i.e. I_{SC}, and for the I_0, i.e. for V_{OC}. The reduction of recombination losses to a minimum is challenging. The general way to minimize recombination losses in solar cells is (i) the increase of the diffusion lengths of minority charge carriers, (ii) the reduction of the surface recombination velocity and (iii) the decrease of the absorber thickness by light trapping.

The application of Equations (4.26), (4.33), (4.43), (4.45), (4.48) and (4.49) opens the opportunity of analyzing general trends in the dependencies of solar cell parameters on parameters of materials and combinations of materials. However, detailed simulations of the complete solar cell including the local morphology and the architecture of the solar cell are required for the optimization of a certain type of solar cell under given conditions taking into account the detailed parameters of implemented materials, combinations of materials and technologies.

Figure 4.20 gives an example for the dependence of I_{SC} and V_{OC} on the surface recombination velocity for different diffusion lengths of electrons

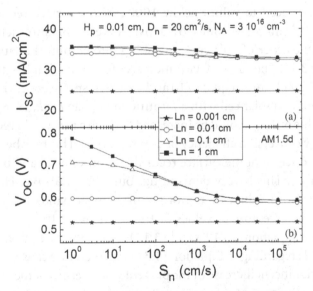

Figure 4.20. Dependence of I_{SC} and V_{OC} at air mass (AM) 1.5d on the surface recombination velocity for a c-Si solar cell with a very thin n-type and a 100 μm thick p-type doped region for different diffusion lengths of electrons photogenerated in the p-type doped region. Resistive losses and losses due to reflection are not considered. An optical path equal to the absorber thickness is taken into account.

photogenerated in the p-type doped absorber at fixed H_p and very thin H_n. Light trapping is not taken into account. The values of I_{SC} and V_{OC} are 25 mA/cm^2 and 0.522 V, respectively, independent of s_n for a diffusion length of only 10 μm. A diffusion length of 100 μm is significantly larger or close to the absorption length for most photons in the sun spectrum with energy above E_g of c-Si. An increase of L_n to 100 μm leads to an increase of I_{SC} and V_{OC} to 34 mA/cm^2 and 0.6 V for low values of s_n and of about 33 mA/cm^2 and 0.58 V for high values of s_n, respectively, i.e. the influence of surface recombination becomes remarkable but is still weak. It is worth noting that the solar energy conversion efficiency is about 16% for c-Si solar cells with related parameters.

An increase of L_n to 1 mm leads to saturation of I_{SC} at 36 mA/cm^2 for low values of s_n while I_{SC} remains at about 33 mA/cm^2 for high values of s_n. At the same time V_{OC} increases only slightly to about 0.59 V for high values of s_n. The V_{OC} reaches 0.62 V for s_n equal to 10^3 cm/s and L_n equal to 1 mm. This means that an increase of the L_n to values above 100 μm is not useful for c-Si solar cells if the surface recombination velocity cannot be reduced to values below 1000 cm/s. A surface recombination velocity of 10^3 cm/ corresponds to a density of recombination active surface defects of the order of 10^{11} cm^{-2} if taking into account Equation (3.26′) and Figure 3.11. The value of V_{OC} increases for a diffusion length of 1 mm to 0.70 and 0.71 V for s_n equal to 10 and 1 cm/s, respectively, corresponding to the record c-Si solar cell with a pn-junction (Zhao *et al.*, 1998).

The value of V_{OC} should be increased to reach a new world record of the solar energy conversion efficiency of c-Si solar cells. This may be reached by a further reduction of the surface recombination velocity and by a further increase of L_n. This is very demanding, but seems possible. However, an increase of L_n to values of the order of 1 cm demands a reduction of the density of acceptors to values of the order of $3 \cdot 10^{15}$ cm^{-3} if taking into account Equations (3.12′) and (3.15). But a decrease of N_A causes an increase of I_0 regarding to Equation (4.34) and therefore causes a reduction of V_{OC}. A significant increase of the present world record of the solar energy conversion efficiency of c-Si solar cells can therefore only be achieved by reducing H_p to values of the order of only 10 μm. This is possible, in principle, by implementing strategies of advanced light trapping into solar cells (see Figure 2.14).

4.5 Summary

A photogenerated charge carrier can pass an ideal charge-selective contact only in the direction in which the potential energy is reduced for the given charge carrier. This condition should hold for both photogenerated electrons and holes, which are separated to the opposite sides of an ideal charge-selective contact.

At the contact between a p-type and an n-type doped semiconductor layer (pn-junction), the potential energy is reduced at the side where the charge carriers are the majority. This means that photogenerated electrons are separated from the p-type doped region into the n-type doped region, and similarly the photogenerated holes are separated from the n-type doped region into the p-type doped region. Therefore ideal charge-selective contacts can be formed by pn-junctions.

Charge-selective contacts such as pn-junctions are sinks for photogenerated minority charge carriers, which reach the sink from neutral regions of the PV absorber by diffusion. Space charge regions are formed at pn-junctions and the corresponding potential drop across the space charge region is the driving force for the diffusion of minority charge carriers towards the pn-junction. Therefore, the potential drop across a pn-junction is the diffusion potential of a pn-junction. The diffusion potential is the upper limit of V_{OC}.

Photogenerated minority charge carriers are collected at a pn-junction. All charge carriers photogenerated in a space charge region of a pn-junction are collected since the drift time is much shorter than the lifetime of photogenerated charge carriers. For charge carriers photogenerated in the neutral bulk of a PV absorber, the efficiency of the collection depends on the distance between the pn-junction and the region in which the minority charge carriers were photogenerated in relation to the diffusion length of the minority charge carriers. Therefore, in first approximation, the quantum efficiency of a solar cell is limited by the reciprocal sum of one and the quotient of the absorption length and the diffusion length. Investigation into the spectral dependence of the quantum efficiency gives the opportunity of obtaining information about the diffusion length of photogenerated minority charge carriers.

The Fermi-energy of photogenerated minority charge carriers is contacted with the Fermi-energy of the majority charge carriers at the opposite side of a *pn*-junction, and the Fermi-energy of majority charge carriers is contacted with external leads in solar cells. Therefore the electric potential between the two external leads of a solar cell corresponds to the difference between the Fermi-energies of the minority charge carriers in the *p*-type and *n*-type doped layers forming the *pn*-junction.

The most important characteristic of a charge-selective contact is the I_0 which limits V_{OC}. A diode equation related to bulk recombination follows from the dependence of the density of minority charge carriers on an external potential and diffusion of minority charge carriers towards the *pn*-junction. The value of I_0 depends on the squared density of intrinsic charge carriers (i.e. on temperature, E_g and effective masses of electrons and holes), on the densities of acceptors and donors at the *p*-type and *n*-type doped regions of the *pn*-junction and on the diffusion constants and diffusion lengths of minority charge carriers in the *p*-type and *n*-type doped regions. The ideality factor of a diode limited by bulk recombination is equal to one.

Shockley–Read–Hall recombination via deep defect states at a *pn*-junction also limits the current density across a *pn*-junction. The dependence of the SRH recombination current on the external potential is described by a diode equation with a diode saturation current density $I_{0,SRH}$. The value of $I_{0,SRH}$ is proportional to the density of intrinsic charge carriers, the extension of the region with a high SRH recombination rate and the inverse SRH lifetime. The ideality factor of a diode limited by SRH recombination at the *pn*-junction is equal to two.

The two-diode model considers in general recombination in the bulk of the PV absorber and SRH recombination at the charge-selective contact. Two diodes with diode saturation current densities I_0 and $I_{0,SRH}$ and with corresponding ideality factors of 1 and 2 are connected in parallel in the two-diode model. Both recombination processes can be distinguished by the thermal activation energy of the I_0, which is equal to the E_g of the PV absorber for bulk recombination and to half of the E_g for SRH recombination at the charge-selective contact.

Surface recombination at the external surface of the *p*-type and *n*-type doped layers of a solar cell can additionally lead to a reduction of the density of minority charge carriers if the diffusion length of minority charge carriers is equal or less than the thickness of the corresponding *p*-type or *n*-type doped layer. The influence of surface recombination is described by the geometry factor multiplied by I_0. The geometry factor depends on the ratio between the thickness of the *p*-type or *n*-type doped layer and the diffusion length of the corresponding minority charge carriers, as well as on the ratio between the surface recombination velocity and the diffusion coefficient divided by the diffusion length of the corresponding minority charge carriers.

Very high solar energy conversion efficiencies can be reached only for solar cells with low thicknesses of PV absorbers where the diffusion length of minority charge carriers is very high and the surface recombination velocities are extremely low. Related conditions can only be reached with c-Si (see Chapter 7) and GaAs (see Chapter 8) solar cells.

4.6 Tasks

T4.1: Diffusion potential

Calculate the U_D of a *pn*-junction with (a) $N_A = 10^{16}$ and $N_D = 10^{19}\,\text{cm}^{-3}$ for c-Si and (b) $N_A = 10^{18}$ and $N_D = 10^{17}\,\text{cm}^{-3}$ for GaAs at room temperature.

T4.2: Space charge regions of a pn-junction

Calculate the thickness of the space charge regions for a *pn*-junction with (a) $N_A = 10^{16}$ and $N_D = 10^{19}\,\text{cm}^{-3}$ for a c-Si solar cell and (b) $N_A = 10^{18}$ and $N_D = 10^{17}\,\text{cm}^{-3}$ for a GaAs solar cell at room temperature.

T4.3: Maximum electric field at a pn-junction

Calculate the maximum electric field at a *pn*-junction with (a) $N_A = 10^{16}$ and $N_D = 10^{19}\,\text{cm}^{-3}$ for a c-Si solar cell and (b) $N_A = 10^{18}$ and $N_D = 10^{17}\,\text{cm}^{-3}$ for a GaAs solar cell at room temperature.

T4.4: Diode saturation current density of a semi-infinite pn-junction in the limit of Auger recombination

Ascertain the I_0 of a semi-infinite *pn*-junction of a c-Si solar cell with $N_A = 10^{16}$ and $N_D = 10^{19}\,\text{cm}^{-3}$ at room temperature and compare with a c-Si record solar cell.

T4.5: Diode saturation current density of a semi-infinite pn-junction in the limit of radiative recombination

Ascertain the I_0 of a semi-infinite *pn*-junction of a GaAs solar cell with $N_A = 10^{18}$ and $N_D = 10^{17}$ cm^{-3} at room temperature and compare with a record GaAs solar cell.

T4.6: Temperature dependence of V_{OC} of c-Si solar cells

Ascertain the temperature dependence of V_{OC} for a c-Si solar cell with the I_0 in the Auger limit.

T4.7: Diode saturation current density for SRH recombination at the pn-junction

Maximum densities of defects with electronic states in the middle of the band gap were obtained with regard to the minimum lifetime condition. Estimate the $I_{0,SRH}$ via defects at a *pn*-junction for a c-Si solar cell with (a) $N_t = 10^{12}$ cm^{-3} and (b) $N_t = 10^{16}$ cm^{-3} at room temperature.

T4.8: Dependence of V_{OC} on SRH recombination at a pn-junction

Ascertain the dependence of V_{OC} on the diode saturation current densities considering bulk recombination (I_0) and SRH recombination ($I_{0,SRH}$) at a *pn*-junction and plot V_{OC} as a function of $I_{0,SRH}$ for $I_0 = 10^{-26}, 10^{-20}, 10^{-14}$ and 10^{-8} A/cm^2 ($I_{SC} = 0.04$ A/cm^2).

T4.9: Geometry factor of solar cells with a pn-junction

Estimate the geometry factor for the electron I_0 of a c-Si solar cell with $N_A = 10^{16}$ and $N_D = 10^{19}$ cm^{-3}. The thickness of the *p*-type doped region and the surface recombination velocity are $H_p = 150\,\mu$m and $s_n = 10$ cm/s, respectively.

5

Ohmic Contacts for Solar Cells

Ohmic contacts connect a photovoltaic (PV) absorber from the two sides of the charge-selective contact of a solar cell with metal leads connecting to external load. This chapter is devoted to the concept of ideal ohmic contacts and to realization principles of ohmic contacts at metal–semiconductor and semiconductor–semiconductor interfaces. Barriers are formed at metal–semiconductor interfaces depending on the metal work function, the electron affinity, doping of the semiconductor and interfacial defects. High densities of interface defects cause Fermi-level pinning. Tunneling of free charge carriers through very thin barrier layers at highly doped semiconductors is decisive for ohmic contacts in solar cells with very high solar energy conversion efficiencies. The transmission probability is derived for a triangular barrier the width and height of which are given by the extension of the space charge region and by the barrier at the contact, respectively. The tunneling or contact resistance is obtained as a function of the barrier height and of the density of dopants in a highly doped semiconductor. Consequences for the tolerable series resistance and design of solar cells are discussed. Furthermore, tunneling contacts are considered between a highly doped n-type and a highly doped p-type semiconductor for integrated series connection in multi-junction solar cells. Ohmic contacts may be realized as well by defect-assisted tunneling through barrier layers.

5.1 Concept of Ideal Ohmic Contacts

Metal leads connect a solar cell with an external load. A metal–semiconductor contact is formed at the interface between a metal lead and

149

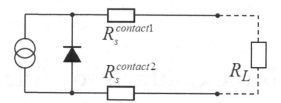

Figure 5.1.   Equivalent circuit of a solar cell with contact resistances and a load.

a photovoltaic (PV) absorber. Metal–semiconductor contacts are crucial for minimizing losses in a solar cell.

Ideal metal–semiconductor contacts are expressed by two ohmic contacts in the equivalent circuit of a solar cell (Figure 5.1). Ohmic means that the voltage drop across the contact is proportional to the current flowing through the contact independent of the direction of the current, i.e. the function of charge-selectivity of a solar cell is not influenced by ohmic metal–semiconductor contacts. As a consequence, ideal ohmic metal–semiconductor contacts do not introduce potential barriers either for mobile electrons or holes. The proportionality factor between the voltage drop across an ohmic contact and the flowing current is known as contact resistance (R_s^{contact1}, R_s^{contact2}).

For ideal ohmic contacts of a solar cell, the Fermi-energies of the two metal leads (E_{Fm1} and E_{Fm2}) are aligned with the Fermi-energies of the majority charge carriers ($E_{\text{Fn}(n)}$ and $E_{\text{Fp}(p)}$) at the two sides of the PV absorber, which are separated by the charge-selective contact (see Chapter 4). Therefore, the potential difference between the external leads of a solar cell (U) is equivalent to the difference between $E_{\text{Fn}(n)}$ and $E_{\text{Fp}(p)}$. With regard to Ohm's law and Equation (4.26), one can write the following condition for ideal ohmic contacts of a solar cell:

$$U = I \cdot R_L = \frac{E_{\text{Fn}(n)} - E_{\text{Fp}(p)}}{q} - I \cdot (R_s^{\text{contact1}} + R_s^{\text{contact2}}) \qquad (5.1)$$

Equation (5.1) becomes equal to Equation (4.26) if the values of $R_s^{\text{contact1,2}}$ are so low so that the voltage drops across the ohmic contacts can be neglected. For reaching very high solar energy conversion efficiencies, the values of R_s^{contact1} and R_s^{contact2} should be lower than the tolerable series resistance of a solar cell (see also Chapter 1).

Ohmic contacts are not charge-selective. This means, in contrast to charge-selective contacts, that a splitting of Fermi-energies is impossible at ohmic contacts. Therefore, the Fermi-energy of the metal (E_{Fm}) is equal to the Fermi-energies of free electrons and holes at ohmic metal–semiconductors contacts (Figure 5.2(a)). The definition of ideal ohmic contacts with n-type or p-type doped semiconductors is given as

$$E_{Fm} \equiv E_{Fn(n)} \equiv E_{Fp(n)} \tag{5.2'}$$

$$E_{Fm} \equiv E_{Fn(p)} \equiv E_{Fp(p)} \tag{5.2''}$$

As a consequence of definition (5.2), the surface recombination velocity is infinitely high at ideal ohmic contacts. In reality, the surface recombination velocity (see Equations (3.26') and (3.26'')) does not exceed the thermal velocity of free charge carriers since there is no physical meaning if the product of the capture cross section and the density of surface states becomes larger than 1. Therefore, the surface recombination velocity at ideal ohmic contacts is about 10^7 cm/s.

Ohmic contacts can also be realized between two semiconductors (Figure 5.2(b)). Ohmic contacts between two semiconductors are called recombination contacts. Recombination contacts are decisive for integrated series connections in multi-junction solar cells (see also Chapters 6 and 8).

Recombination losses at ohmic contacts have to be minimized despite the fact that the surface recombination velocity cannot be reduced. The

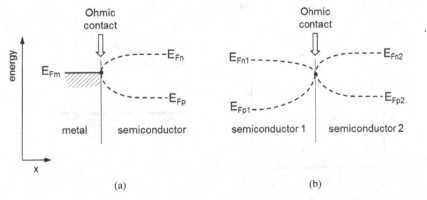

Figure 5.2. Band diagrams of ideal Ohmic contacts between a metal and a semiconductor (a) and between two semiconductors (b).

recombination rate at ohmic contacts is proportional to the density of photogenerated charge carriers. Therefore, recombination losses at ohmic contacts can be reduced by minimizing the density of photogenerated minority charge carriers in front of the ohmic contact. The density of photogenerated minority charge carriers at ohmic contacts is reduced by implementing potential barriers, which reject minority charge carriers in the PV absorber in front of the ohmic contact. Related barriers, for example, can be achieved with a so-called back surface field (see Chapter 7).

Another measure for reducing recombination losses at ohmic contacts in solar cells is related to the drastic reduction of the area of the ohmic contact by factor of the order of 1000 in combination with the excellent surface passivation of the part of the semiconductor surface (see Chapter 3) that is not covered by an ohmic contact (Figure 5.3).

The ohmic contact area is minimized, for example, by using interconnected grid lines at illuminated front contacts and by applying contact dots at back contacts. The distance between ohmic contact dots has to be larger than the thickness of the PV absorber and/or the diffusion length of minority charge carriers. In this case, the probability of photogenerated minority charge carriers reaching the charge-selective contact is much higher than the probability of recombination at the ohmic contact. The ohmic contact dots are connected with a metallization layer at back contacts. The reduction of the ohmic contact area is decisive for reaching very high solar energy conversion efficiencies, especially with crystalline silicon (c-Si) solar cells (see Chapter 7).

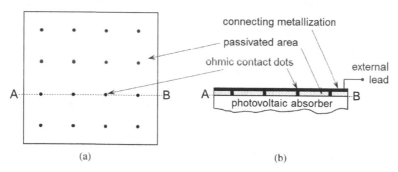

(a)   (b)

Figure 5.3. Bottom view of a solar cell with ohmic contact dots which connect the metallization and passivated areas (a) and a cross-section along line A–B (b).

The minimization of the ohmic contact area as well as the concentration of sunlight demand very low resistances of metal–semiconductor contacts in solar cells of the order of 10^{-4}–10^{-5} Ωcm^2 or even less.

5.2 Ohmic Metal–Semiconductor Contacts

The behavior of metal–semiconductor contacts has been described in numerous textbooks, such as Rhoderick (1978) or Sze (1981). Here only the principal aspects that are important for solar cells are considered.

5.2.1 *Work function, electron affinity and ionization energy*

Fermi-energies of metals (E_{Fm}) in relation to the energies of valence and conduction band edges of semiconductors (E_V and E_C) are important for charge transport across metal–semiconductor interfaces. All characteristic energies can be compared to one common reference energy — the vacuum level (E_{vac}). This is expressed in Figure 5.4.

The work function of a metal (Φ_m) is defined as the difference between the vacuum level and the Fermi-energy of the metal.

$$\Phi_m \equiv E_{vac} - E_{Fm} \tag{5.3}$$

The work function of a metal corresponds to the energy, which is needed to replace a free electron from the metal into the vacuum, for example, by photoemission. In an ultra-high vacuum, the work function of pure metal surfaces depends on the nature of the metal and on the facet

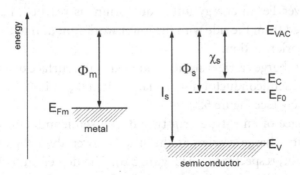

Figure 5.4. Energy diagram of the metal work function (Φ_m) and of the electron affinity (χ_s), the work function (Φ_s) and the ionization energy (I_s) of a semiconductor.

from which electrons are emitted into the vacuum. For example, the values of Φ_m range between 5.1 and 5.47 eV for gold surfaces, between 4.52 and 4.74 eV for silver surfaces, and between 4.06 and 4.26 eV for aluminum surfaces. Calcium has a low work function of 2.87 eV (for more details see Sze, 1981 and references therein).

The work function of a semiconductor (Φ_s) is defined as the difference between the vacuum level and the Fermi-energy of the semiconductor in thermal equilibrium (E_{F0}). The Fermi-energy of a semiconductor in thermal equilibrium — and therefore also Φ_s — depend on the doping of the semiconductor. The ionization energy of a semiconductor (I_s) is the energy needed to transfer an electron from a bond state into vacuum, i.e. the difference between the vacuum level and the valence band edge. The electron affinity of a semiconductor (χ_s) is defined as the difference between the conduction band edge and the vacuum level. The work function of a semiconductor can be calculated if the ionization energy or the electron affinity and the doping of the semiconductor (see Equation (2.37)) are known:

$$\Phi_s = I_s - (E_{F0} - E_V) = \chi_s + (E_C - E_{F0}) \tag{5.4}$$

The electron affinity, for example, is 4.05 eV for c-Si and 4.07 eV for GaAs (Cowley and Sze, 1965).

5.2.2 *Formation of metal–semiconductor contacts*

When a metal and a semiconductor are brought into intimate contact, electrons flow from the side with the higher Fermi-energy to the side with the lower Fermi-energy until equilibrium is reached, i.e. electrons are transferred from the side with the lower work function to the side with the higher work function.

Majority charge carriers are accumulated at the surface of an n-type or p-type doped semiconductor if Φ_m is larger than Φ_s or if Φ_m is lower than Φ_s, respectively (see Figure 5.5).

The surface of an n-type or p-type doped semiconductor is depleted from majority charge carriers if the Φ_m is lower than Φ_s or if Φ_m is higher than Φ_s, respectively (see Figure 5.6). The depletion regions of the n-type or p-type doped semiconductors are charged positively or negatively, respectively, due to the fixed charge of ionized donors or acceptors.

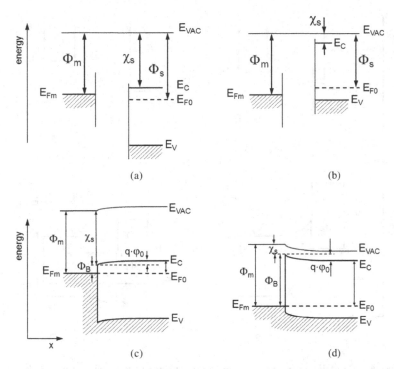

Figure 5.5. Band diagrams of a metal and an *n*-type doped semiconductor with $\Phi_m < \Phi_s$ (a, c) and of a metal and a *p*-type doped semiconductor with $\Phi_m > \Phi_s$ (b, d) before (a, b) and after (c, d) contact formation. Majority charge carriers are accumulated at the semiconductor surfaces (c, d).

Charging the semiconductor near the metal–semiconductor contact leads to the formation of a space charge region in a similar manner to the formation of a space charge region at a *pn*-junction (see Chapter 4). The distribution of the electric field and of the electrostatic potential in the space charge region can be calculated by integrating the Poisson Equation (4.1) once or twice, respectively.

The transfer of electrons from the metal into the semiconductor requires energy equal to or larger than the difference between the electron affinity of the semiconductor and the metal work function. The energetic barrier for electron transfer across metal–semiconductor contacts (Φ_B) is given by Φ_m and χ_s (the Schottky limit)

$$\Phi_B = \Phi_m - \chi_s \tag{5.5}$$

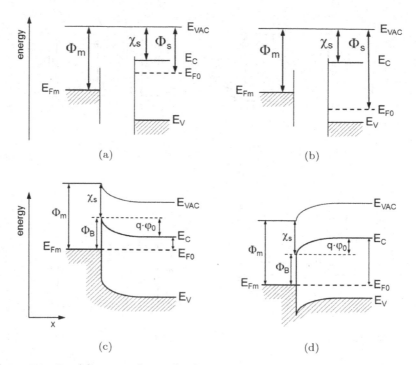

Figure 5.6. Band diagrams of a metal and an n-type doped semiconductor with $\Phi_m > \Phi_s$ (a, c) and another p-type doped semiconductor with $\Phi_m < \Phi_s$ (b, d) before (a, b) and after (c, d) contact formation. The semiconductor surfaces are depleted from majority charge carriers (c, d).

The conduction or valence bands of depleted surfaces of n-type or p-type doped semiconductors bend upwards or downwards, respectively. The so-called surface band bending (φ_0) is positive or negative for depleted surfaces of n-type or p-type doped semiconductors, respectively. The surface band bending is equal to the difference between Φ_B and the difference between the conduction band edge and the Fermi-energy in the bulk semiconductor ($E_C - E_{F0}$).

$$q \cdot \varphi_0 = \Phi_B - (E_C - E_{F0}) \qquad (5.6)$$

For ohmic contacts, the potential energy of majority charge carriers should decrease towards the contact such that majority charge carriers can flow from the semiconductor into the metal. This is the case for metal–semiconductor contacts which are in accumulation (Figures 5.5(c) and 5.5(d)). The surface band bending at metal–semiconductor contacts that are in depletion corresponds to the increase of the potential energy of majority charge carriers moving from the bulk of the semiconductor towards the metal–semiconductor contact (Figures 5.6(c) and 5.6(d)). Metal–semiconductor contacts with a barrier for majority charge carriers, i.e. $q \cdot \varphi_0$, are called Schottky contacts (Schottky, 1938).

Metals with a relatively low work function close to the electron affinity of the semiconductor are required for the formation of an ohmic contact with a given n-type doped semiconductor. In contrast, metals with a much larger metal work function close to the ionization energy of the semiconductor are needed for the formation of an ohmic contact with the same semiconductor, p-type doped.

The barrier height depends on the nature of the contacting metal and semiconductor in the Schottky limit. However, interfacial layers at metal–semiconductor contacts can strongly influence the value of Φ_B. Interfacial layers at metal–semiconductor contacts are caused, for example, by inter-diffusion of atoms across the metal–semiconductor interface, by partial replacement of atoms at the semiconductor surface with metal atoms or by residual contamination, which was not removed before contact formation. The thickness of related interfacial layers can be of the order of several mono-layers of atoms.

Interfacial layers introduce defect states at the surface of the semicon-ductor very close to the metal contact (see Figure 5.7). An electric dipole is formed at the interface dipole layer disturbing the dependence of the Φ_B on Φ_m:

$$\Phi_B = \Phi_m - \chi_s - \Delta \tag{5.7}$$

Defect states are part of metal–semiconductor contacts. Occupied donor interface defect states are neutral, whereas unoccupied donor inter-face defect states are positively charged. Occupied or unoccupied acceptor interface defect states are negatively charged or neutral, respectively.

5.2.3 Fermi-level pinning at metal–semiconductor contacts

A variation of the metal work function causes a shift of the Fermi-energy at the semiconductor surface (E_{FS}) and therefore a change of the occupation of defect states, i.e. the electrostatic charge at the semiconductor surface is

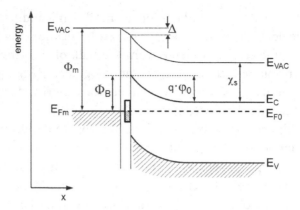

Figure 5.7. Band diagram of a metal–semiconductor contact with an interfacial layer and defects at the semiconductor surface.

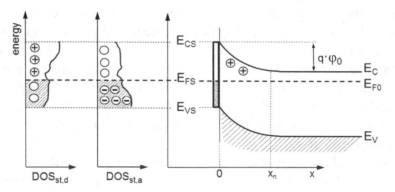

Figure 5.8. Band diagram and schematic densities of surface states at a semiconductor surface with occupied and unoccupied donor and acceptor surface states.

changed due to a change of E_{FS}. The change of the electrostatic charge at surface states is compensated by a change of the electrostatic charge in the depletion region of the semiconductor.

Defect states with a high density can pin the Fermi-energy at a semiconductor surface. Fermi-level pinning means that E_{FS} cannot be shifted in relation to the conduction or valence band edges at the surface of the semiconductor (E_{CS} and E_{VS}, respectively). In this case, the barrier height at a metal–semiconductor contact is independent of the metal work function (the so-called Bardeen limit; Bardeen, 1947). In the following, the Fermi-level pinning will be analyzed for a free semiconductor surface with donor and acceptor surface states (Figure 5.8).

With regard to charge neutrality at a free semiconductor surface, the sum of the charge in the space charge region (Q_{sc}), of the charge in donor ($Q_{st,d}$) and of the charge in acceptor ($Q_{st,a}$) states is equal to 0.

$$0 = Q_{sc} + Q_{st,d} + Q_{st,a} \tag{5.8}$$

Following Equations (4.5) and (4.6), the charge in the space charge region, for example, for an n-type semiconductor is

$$Q_{sc,n} = \sqrt{\frac{2 \cdot \varepsilon_r \cdot \varepsilon_0}{q}} \cdot N_D \cdot \varphi_0 \tag{5.9'}$$

The band bending at the surface and the space charge of a p-type doped semiconductor in depletion are negative

$$Q_{sc,p} = -\sqrt{\frac{2 \cdot \varepsilon_r \cdot \varepsilon_0}{q}} \cdot N_D \cdot |\varphi_0| \tag{5.9''}$$

In thermal equilibrium, the charge in occupied acceptor and unoccupied donor states follows from integration over the densities of acceptor ($DOS_{st,a}$) and donor ($DOS_{st,d}$) surface states from the valence band edge at the surface (E_{VS}) to the Fermi-energy at the surface (E_{FS}) surface, and from E_{FS} to the conduction band edge at the surface (E_{CS}), respectively.

$$Q_{st,a} = \int_{E_{VS}}^{E_{FS}} DOS_{st,a}(E) \cdot dE \tag{5.10'}$$

$$Q_{st,d} = \int_{E_{FS}}^{E_{CS}} DOS_{st,d}(E) \cdot dE \tag{5.10''}$$

For qualitative analysis, it is sufficient to assume constant distributions of $DOS_{st,a}$ and $DOS_{st,d}$. The densities of acceptor and donor surface defects are $N_{st,a}$ and $N_{st,d}$, respectively.

$$Q_{st,a} = -N_{st,a} \cdot \frac{E_{FS} - E_{VS}}{E_g} \tag{5.11'}$$

$$Q_{st,d} = N_{st,d} \cdot \frac{E_{CS} - E_{FS}}{E_g} \tag{5.11''}$$

The differences E_{FS}–E_{VS} and E_{CS}–E_{FS} can be expressed by the surface band bending and by the Fermi-energy of majority charge carriers in the bulk of an *n*-type doped semiconductor

$$E_{FS} - E_{VS} = E_g - (E_C - E_{F0}) - q \cdot \varphi_0 \qquad (5.12')$$

$$E_{CS} - E_{FS} = (E_C - E_{F0}) + q \cdot \varphi_0 \qquad (5.12'')$$

Similarly, the differences E_{FS}–E_{VS} and E_{CS}–E_{FS} can be expressed for a *p*-type doped semiconductor by

$$E_{FS} - E_{VS} = -[(E_V - E_{F0}) + q \cdot \varphi_0] \qquad (5.12''')$$

$$E_{CS} - E_{FS} = E_g + (E_V - E_{F0}) + q \cdot \varphi_0 \qquad (5.12'''')$$

The differences between the conduction and valence band edges and the Fermi-energy in the bulk of the semiconductor can be calculated by using Equation (2.37) and by taking into account complete ionization of donors or acceptors.

$$E_C - E_{F0} = k_B \cdot T \cdot \ln \frac{N_C}{N_D} \qquad (5.13')$$

$$-(E_V - E_{F0}) = k_B \cdot T \cdot \ln \frac{N_V}{N_A} \qquad (5.13'')$$

Equation (5.8) can be transformed for *n*-type doped semiconductors.

$$\sqrt{\frac{2 \cdot \varepsilon_r \cdot \varepsilon_0 \cdot N_D}{q}} \cdot \sqrt{\varphi_0} + \frac{q \cdot \varphi_0 + (E_C - E_{F0})}{q} \cdot N_{st,d}$$

$$- \frac{E_g - q \cdot \varphi_0 - (E_C - E_{F0})}{q} \cdot N_{st,a} = 0 \qquad (5.14)$$

An equivalent expression can be obtained for *p*-type doped semiconductors.

Equation (5.14) has the following form:

$$0 = \varphi_0^2 - (2 \cdot B + A) \cdot \varphi_0 + B^2 \qquad (5.15')$$

with

$$A = \frac{2 \cdot \varepsilon_r \cdot \varepsilon_0 \cdot N_D}{q} \cdot \left[\frac{E_g}{q \cdot (N_{st,a} + N_{st,d})} \right]^2 \qquad (5.15'')$$

and

$$B = \frac{N_{st,a}}{N_{st,a} + N_{st,d}} \cdot \frac{E_g}{q} - \frac{E_C - E_{F0}}{q} \qquad (5.15''')$$

The solution of Equation (5.15′) in a physical meaning is

$$\varphi_0 = \frac{2 \cdot B + A}{2} - \sqrt{\left(\frac{2 \cdot B + A}{2}\right)^2 - B^2} \qquad (5.16)$$

Figure 5.9 shows the dependence of the surface band bending of an *n*-type doped semiconductor as a function of $N_{st,d}$ for different values of $N_{st,a}$. If $N_{st,d} = N_{st,a} = N_{st}$, the surface band bending can be neglected for values less than $5 \cdot 10^{10}$ cm^{-2}, where φ_0 increases for larger values of $N_{st,d}$ and saturates at large values of $N_{st,d}$ above 10^{13} cm^{-2}. The value of E_{FS} saturates at the middle of E_g (the band gap) for N_{st} larger than 10^{13} cm^{-2}, i.e. the Fermi-energy is pinned at midgap. For comparison, a density of

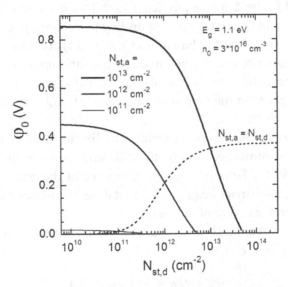

Figure 5.9. Dependence of the surface band bending at a free surface of an *n*-type doped semiconductor on the density of donor surface defects for constant values of the density of acceptor surface defects ($N_{st,a} = 10^{11}$, 10^{12} and 10^{13} cm^{-2}, thin, medium and thick solid lines, respectively) and for $N_{st,d} = N_{st,a}$ (dashed line). The Fermi-energy at the surface is pinned if $N_{st,a}$ and/or $N_{st,d}$ are larger than about 10^{13} cm^{-2}. The surface band bending should be as low as possible for ohmic contacts.

surface states larger than 10^{13} cm^{-2} was obtained for metal contacts with c-Si, gallium phosphide (GaP) and GaAs (Cowley and Sze, 1965). The ratio between N_D (the density of donors) and N_{st}^2 is important for Fermi-level pinning (see also task T5.1 at the end of this chapter).

For an n-type doped semiconductor and for large values of $N_{st,a}$, the value of E_{FS} can be pinned at energies closer to E_{VS} ($N_{st,a} \gg N_{st,d}$, large surface band bending) or closer to E_{CS} ($N_{st,a} \ll N_{st,d}$, very low surface band bending). Similarly, E_{FS} can be pinned at energies closer to E_{CS} ($N_{st,d} \gg N_{st,a}$, large surface band bending) or closer to E_{VS} ($N_{st,d} \ll N_{st,a}$, very low surface band bending) for a p-type doped semiconductor.

The ratio between the densities of donor and acceptor surface states depends on chemical reactions at metal–semiconductor contacts. The barrier height, for example, can sensitively depend on the thermodynamic heat of reaction (Brillson, 1978). The barrier height at metal–semiconductor contacts is less for metals or metal alloys, which reactive at the semiconductor surface.

A shift of E_{FS} by ΔE_{FS} is equivalent to a shift of $q \cdot \varphi_0$ by $q \cdot \Delta\varphi$. Changes of Q_{st} (ΔQ_{st}) and Q_{sc} (ΔQ_{sc}) corresponding to a change in E_{FS} and φ_0 can be calculated with respect to Equations (5.9)–(5.11) and (5.16)–(5.18). For example, a change of E_{FS} may be caused by the variation of the metal work function at the metal–semiconductor contact. A change of E_{FS} is impossible if ΔQ_{st} is larger than the corresponding value of ΔQ_{sc}, i.e. E_{FS} is pinned (see also task T5.2).

In the case of Fermi-level pinning, the barrier height at the metal–semiconductor contact is given by the difference between the conduction band edge and the Fermi-energy at the surface of the semiconductor. In this instance, the barrier height at the metal–semiconductor contact does not depend on the nature of the metal.

The barrier at a metal–semiconductor contact can be described by the following empirical equation:

$$\Phi_B = \Phi_0 + S \cdot (\Phi_m - \chi_s) \qquad (5.17')$$

The value of Φ_0 corresponds to the barrier height in the Bardeen limit and is equal to 0 in the Schottky limit. The empirical S-parameter in Equation (5.17$'$) is equal to 0 for the Bardeen limit and equal to 1 for the Schottky limit. PV absorbers with covalent bonds are close to the Bardeen limit, for example, S is about 0.05 for c-Si (Kurtin *et al.*, 1970). The value of S increases for compound semiconductors with increasing ionic character of bonding, for example, S is about 0.25 for cadmium telluride (CdTe), and S is practically equal to 1 for ionically bonded semiconductors or insulators such as zinc oxide (ZnO) (Kurtin *et al.*, 1970).

The Fermi-energy at a metal–semiconductor contact is often close to the middle of the band gap of the semiconductor in the Bardeen limit, i.e. the semiconductor is in depletion. A depletion region near metal–semiconductor contacts is a barrier layer (Schottky barrier (Schottky, 1938)) for majority charge carriers (Figures 5.6(c) and 5.6(d)). Schottky barriers are not ohmic but charge-selective. The analysis of charge transport across Schottky barriers (see, for example, Sze, 1981) is beyond the scope of this book since Schottky barriers are usually not applied for solar cells due to the limitation of the open circuit voltage (V_{OC}) — which cannot exceed band bending — and to a very high surface recombination velocity at metal–semiconductor contacts. Additional measures are required for the realization of ohmic metal–semiconductor contacts close to the Bardeen limit if depletion at the semiconductor surface cannot be avoided.

5.3 Tunneling Ohmic Contacts

5.3.1 *Tunneling at metal–semiconductor contacts*

Space charge regions at charges-selective contacts become very thin for very high densities of donors or acceptors and low values of band bending. For example, the width of the space charge region is about 3 nm for a density of donors of about $5 \cdot 10^{19}$ cm^{-3} and for a band bending of 0.3 V (Figure 4.5). Tunneling of free charge carriers through very thin barrier layers is key for the formation of tunneling ohmic contacts (see, for example, Yu, 1970).

The density of free charge carriers is larger than the effective density of states of the respective band in very highly doped semiconductors. This means that the E_{F0} is within the valence or conduction bands. Semiconductors for which $E_{F0} < E_V$ or $E_{F0} > E_C$ are called degenerated. The doping type of degenerated semiconductors is denoted by an additional plus (n^+-type or p^+-doping type).

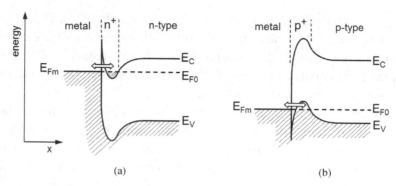

Figure 5.10. Band diagrams of tunneling metal–semiconductor contacts with *n*-type (a) and *p*-type (b) doped semiconductors. The arrows mark the tunneling transitions.

For degenerated semiconductors, the width of the space charge region near a metal contact can be calculated by modifying Equations (4.5′) and (4.5″).

$$x^{n+}_{SCR} = \sqrt{\frac{2 \cdot \varepsilon_r \cdot \varepsilon_0}{q \cdot N_D^+} \cdot \Phi_B} \tag{5.17″}$$

$$x^{p+}_{SCR} = \sqrt{\frac{2 \cdot \varepsilon_r \cdot \varepsilon_0}{q \cdot N_A^+} \cdot (E_g - \Phi_B)} \tag{5.17‴}$$

Figure 5.10 shows the band diagrams for tunneling metal–semiconductor contacts with *n*-type or *p*-type doped semiconductors.

For the formation of an ohmic contact with majority charge carriers in an *n*-type doped semiconductor, a highly *n*-type doped (*n*$^+$-type) layer is prepared on top of the *n*-type doped layer before the metal contact layer is deposited (for more details see Sze, 1981). The potential energy of electrons in the conduction band decreases from the *n*-type to the *n*$^+$-type doped region, and electrons can tunnel from the *n*$^+$-doped region into the metal contact. In a similar manner, an ohmic contact with holes in *p*-type doped semiconductors is formed by implementing a *p*$^+$-doped layer between the metal and the *p*-type semiconductor. The potential energy of holes in the valence band decreases towards the *p*$^+$-type doped region. Electrons can tunnel from the metal contact into unoccupied states in the valence band of the *p*$^+$-type doped region and recombine there with collected holes.

5.3.2 *Transmission probability of a tunneling barrier*

Quantum mechanical tunneling (see, for example, Griffiths, 2004) becomes possible due to the fact that an electron with certain total energy (E) can be considered as a wave with a certain wavelength, the de Broglie wavelength (λ_B)

$$\lambda_B = \frac{\hbar}{2\pi \cdot \sqrt{2 \cdot m_e^* \cdot E}} \qquad (5.18)$$

The Planck constant ($\hbar$) is equal to $1.05 \cdot 10^{-34}$ Js. Incidentally, the Planck constants $\hbar$ and h differ by a factor of 2π, and $\hbar$ is traditionally used in equations describing tunneling.

The total energy of free electrons is of the order of hundreds of meV in PV absorbers. The de Broglie wavelength, for example, is about 1.2 nm or 3.8 nm for electrons with energy of 1 eV or 0.1 eV, respectively, i.e. the de Broglie wavelength of free electrons is of the same order as the width of a space charge region of a highly doped semiconductor. Therefore, an electron at the barrier can be treated as a wave. Figure 5.11 depicts the particle and wave pictures employed to describe charge transport and tunneling of a free electron at a rectangular barrier.

The wave function of the electron (ψ) with an effective mass (m_e^*) can be described by the one-dimensional stationary Schrödinger equation with total energy (E) and potential energy (U_{pot}):

$$\frac{\partial^2 \psi(x)}{\partial x^2} + \frac{2 \cdot m_e^*}{\hbar^2} \cdot (E - \Phi_B) = 0 \qquad (5.19)$$

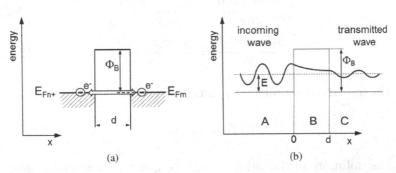

Figure 5.11. Particle (a) and wave (b) pictures of an electron at a rectangular barrier. Regions A, B and C denote the regions of propagation of the incoming wave, of the wave in the barrier and of the transmitted wave, respectively.

A tunneling barrier is characterized by the transmission probability. For simplicity, electrons are considered as plane waves for calculating the transmission probability of electrons ($T_{tr,e}$). The product of the wave function and the conjugate complex wave function describes the probability of finding an electron in a certain unit of space. The Schrödinger equation can be solved for the regions of the incoming wave (index A), of the wave in the barrier (index B) and of the transmitted wave (index C) with an approach for solving the Schrödinger equation, which takes into account reflected waves in the regions of the incoming wave and of the wave in the barrier.

$$\psi_A = A_{in} \cdot \exp(i \cdot k_A \cdot x) + A_{refl} \cdot \exp(-i \cdot k_A \cdot x) \tag{5.20'}$$

$$\psi_B = B_{in} \cdot \exp(i \cdot k_B \cdot x) + B_{refl} \cdot \exp(-i \cdot k_B \cdot x) \tag{5.20''}$$

$$\psi_C = C \cdot \exp(i \cdot k_C \cdot x) \tag{5.20'''}$$

The wave vectors are the same for regions A and C (k_A equal to k_C) and the wave vector is imaginary in the barrier region. The coefficient A_{in} can be set to 1, the coefficients A_{refl}, B_{in}, B_{refl} and C can be calculated by considering continuity of the wave function and of the first derivative of the wave function

$$\psi_A(x = 0) = \psi_B(x = 0) \tag{5.21'}$$

$$\psi_B(x = d) = \psi_C(x = d) \tag{5.21''}$$

$$\frac{\partial \psi_A(x = 0)}{\partial x} = \frac{\partial \psi_B(x = 0)}{\partial x} \tag{5.21'''}$$

$$\frac{\partial \psi_B(x = d)}{\partial x} = \frac{\partial \psi_C(x = d)}{\partial x} \tag{5.21''''}$$

The transmission probability can be written as

$$T_{tr} = C \cdot C^* \tag{5.22}$$

The solution of Equation (5.22) is given in textbooks (see, for example, Griffith, 2004). The commonly used approximation of the transmission probability for a rectangular barrier is given by the following equation:

$$T_{tr,e} = \exp\left(-2 \cdot \sqrt{2 \cdot m_e^* \cdot (\Phi_B - E)} \cdot \frac{d}{\hbar}\right) \qquad (5.23)$$

The barrier is usually distributed across a space charge region depending on doping, distribution of dopants near the metal contact, on metallurgical imperfections at the interface, roughness at the interface and other factors. The transmission probability can be calculated by integrating over the function of the barrier distribution in a space charge region ($U_{pot}(x)$).

$$T_{tr,e} = \exp\left(-\frac{2}{\hbar} \cdot \int_0^{x_{SCR}} \sqrt{2 \cdot m_e^* \cdot (U_{pot}(x) - E)} \cdot dx\right) \qquad (5.24)$$

The barrier of a space charge region at a metal contact with a highly doped n-type semiconductor can be approximated by a triangular shape shown in Figure 5.12.

$$U_{pot}(x) - E = \Phi_B \cdot \left(1 - \frac{x}{x_{SCR}}\right) \qquad (5.25)$$

The electron transmission probability of a triangular barrier is obtained by considering Equations (5.24), (5.25) and (5.17''):

$$T_{tr,e} = \exp\left(-\frac{8}{3} \cdot \sqrt{\frac{m_e^* \cdot \varepsilon_r \cdot \varepsilon_0}{N_D^+}} \cdot \frac{\Phi_B}{q \cdot \hbar}\right) \qquad (5.26)$$

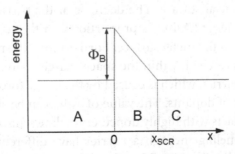

Figure 5.12. Triangular barrier for approximation of the space charge region at an ohmic tunneling metal–semiconductor contact.

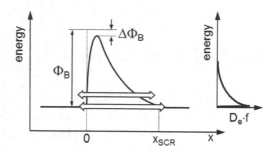

Figure 5.13. Barrier at a tunneling ohmic metal–semiconductor contact considering barrier lowering (caused by image charge) and energy distribution of electrons crossing the barrier.

To calculate the hole transmission probability of a triangular barrier at a metal contact with a highly *p*-type doped semiconductor, N_D^+, m_e^* and Φ_B in Equation (5.26) are replaced by N_A^-, m_h^* and $E_g - \Phi_B$.

Equation (5.26) is very useful for estimating the order of magnitude of the transmission probability at tunneling ohmic metal–semiconductor contacts. However, the values of T_{tr} are often underestimated due to barrier lowering and the energy distribution of free electrons (Figure 5.13). Furthermore, the real effective mass of the electron tunneling through the barrier and the exact shape of the barrier following from the solution of the Poisson equation have to be considered for a more precise analysis of T_{tr}.

A lowering of the barrier is caused by the image charge arising at the metal side of the contact when an electron is moving towards the barrier (image force lowering, also called the Schottky effect, see, for example, Sze, 1981). The image force increases with decreasing distance between the electron and the metal surface. The decrease of the barrier caused by the image potential energy ($\Delta\Phi_B$) is proportional to the reciprocal distance of the electron from the metal surface. Furthermore, the maximum of the reduced barrier is reached within the space charge region. The additional reduction of the barrier, which is caused by the image force, increases with increasing density of dopants. The value of $\Delta\Phi_B$ can be about 10–20% of Φ_B for metal contacts with highly doped crystalline semiconductors.

Electrons tunneling through a barrier have different kinetic energy levels. The energy distribution of free electrons is given by the product of the density of states and the Fermi-function (see Chapter 2). The width of the barrier is reduced and therefore T_{tr} is increased for electrons tunneling at higher energy. This so-called thermionic field emission enhanced tunneling becomes more important with decreasing density of dopants in highly doped semiconductors.

5.3.3 Contact resistance of a triangular tunneling barrier

The contact resistance (R_C) of a solar cell has to be compared with the tolerable series resistance (see Chapter 1), which is, for example, of the order of $10^{-1}\,\Omega\text{cm}^2$ or $10^{-4}\,\Omega\text{cm}^2$ for solar cells illuminated at air mass (AM)1.5 or at concentrated sunlight (concentration factor 1000), respectively. The reduction of the ohmic contact area at finger grids or at contact dots has to be taken into account as well. Therefore, the contact resistance of ohmic contacts should be of the order of $10^{-6}\,\Omega\text{cm}^2$ or even less for very efficient solar cells operated at concentrated sunlight.

The inverse contact resistance is defined as the derivative of the current density over the potential drop across the contact at 0 potential.

$$\frac{1}{R_C} = \frac{dI}{dU}\bigg|_{U \to 0} \tag{5.27}$$

The current across an ohmic contact, which can be supported by tunneling, is proportional to the transmission probability of the barrier, to the density of free charge carriers at the side of the highly doped semiconductor, and to the thermal velocity of free charge carriers. The barrier is reduced under forward bias or under illumination by the product of the elementary charge and the potential drop across the barrier. The tunneling current of electrons can be expressed by the equation

$$I_{\text{tunn},e} = q \cdot N_D^+ \cdot v_{\text{th}} \cdot \exp\left(-\frac{8}{3} \cdot \sqrt{\frac{m_e^* \cdot \varepsilon_r \cdot \varepsilon_0}{N_D^+}} \cdot \frac{\Phi_B - q \cdot U}{q \cdot \hbar}\right) \tag{5.28}$$

The resulting contact resistance of a tunneling ohmic contact is

$$R_{c,\text{tunn},e} = \frac{3 \cdot \hbar}{8 \cdot q \cdot v_{\text{th}}} \cdot \sqrt{\frac{1}{N_D^+ \cdot m_e^* \cdot \varepsilon_r \cdot \varepsilon_0}}$$

$$\times \exp\left(\frac{8}{3} \cdot \sqrt{\frac{m_e^* \cdot \varepsilon_r \cdot \varepsilon_0}{N_D^+}} \cdot \frac{\Phi_B}{q \cdot \hbar}\right) \tag{5.29}$$

It is no mean feat to obtain correct values for the effective mass of the tunneling charge carrier and for the dielectric constant, taking into account the extremely short duration of the tunneling process. A good approximation is reached for metal contacts with c-Si by taking the static dielectric constant of c-Si. Equation (5.29) can be simplified if approximating the effective electron or hole masses by the mass of the free electron in vacuum ($m_e = 9.1 \cdot 10^{-31}$ kg) and the relative dielectric constant by the relative dielectric constant of c-Si.

$$R_{c,\text{tunn},e} \approx \frac{\Omega \text{cm}^2}{1.26 \cdot 10^9} \cdot \sqrt{\frac{10^{19}\text{cm}^{-3}}{N_D^+}} \cdot \exp\left(79 \cdot \sqrt{\frac{10^{19}\text{cm}^{-3}}{N_D^+}} \cdot \frac{\Phi_B}{eV}\right) \qquad (5.30)$$

For getting $R_{c,\text{tunn},h}$, the values of N_D^+, m_e^* and Φ_B in Equation (5.30) should be replaced by the values of N_A^-, m_h^* and $E_g - \Phi_B$. The resistance of the tunneling contact is plotted in Figure 5.14 as a function of the density of donors.

Equation (5.30) is an approximation which does not take into account barrier lowering, the real barrier distribution in space, local morphology

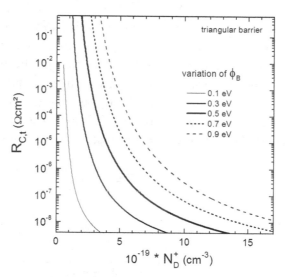

Figure 5.14. Resistance of an electron tunneling contact as a function of the density of donors for different values of Φ_B. The space charge region is approximated by a triangular barrier. The effective electron mass and the relative dielectric constant are set to the free electron mass and 12, respectively.

or the correct values of the effective mass and dielectric constant. The consideration of these factors demands sophisticated models and can lead to relatively large differences with the approximation given by Equation (5.30). Nevertheless, Equation (5.30) gives a clear idea about the orders of magnitude of tunneling contact resistances and their dependence on barrier height and density of dopants.

Small changes in the density of dopants or in the barrier height can cause large changes in the contact resistance. For example, the resistance of the tunneling contact decreases from about 10^{-1} to 10^{-6} Ωcm^2 if increasing the density of dopants from $2 \cdot 10^{19}$ to $5 \cdot 10^{19}$ cm^{-3} for a barrier of 0.5 eV. Therefore, the precise control of doping as well as the conditioning of the metal–semiconductor interface is very important for solar cells.

The exponential dependence of the tunneling resistance on the barrier height shows the importance of the choice of the metal or metal alloy for the contact, as well as the importance of the control of interface states at ohmic contacts. Sophisticated technologies of contact formation including preparation of semiconductor surfaces, metal deposition and activation of chemical reactions at metal–semiconductor interfaces have been developed especially for c-Si and III–V semiconductors with respect to requirements for electronic and opto-electronic applications. For example, $Au_{0.88}Ge_{0.12}$ is used for the formation of ohmic contacts at n-type doped GaAs (Sze, 1981).

Ohmic contacts can be realized with tunneling contacts between metals and highly doped semiconductors. The ohmic contact area can be strongly reduced for related tunneling contacts without remarkable power losses in the solar cell. In principle, there is no limitation of tunneling ohmic contacts for the implementation of (i) contact fingers at front contacts and (ii) contact dots at back contacts of solar cells (see Figure 5.3), and further, (iii) very low contact resistances allow operation of solar cells at highly concentrated sunlight.

5.3.4 *Tunneling resistance at degenerated pn-heterojunctions*

The monolithic series connection of stacked solar cells plays an important role for the realization of very high solar energy conversion efficiencies with multi-junction solar cells (see Chapter 6). A monolithic series

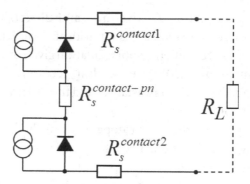

Figure 5.15. Equivalent circuit of two stacked solar cells with load, contact resistances between both solar cells and between metal leads and semiconductors.

connection of stacked solar cells means the connection of the Fermi-energies of majority charge carriers in n-type and p-type doped semiconductors by an ohmic contact so that the photovoltages of both solar cells are added at the load resistance (see the equivalent circuit in Figure 5.15). The ohmic contact between the n-type and p-type doped semiconductors should not influence the function of charge-selectivity of both connected solar cells. In multi-junction solar cells, the n-type and p-type doped regions of recombination contacts consist of semiconductors with different band gaps, E_{g1} and E_{g2} (i.e. a pn-heterojunction; see also Chapter 8).

The contact between n-type and p-type doped semiconductors leads to the formation of space charge regions at both sides of the contact (see Chapter 4). The space charge regions are very thin for very high densities of donors and acceptors at both the n-type and p-type doped sides of the pn-heterojunction. The Fermi-energies in the bulk of highly doped n-type or p-type doped semiconductors are within the conduction or valence bands, respectively. Such semiconductors form degenerated pn-heterojunctions. Degenerated pn-heterojunctions are applied for integrated series connection in multi-junction solar cells with very high solar energy conversion efficiency (see Chapter 8).

The wave functions of electrons and holes at the highly doped n-type and highly doped p-type sides of a degenerated heterojunction overlap within the space charge region. Therefore it is reasonable to assume that tunneling recombination of an electron from the highly doped n-type side with a hole from the highly doped p-type side takes place within the barrier.

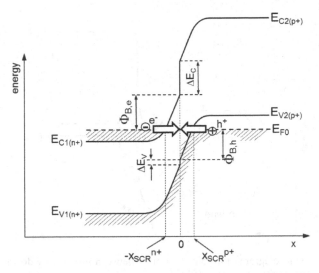

Figure 5.16. Band diagram of a degenerated *pn*-heterojunction forming a tunneling recombination contact.

Figure 5.16 shows a degenerated *pn*-heterojunction. Heterojunctions between two semiconductors are characterized by the offsets of the conduction and valence band edges (ΔE_C and ΔE_V, see also Chapter 8). The barrier heights for the tunneling of electrons ($\Phi_{B,e}$) and holes ($\Phi_{B,h}$) correspond to the respective potential drops across the space charge regions (see Chapter 4) at the *n*-type and *p*-type doped regions (see also task T5.5 at the end of this chapter).

For the example shown in Figure 5.16, the barrier heights for tunneling recombination of electrons and holes can be obtained from the following equations:

$$E_{g2} = \Phi_{B,e} + \Phi_{B,h} + \Delta E_C \qquad (5.31')$$

$$E_{g1} = \Phi_{B,e} + \Phi_{B,h} + \Delta E_V \qquad (5.31'')$$

The values of $\Phi_{B,e}$ and $\Phi_{B,h}$ depend on the densities of donors and acceptors and are connected via the condition of charge neutrality. The shapes of the barriers at degenerated *pn*-heterojunctions can be well approximated by a triangular barrier shape similarly to tunneling metal–semiconductor contacts. Therefore, as for tunneling metal–semiconductor contacts, Equation (5.29) can be used for calculating the contact resistance

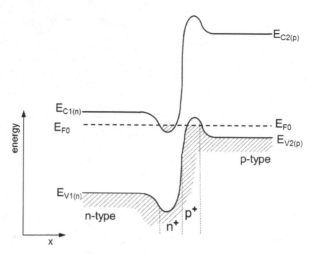

Figure 5.17. Band diagram of an ohmic contact between moderately doped n-type and p-type semiconductors.

of degenerated pn-heterojunctions. The tunneling or recombination resistance can be calculated by adding the tunneling resistance of electrons and holes. For this purpose Φ_B is replaced by $\Phi_{B,e}$ or $\Phi_{B,h}$ in Equation (5.29) and the corresponding values of m_e^*, m_h^*, N_A^+, N_D^+ and ε_r (the relative dielectric constant) are taken. The contact resistance of degenerated pn-heterojunctions is limited by the space charge region at the side with the lower density of dopants due to the exponential dependence of the transmission probability on the thickness of the barrier layer.

The band diagram of an ohmic contact between moderately doped n-type and p-type semiconductors is shown in Figure 5.17. The structure of the complete ohmic contact consists of a stack of $n/n^+/p^+/p$-type doped semiconductors. It is interesting to remark that the potential energy of minority charge carriers increases towards the tunneling pn-contact so that the density of minority charge carriers and the recombination rate are reduced (back surface field, see also Chapter 7).

5.3.5 *Tunneling via defect states near metal–semiconductor contacts*

Sometimes, a PV absorber cannot be highly doped, and a suitable metal cannot be found for realizing an ohmic contact. In such a case, it is useful to incorporate deep defect states in the absorber near the metal–semiconductor contact in order to obtain ohmic character of the contact. The point is that deep defect states can be occupied by electrons or holes and that wave functions of occupied and unoccupied defect states can overlap in space. Charge carriers can be transferred between localized states if there is an overlap between the wave functions of the states in two neighboring defects. Therefore, mobile charge carriers can penetrate a barrier by tunneling via localized defect states if the distance between localized defect states is of the order of the tunneling lengths.

The average distance between neighboring defects (Δx) is $(N_t)^{-3}$. For example, for defect densities of 10^{20} or $10^{19}\,\mathrm{cm}^{-3}$, Δx is about 2.2 or 4.6 nm, respectively, which is of the order of tunneling lengths. The presence of deep defect states in the space charge region reduces the distance for tunneling and the barrier. For example, if one defect state participating in tunneling is located in the middle of the space charge region, the distance for tunneling is reduced to $x_{SCR}/2$ and the barrier is reduced to $\Phi_B/2$. The reduction of x_{SCR} by two times is equivalent to the increase of $N_{d(a)}$ by 4 times (see Equation (4.5)). In this case, the tunneling resistance can be estimated by using Equation (5.30). With regard to Figure 5.14, tunneling assisted by defect states can reduce $R_{c,\mathrm{tunn}}$ by many orders of magnitude.

Figure 5.18 gives an impression about defect-assisted electron tunneling through a space charge region. First, transferred electrons fall from

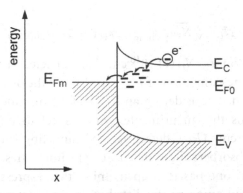

Figure 5.18. Band diagram of an ohmic metal–semiconductor contact with a large density of deep defect states in the *n*-type doped semiconductor near the contact.

the conduction band of the PV absorber into trap states closest to the end of the space charge region. Then the electrons tunnel towards the metal surface via several defect states. Successive tunneling via localized defect states is also called hopping.

For multiple trapping, the tunneling rate from an initial localized state (denoted by index i) to a final localized state (denoted by index f) can be described by Miller–Abrahams tunneling (Miller and Abrahams, 1960) in the simplest approximation. The initial and final localized states are characterized by their distance (R_{if}) and by their energies (E_i and E_f).

The system has only two parameters for Miller–Abrahams tunneling: the maximum tunneling rate ($\omega_{0,\text{tunn}}$) and the reciprocal tunneling length (α_{tunn}). The maximum tunneling rate corresponds to a vibration frequency of a bond configuration at which charge carriers are localized and is of the order of 10^{12} s^{-1}. The tunneling length is of the order of 0.5–1.5 nm. The tunneling step is thermally activated if the energy of the electron is larger in the final than in the initial state.

$$\omega_{\text{tunn}} = \omega_{0,\text{tunn}} \cdot \exp\left(-2 \cdot \alpha_{\text{tunn}} \cdot R_{if}\right) \cdot \exp\left(-\frac{(E_i - E_f) - |E_i - E_f|}{2 \cdot k_{\text{B}} \cdot T}\right)$$

$$(5.32)$$

A rather crude approximation can be obtained for the current density supported by trap-assisted tunneling through a barrier. For this purpose, a narrow distribution of R_{if} and directed tunneling are assumed. The average displacement per hopping step then corresponds to a value close to Δx and it can be written

$$I_{\text{tunn}} \sim q \cdot N_D \cdot \sqrt[3]{N_t} \cdot \omega_{0,\text{tunn}} \cdot \exp\left(-2 \cdot \alpha_{\text{tunn}} \cdot \sqrt[3]{N_t}\right) \qquad (5.33)$$

Densities of donors (N_D) of 10^{18} cm^{-3} and of defects (N_t) of 10^{20} cm^{-3} are assumed for the estimation of I_{tunn}. An I_{tunn} of the order of 10 mA/cm^2 may be supported if considering an α_{tunn} of 1.0 nm and four tunneling steps. This means that in principle, with respect to possible values of short-circuit currents (I_{SC}), defect-assisted tunneling can be applied for contacting PV absorbers illuminated at light intensities of the order of AM1.5. However, one has to keep in mind that expression (5.33) does not consider an inhomogeneous distribution of defect states or a change of the distribution of the electric potential due to charging of trap states. Furthermore, the controlled incorporation of defects near metal–semiconductor interfaces may be challenging.

5.4 Summary

Ohmic contacts are the interface connecting the active part of a solar cell, i.e. the PV absorber and the charge-selective contact, with metal leads connected to the external load. Ohmic contacts are passive, i.e. they should not contribute to photogeneration and charge-separation. Ideal ohmic contacts are also called recombination contacts. Fermi-level splitting is impossible at ideal ohmic contacts.

The surface recombination velocity is maximal at ohmic contacts in general. For reducing the surface recombination rate at ohmic contacts in solar cells, the density of photogenerated charge carriers has to be reduced for minority charge carriers. This can be done (i) by implementing barriers for minority charge carriers, for example, with back surface fields, (ii) reducing the area of ohmic contacts in addition to passivation of the surface area that is not covered by ohmic contacts or (iii) increasing the distance between the charge-selective contact and the ohmic contact to a value larger than the diffusion length of minority charge carriers. The implementation of a back surface field and the reduction of the ohmic contact area are decisive for reaching very high solar energy conversion efficiencies, for example, with c-Si (see Chapter 7) or GaAs (see Chapter 8) solar cells. The increase of the distance between the charge-selective and ohmic contacts to values larger than the diffusion length of minority charge carriers is important for thin-film solar cells based, for example, on chalcopyrite or CdTe thin-film absorber layers (see Chapter 9).

The electronic properties of an interface between a metal and a semiconductor depend on the metal work function, the electron affinity, doping of the semiconductor and on defect formation or chemical reactions at the metal–semiconductor interface. At ideal ohmic contacts, no barriers are formed for the transport of majority charge carriers from the semiconductor into the metal. This is the case if the metal work function ranges between the electron affinity and the Fermi-energy of an *n*-type doped semiconductor or between the ionization energy and the Fermi-energy of a *p*-type doped semiconductor. Therefore a metal or metal alloy with a reliable work function has to be chosen for forming an ohmic contact with a given doped semiconductor.

Barriers for the transport of majority charge carriers, so-called Schottky barriers, are often formed at metal–semiconductor interfaces and the barrier heights are sometimes independent of the metal work function. In the extreme cases (i) the barrier height at the metal–semiconductor interface correlates exactly with the metal work function (Schottky limit) or (ii) the barrier height does not correlate at all with the metal work function (Bardeen limit). The Bardeen limit is caused by a very high density of electronic states in the forbidden band gap of the semiconductor at the metal–semiconductor interface. The Fermi-energy at a semiconductor surface is pinned at a certain value if the density of surface defects exceeds values of the order of 10^{13} cm^{-2}.

A transport barrier for majority charge carriers from a semiconductor towards a metal–semiconductor contact is characterized by the surface band bending at the semiconductor surface. A barrier layer against the transport of majority charge carriers acts as a charge-selective contact in a semiconductor by depleting majority charge carriers in the barrier region. The thickness of a surface space charge region must be drastically reduced for enabling tunneling transport of majority charge carriers from the semiconductor across the metal–semiconductor interface into the metal.

The width of a surface space charge region can be reduced to a few nm for highly doped semiconductors with densities of majority charge carriers larger than 10^{19} cm^{-3}. Therefore, very thin and highly doped semiconductor layers are deposited at semiconductor absorbers to form ohmic metal–semiconductor contacts.

The barrier layer at a contact between a metal and a highly doped semiconductor can be approximated by a triangular barrier, the width and the height of which are given by the extension of the space charge region and by the barrier height at the metal–semiconductor contact, respectively. The transmission probability of the barrier layer increases exponentially with the increasing square root of the density of majority charge carriers and with decreasing barrier height. The reduction of the effective barrier by image force, the precise shape of the barrier layer, the distribution of the kinetic energy of mobile charge carriers and local fluctuations of the surface band bending can be considered in more detailed models.

The tunneling or contact resistance of a space charge region at a metal–semiconductor contact can be as low as $10^{-6}\,\Omega\text{cm}^2$ for low barrier heights of about 0.3 eV and for very high densities of majority charge carriers of about $3 \cdot 10^{19}\,\text{cm}^{-3}$. A contact resistance of about $10^{-6}\,\Omega\text{cm}^2$ is about 5 orders of magnitude lower than the tolerable series resistance of a solar cell illuminated at AM1.5. Therefore, the area of the ohmic contact can be strongly reduced in order to reduce the influence of surface recombination at the ohmic contact on the performance of solar cells. Additionally, solar cells with related ohmic contacts can be illuminated at concentrated sunlight without remarkable resistive losses at the ohmic contacts.

Very low contact resistances are also reached for tunneling between highly n-type doped and highly p-type doped semiconductors. Tunneling or recombination contacts are very important for the realization of integrated series connection in multi-junction solar cells with the highest solar energy conversion efficiency at concentrated sunlight (see Chapter 8).

Sometimes defect-assisted tunneling through barrier layers may be useful in the reduction of the resistance at ohmic contacts, for example, in the case of a relatively large barrier height.

5.5 Tasks

T5.1: Surface band bending in presence of surface defects

Ascertain the surface band bending at n-type doped semiconductors with equal densities of donor and acceptor surface defects of 10^{13} and $10^{11}\,\text{cm}^{-2}$. The densities of donor and acceptor surface states are constant over the E_g. The densities of donors, the effective densities of states at the conduction band edge (N_C) and E_g of the semiconductors are (a) $N_D = 10^{16}\,\text{cm}^{-3}$, $N_C = 3 \cdot 10^{19}\,\text{cm}^{-3}$ and $E_g = 1.1\,\text{eV}$ (c-Si) and (b) $N_D = 10^{17}\,\text{cm}^{-3}$, $N_C = 5 \cdot 10^{17}\,\text{cm}^{-3}$ and $E_g = 1.42\,\text{eV}$ (GaAs). The relative dielectric constant is 12.

T5.2: Fermi-level pinning

Is it possible to shift the Fermi-energy at a semiconductor surface by 0.1 eV for a density of surface defects of (a) $10^{11}\,\text{cm}^{-2}$ and (b) $10^{13}\,\text{cm}^{-2}$? The densities of donor and acceptor surface states are equal and constant over the E_g. The values of N_C, N_D, ε_r and E_g are $3 \cdot 10^{19}$ cm^{-3}, 10^{16} cm^{-3}, 12 and 1.1 eV, respectively.

T5.3: Engineering of surface band bending with interface defects

Calculate the surface band bending for n-type ($N_D = 10^{18}$ cm^{-3}) and p-type ($N_A = 10^{19}$ cm^{-3}) doped semiconductors with different densities of donor and acceptor surface defects of (a) $N_{st,a} = 3 \cdot 10^{13}$ and $N_{st,d} = 8 \cdot 10^{13}$ cm^{-2} and (b) $N_{st,a} = 10^{14}$ and $N_{st,d} = 2 \cdot 10^{13}$ cm^{-2}. The densities of donor and acceptor surface states are constant over the band gap. The values of N_C, N_V, ε_r and E_g are $1 \cdot 10^{18}$ cm^{-3}, $1 \cdot 10^{19}$ cm^{-3}, 12 and 1.42 eV, respectively. Discuss the relevance of the result for the formation of ohmic contacts.

T5.4: Transmission probability of tunneling contacts with inorganic and organic semiconductors

Inorganic and organic semiconductors can be highly doped while the effective electron mass can be very low for inorganic semiconductors (for example, $m_e^* = 0.066 \cdot m_e$) and the dielectric constant can be much lower for organic (for example, $\varepsilon_r = 3$) than for inorganic (for example, $\varepsilon_r = 12$) semiconductors. Compare the transmission probabilities of tunneling contacts with an inorganic and with an organic semiconductor for a barrier height of 0.4 eV and a density of donors of $5 \cdot 10^{19}$ cm^{-3}.

T5.5: Recombination contact between highly doped semiconductors

Estimate the tunneling resistance between two highly doped semiconductors with band gaps of 1.32 and 1.84 eV. The semiconductor with $E_g = 1.32$ eV is n-type doped ($N_D^+ = 5 \cdot 10^{19}$ cm^{-3}) and the semiconductor with $E_g = 1.84$ eV is p-type doped ($N_A^+ = 2 \cdot 10^{20}$ cm^{-3}). The effective electron and hole masses are 0.07 and 0.4 times the mass of the free electron, respectively. The dielectric constant and the conduction band offset are 12 and 0.3 eV, respectively. Discuss consequences for multi-junction solar cells under concentrated sunlight.

T5.6: Schottky-barriers for charge-separation and ohmic contacts

Schottky-barriers can be used for charge separation. Discuss advantages and disadvantages in comparison with a *pn*-junction and separate ohmic contacts.

6

Maximum Efficiency of Solar Cells

The diode equation of an ideal solar cell is obtained following the balance between thermal generation and radiative recombination (Shockley and Queisser, 1961; Würfel, 2005). The diode saturation current density of an ideal solar cell depends only on the band gap of the photovoltaic (PV) absorber and its temperature. The Shockley–Queisser limit describes the dependence of the solar energy conversion efficiency (η) of an ideal solar cell on the band gap (E_g) of its photovoltaic absorber illuminated at air mass (AM)1.5 and 25°C. The maximum value of η is 32% for an E_g between 1.1 and 1.5 eV. The reduction of thermalization losses by spectral splitting is analyzed for multi-junction solar cells with several E_g. Optimum combinations of E_g are derived from the current-matching condition of multi-junction solar cells. Maximum efficiencies are calculated for tandem solar cells, triple- and quadruple-junction solar cells. The maximum efficiency is further increased under illumination with concentrated sunlight, for example, up to 66% for a quadruple-junction solar cell illuminated with sunlight concentrated 1000 times. The increase of the quantum efficiency by down-conversion and impact ionization and the influence of up-conversion on the maximum efficiency are discussed.

6.1 Diode Equation of the Ideal Single-Junction Solar Cell

6.1.1 *Thermal generation and radiative recombination rate constant*

Photons absorbed in a photovoltaic absorber are converted into free electrons and holes. A minimum of recombination of free electrons and holes is desired for a maximum energy conversion efficiency of solar cells.

One absorbed photon generates one electron–hole pair. The quantum efficiency (see Chapter 1) of an ideal photovoltaic absorber is equal to 0 for photons with energies (E_{ph}) less than the forbidden E_g (band gap) and equal to 1 for all incoming photons with energies equal to or larger than E_g.

$$QE(E_{ph}) = \begin{cases} 0 & (E_{ph} < E_g) \\ 1 & (E_{ph} \geq E_g) \end{cases} \tag{6.1}$$

Equation (6.1) describes the condition of an ideal photovoltaic (PV) absorber with a continuum of states in the valence and conduction bands at energies below the valence band edge and above the conduction band edge, respectively. Condition (6.1) requires that (i) losses caused by reflection are minimized to 0, (ii) optical transmission losses are reduced to 0, i.e. the product of the absorption coefficient and of the optical path is much larger than 1 for all photon energies equal to or larger than E_g (see Chapter 2), and (iii) there are no collection losses, i.e. diffusion lengths of minority charge carriers are much larger than the absorption length at any wavelength of absorbed light (see Chapter 4).

The thermal generation rate (G_0) describes the generation of free electrons and holes due to absorption of blackbody radiation emitted by the photovoltaic absorber (see Chapter 2). The thermal generation rate is only a function of E_g for ideal absorbers.

$$G_0(E_g) = \int_{E_g}^{\infty} \Phi_0(E_{ph}) \cdot dE_{ph} \tag{6.2}$$

The photon flux of blackbody radiation ($\Phi_0(E_{ph})$) follows directly from the Planck equation while T_0 is the temperature of the ideal PV absorber under operation of the solar cell.

$$\Phi_0(E_{ph}) = \frac{2 \cdot \Omega_s}{h^3 \cdot c^2} \cdot \frac{(E_{ph})^2}{\exp\left(\frac{E_{ph}}{k_B \cdot T_0}\right) - 1} \tag{6.3}$$

For an ideal PV absorber, the space angle (Ω_s) in Equation (6.3) is equal to the full space angle (4π).

The density of free electrons and holes is balanced by all possible recombination processes in thermal equilibrium. This principle is also called the detailed-balance principle. Figure 6.1(a) shows thermal generation and radiative recombination.

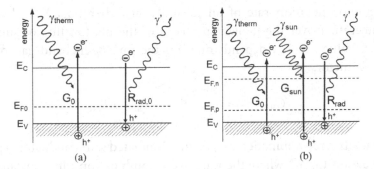

Figure 6.1. Energy diagram of an ideal PV absorber in thermal equilibrium (a) and under illumination (b). Photons emitted by blackbody radiation, photons emitted by the sun and photons emitted by radiative recombination of the PV absorber are denoted by γ_{therm}, γ_{sun} and γ', respectively.

The Shockley–Read–Hall (SRH) and surface recombination rates (see Chapter 3) are minimized to 0 for ideal photovoltaic absorbers. Furthermore, the densities of free electrons and/or holes are considered low in thermal equilibrium and consequently Auger recombination can be neglected in ideal PV absorbers. Therefore, the thermal generation rate is balanced only by radiative recombination in an ideal PV absorber ($R_{rad,0}$)

$$G_0\left(E_g, T_0\right) = R_{rad,0} \qquad (6.4)$$

The radiative recombination rate constant of an ideal PV absorber (B_{ideal}) can be calculated using the definition of the thermal generation rate of an ideal PV absorber (Equation (6.2)) and the density of intrinsic charge carriers (see Equation (2.36)):

$$B_{ideal} = \frac{G_0\left(E_g, T_0\right)}{n_i^2} \qquad (6.5)$$

For ideal PV absorbers, the radiative recombination rate constant depends only on the operation temperature and on the forbidden band gap, which is the only material parameter of an ideal PV absorber.

6.1.2 *Radiative recombination rate of an illuminated ideal absorber*

The densities of free electrons and holes of a PV absorber increase under illumination due to absorption of photons from sunlight (Figure 6.1(b)).

The photogeneration rate of an ideal PV absorber (G_{sun}) is obtained by integrating over the known spectrum of the photon flux of sunlight (Φ_{sun}). Value of G_{sun} depends only on the forbidden band gap of the semiconductor.

$$G_{sun}(E_g) = \int_{E_g}^{\infty} \Phi_{sun}(E_{ph}) \cdot dE_{ph} \qquad (6.6)$$

The radiative recombination rate of an illuminated semiconductor is given by Equation (3.11′) where the radiative recombination rate constant will be substituted by the expression in Equation (6.5) for ideal PV absorbers. Therefore, it can be written

$$R_{rad} = G_0(E_g, T_0) \cdot \frac{n \cdot p}{n_i^2} \qquad (6.7)$$

The densities of free electrons and holes (n and p, respectively) are given by the Boltzmann equations (Equations (2.30′) and (2.30″)). Under illumination, the Fermi-energies split into separate Fermi-energies for free electrons (E_{Fn}) and free holes (E_{Fp}). Following the Boltzmann statistics of free electrons and holes in a semiconductor and following the definition of the density of intrinsic charge carriers, the radiative recombination rate of an illuminated ideal photovoltaic absorber is given by the following expression:

$$R_{rad} = G_0(E_g, T_0) \cdot \exp\left(\frac{E_{Fn} - E_{Fp}}{k_B \cdot T_0}\right) \qquad (6.8)$$

Equation (6.8) means that the radiative recombination rate of an illuminated ideal PV absorber depends only on the thermal generation rate, the operation temperature of the solar cell and the Fermi-level splitting ($E_{Fn} - E_{Fp}$).

For ideal solar cells, the Fermi-level splitting in the ideal PV absorber is conserved across the charge-selective contact (see Chapter 4) and between the two ohmic contacts (see Chapter 5). Therefore, the potential at the external leads is equal to the Fermi-level splitting in the ideal PV absorber divided by the elementary charge. The radiative recombination rate can be written (for more details, see, Würfel, 2005):

$$R_{rad} = G_0(E_g, T_0) \cdot \exp\left(\frac{q \cdot U}{k_B \cdot T_0}\right) \qquad (6.9)$$

The radiative recombination rate of an illuminated ideal solar cell depends only on the thermal generation rate of the ideal PV absorber and on the potential drop at the external contacts of the solar cell.

6.1.3 Diode saturation current of an ideal single-junction solar cell

The generation and recombination rates in Equations (6.2), (6.4), (6.6) and (6.9) correspond to particle fluxes through a unit area within a unit time. Therefore, thermal generation, photogeneration and recombination rates can be ascribed directly to the respective current densities.

The difference between the sum of the thermal and photogeneration rates and the recombination rate of an ideal PV absorber is related to the current density through a load connected with an ideal solar cell. The electric current has the opposite direction of the direction of the flowing electrons. Therefore, the following equation can be written for the current density of an ideal solar cell:

$$I = q \cdot R_{\text{rad}} - (q \cdot G_0 + q \cdot G_{\text{sun}}) \tag{6.10}$$

Equation (6.10) can be transformed by taking into account Equation (6.9)

$$I = q \cdot G_0 \cdot \left[\exp\left(\frac{q \cdot U}{k_B \cdot T_0} \right) - 1 \right] - q \cdot G_{\text{sun}} \tag{6.11'}$$

$$I_0 = q \cdot G_0 \tag{6.11''}$$

$$I_{\text{SC}}^{\text{max}} = q \cdot G_{\text{sun}} \tag{6.11'''}$$

Equation (6.11') is the diode equation (see Chapter 1) of an ideal single-junction solar cell with an ideal photovoltaic absorber. The short-circuit current density is identical to the maximum short-circuit current density considered for the analysis of the ultimate solar energy conversion efficiency (see Chapter 2, Figure 2.3). The diode saturation current density (I_0) of the ideal solar cell is given by the thermal generation rate

$$I_0 = \frac{2 \cdot q \cdot \Omega_S}{h^3 \cdot c^2} \cdot \int_{E_g}^{\infty} \frac{(E_{\text{ph}})^2}{\exp\left(\frac{E_{\text{ph}}}{k_B \cdot T_0} \right) - 1} \cdot dE_{\text{ph}} \tag{6.12}$$

Equation (6.12) allows for the calculation of the I_0 density of ideal solar cells as a function of the forbidden band gap of the ideal PV absorber and

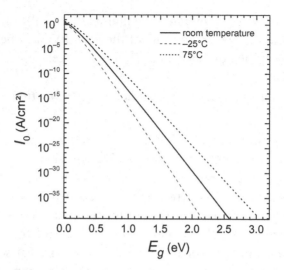

Figure 6.2. Dependence of the diode saturation current density of an ideal solar cell on the forbidden band gap of the ideal PV absorber at different temperatures.

of the operation temperature of the ideal solar cell. Figure 6.2 shows a plot of I_0 as a function of E_g.

For example, an ideal solar cell has a diode saturation current density of about 10^{-15} A/cm^2 at room temperature for an ideal PV absorber with a band gap equal to 1.1 eV.

For band gaps larger than about 0.3 eV, the dependence of I_0 on E_g fits excellently into the following equation:

$$I_0 = \frac{2 \cdot q \cdot \Omega_S \cdot k_B \cdot T_0}{h^3 \cdot c^2} \cdot (E_g)^2 \cdot \exp\left(-\frac{E_g}{k_B \cdot T_0}\right) \qquad (6.13)$$

Equation (6.13) gives the opportunity to calculate with ease the I_0 density of ideal solar cells. Equation (6.13) shows also that the I_0 density of ideal solar cells is thermally activated with activation energy equal to E_g.

The open-circuit voltage (V_{OC}) is limited by I_0 (see Chapter 1, Equation (1.19)). Figure 6.3 shows the dependence of V_{OC} on the band gap of an ideal solar cell illuminated at AM1.5.

The values of V_{OC} are significantly below the band gaps divided by the elementary charge. Therefore, the solar energy conversion efficiency of ideal solar cells is significantly lower than the ultimate efficiency (see Chapter 2). Furthermore, V_{OC} becomes very low when I_0 increases up to the order of

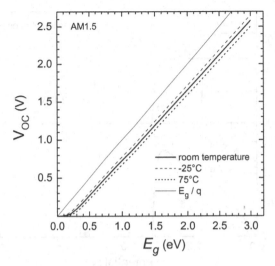

Figure 6.3. Dependence of the open-circuit voltage of an ideal solar cell on the forbidden band gap of the ideal PV absorber at different temperatures.

I_{SC} (short-circuit current), which is the case for band gaps less than about 0.2–0.3 eV. For this reason, band gaps of this region are not useful for PV solar energy conversion.

6.2 Maximum Efficiency of Ideal Single-Junction Solar Cells

6.2.1 *The Shockley–Queisser limit of solar energy conversion efficiency*

An ideal single-junction solar cell consists of an ideal photovoltaic absorber with one forbidden band gap and ideal charge-selective and ohmic contacts. The theoretical maximum energy conversion efficiency of ideal single-junction solar cells was calculated for the first time by Shockley and Queisser (1961) starting from the detailed-balance principle. The maximum efficiency of solar cells is generally obtained at a temperature of 25°C.

The current–voltage characteristics of ideal single-junction solar cells can be calculated by using Equation (6.11') with the diode saturation current density and the maximum short-circuit current density given by Equations (6.11'), (6.12) and (2.2), respectively. Figure 6.4 shows current–voltage characteristics and the corresponding power–voltage characteristics of ideal single-junction solar cells with different band gaps illuminated at

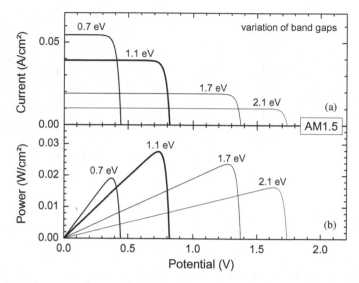

Figure 6.4. Current–voltage and power–voltage characteristics of ideal single-junction solar cells with different band gaps operated at room temperature and illuminated at AM1.5d.

AM1.5d. The short-circuit current densities decrease and the open-circuit voltages increase with increasing band gap. A maximum power is reached at 0.029 W/cm^2 for a band gap of 1.1 eV, which corresponds to an energy conversion efficiency of roughly 32%.

The resulting energy conversion efficiencies of ideal single-junction solar cells are plotted in Figure 6.5 as a function of the forbidden band gap (25°C, AM1.5d). This dependence is also known as the Shockley–Queisser limit. The solar energy conversion efficiency is about 32–33% in the maximum and above 30% for ideal PV absorbers with band gaps between 1.0 and 1.6 eV.

The maximum solar energy conversion efficiency decreases towards lower band gaps and reaches values below 1% for band gaps less than 0.3 eV. Therefore, semiconductors with band gaps below 0.3 eV are not used for PV solar energy conversion.

The maximum solar energy conversion efficiency is still about 21% for a band gap of 2 eV and decreases to about 11% for a band gap of 2.5 eV. For band gaps above 3.2 eV, the maximum solar energy conversion efficiency is less than 2%.

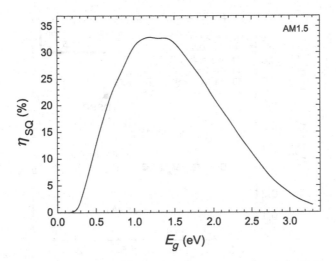

Figure 6.5. Dependence of the maximum solar energy conversion efficiency for ideal single-junction solar cells as a function of the band gap of ideal PV absorbers operated at 25°C and illuminated at AM1.5d (Shockley–Queisser limit).

It is useful to compare the solar energy conversion efficiency of record solar cells (Green *et al.*, 2016) with the Shockley–Queisser limit of ideal solar cells with the same band gaps. Normalized differences between the efficiency in the Shockley–Queisser limit and the efficiency of record solar cells are plotted in Figure 6.6 for some solar cells based on single crystalline absorbers of crystalline silicon (c-Si) (wafer based technology, see also Chapter 7) and of the III–V semiconductor family (epitaxy-based technology, GaAs, InP, see also Chapter 8), based on thin-film solar cells such as Cu(In,Ga)Se$_2$, CdTe, CuInS$_2$ and a-Si:H (absorber deposition on foreign substrates, see also Chapter 9) and based on nanocomposite absorbers with quantum dots, dye molecules (DSSC — dye-sensitized solar cell), CH$_3$NH$_3$PbI$_3$ perovskite-sensitized or organic polymers (see Chapter 10).

The lowest deviations between the Shockley–Queisser limit and a world record solar cell are obtained for advanced single crystalline absorbers. The deviations are about 14%, 23% and 32% for gallium arsenide (GaAs, $E_g = 1.42\,\text{eV}$), c-Si ($E_g = 1.1\,\text{eV}$) and indium phosphide (InP, $E_g = 1.34\,\text{eV}$) absorbers, respectively. The deviations increase for thin-film solar cells with an increasing band gap from 38% for copper indium gallium

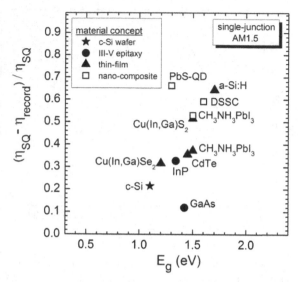

Figure 6.6. Deviation between the solar energy conversion efficiency in the Shockley–Queisser limit and record solar cells with the corresponding band gap for the different classes of materials (the efficiencies of record solar cells are taken from Green *et al.* (2016) and, in the case of $CH_3NH_3PbI_3$ (nanocomposite), Burschka *et al.* (2013)).

diselenide (Cu(In,Ga)Se$_2$, E_g = 1.2 eV), to about 40% for cadmium telluride (CdTe, E_g = 1.4 eV) and to about 56% for copper indium disulfide (CuInS$_2$, E_g = 1.5 eV) or 64% for hydrogenated amorphous silicon (a-Si:H, E_g = 1.7 eV).

The comparison between the Shockley–Queisser limit and the efficiency of record solar cells gives an impression about limitations imposed by materials and of achievements in semiconductor and solar cell technologies. For example, optical, resistive and recombination losses have been reduced to a minimum for solar cells with a GaAs absorber (see Chapter 8). Silicon crystals can be produced in large quantities and of the highest quality so that high solar energy conversion efficiencies are reached even in mass production (see Chapter 7). Fundamental materials limitations by disorder, for example in amorphous silicon, can be reduced but not eliminated by technological optimization (see Chapter 9). Therefore, the Shockley–Queisser limit does not give a realistic limit for related materials. The second part of the book (Chapters 7–10) gives deeper insight into materials concepts for different classes of PV absorbers and solar cells.

6.2.2 *Role of temperature and concentrated sunlight*

The Shockley–Queisser limit is defined for thermal generation at 25°C and photogeneration at AM1.5. The solar energy conversion efficiency increases with decreasing operation temperature of the solar cell due to the thermal activation of the diode saturation current density and/or with increasing concentration of the sunlight. The upper limit of the solar energy conversion efficiency is given by the ultimate solar energy conversion efficiency (see Chapter 2). Figure 6.7 compares The Shockley–Queisser limit with the ultimate efficiency and the dependence of the solar energy conversion efficiency on the band gap of ideal solar cells operated at 75°C and −25°C at AM1.5 and at concentrated sunlight with a concentration factor of 1000 by keeping the temperature at 25°C.

The ultimate efficiency has a maximum of about 48% at a band gap of about 1.0 eV. The efficiency in the Shockley–Queisser limit is 31% for an ideal solar cell with a band gap of 1.0 eV. The efficiency of the same ideal solar cell decreases to about 27% or it increases to about 34% if the temperature increases to 75°C or decreases to −25°C, respectively. On the

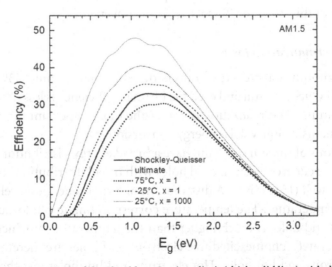

Figure 6.7. Comparison of the Shockley–Queisser limit (thick solid line) with the ultimate efficiency (thin solid line) and with the dependence of the maximum solar energy conversion efficiency on the band gap for ideal single-junction solar cells operated at AM1.5 at 75°C (thick dotted line) and −25°C (short dashed line) and operated at concentrated sunlight with a concentration factor of 1000 at 25°C (thin dotted line).

other hand, the efficiency of an ideal solar cell with a band gap of 1.0 eV increases to about 40% if the sunlight is concentrated by 1000 times and if the operation temperature is kept at 25°C.

Relative changes of the efficiency are much more pronounced at lower band gaps. For example, the efficiency of an ideal solar cell with E_g equal to 0.3 eV operated at room temperature increases from about 1% at AM1.5 to about 10% at concentrated sunlight with a concentration factor of 1000. In contrast the efficiency increases only from about 21% at AM1.5 to 24% at sunlight concentrated with a factor 1000 for an ideal solar cell with a band gap of 2.0 eV. Therefore, the range of useful PV absorbers can be extended to lower band gaps under illumination with concentrated sunlight.

The maximum concentration factor of sunlight is limited due to the extension of the sun as a light source. An absorber cannot emit photons at higher energy than it receives them, i.e. the maximum temperature of an absorber cannot be above the temperature at the surface of the sun (5800 K). The maximum concentration factor of sunlight can be calculated from the balance of energy fluxes that are received and emitted by an absorber. The maximum concentration factor of sunlight on earth is equal to 46,200 (see also task T6.2 at the end of this chapter). However, there is no practical meaning of this value for PV solar energy conversion.

6.2.3 Thermalization losses

The maximum solar energy conversion efficiency is only 33% in the Shockley–Queisser limit and even the ultimate efficiency is less than 50% in the maximum. Thermalization is one of the most important fundamental loss mechanisms in PV solar energy conversion.

Photons of the sun spectrum have a broad energy distribution and can therefore excite free electrons and holes with broad distributions of kinetic energy as well (Figure 6.8). A distribution of kinetic energies is related to a certain temperature. The temperature corresponding to photoexcited free electrons and holes is much higher than the temperature at which a solar cell is operated. Photoexcited free electrons and holes are therefore called hot electrons and hot holes. Hot electrons and hot holes loss their excess kinetic energy within a very short time due to interactions with atoms in the semiconductor. The cooling process of hot charge carriers is called thermalization and can last as long as tens of picoseconds. The excess kinetic

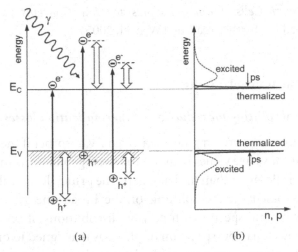

Figure 6.8. Photoexcitation of free electrons and holes at different kinetic energies denoted by the open double-ended arrows (a) and schematic energy distributions of excited and thermalized free electrons and holes (b).

energy of hot electrons and holes is transferred into vibrations of atoms in the photovoltaic absorber resulting in an increase of the temperature of the solar cell.

Free electrons and holes behave like an ideal gas. The kinetic energy of a particle in an ideal gas ($W_{\text{ideal gas}}$) is given by the following equation:

$$W_{\text{ideal gas}} = \frac{3}{2} \cdot k_B \cdot T \tag{6.14}$$

Photons of the sun spectrum can heat up photoexcited electrons and holes up to the temperature at the surface of the sun. In this case, photo-excited electrons and holes get a maximum kinetic energy. For the estimation of thermalization losses in solar cells the value of $W_{\text{ideal gas}}$ has to be doubled since excited electrons and holes obtain high temperatures. Therefore, the maximum kinetic energy of excited electrons and holes is 1.5 eV ($k_B \cdot T_S$ is about 0.5 eV). This value has to be compared with the band gap of a given PV absorber. For example, the thermalization losses are about 50% for a solar cell with a semiconductor having a band gap equal to 1.5 eV. For comparison, the losses are about 58% or 68% for the ultimate efficiency or for the Shockley–Queisser limit, respectively, for a band gap of 1.5 eV. Thermalization losses increase with decreasing band gap.

Physics of Solar Cells. From Principles to New Concepts by Würfel is recommended for further reading (Würfel, 2005).

6.3 Multi-Junction Solar Cells

6.3.1 *Spectral splitting for reduction of thermalization losses*

Thermalization losses are very low for a given PV absorber if the energies of absorbed photons have values between the band gap of the semiconductor and energies little larger than the band gap. The principle of multi-junction solar cells is based on the splitting of the broad energy spectrum of sunlight into several spectra with narrow distributions of photon energy (Figure 6.9). Each narrow spectrum of photons is assigned to one ideal PV absorber with a band gap equal to the lowest photon energy in the given narrow spectrum. The photons of each narrow spectrum are absorbed by the semiconductor with optimum band gap and converted into electric power at maximum efficiency with a separate solar cell. All separate solar

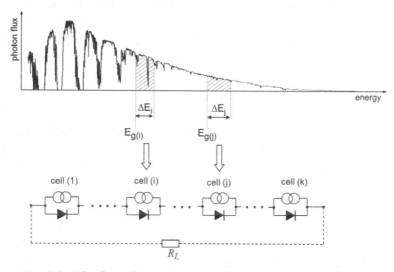

Figure 6.9. Principle of a multi-junction solar cell including spectral splitting of the sun spectrum into separate narrow spectra, assignment of an optimum band gap for each narrow spectrum and series connection of separate solar cells to one multi-junction solar cell. The PV absorber with the lowest band gap (bottom cell) has the index 1, the semiconductor with the highest band gap (top cell) the index k.

cells with different absorber are connected in series so that a multi-junction solar cell can be treated like one solar cell connected to one load resistance.

Figure 6.9 depicts the principle of an ideal multi-junction solar cell with k band gaps while $E_{g(1)}$ and $E_{g(k)}$ denote the lowest and highest values, respectively. The solar cells with the lowest and highest band gaps are called bottom and top solar cells, respectively.

Figure 6.10 shows an example for the band diagram of a multi-junction solar cell with four different band gaps in thermal equilibrium. A multi-junction solar cell with four different band gaps contains four pn-homo-junctions as charge-selective contacts (see Chapter 4) — one for each separate solar cell — and three degenerated pn-hetero-junctions as ohmic recombination contacts (see Chapter 5) for connecting the solar cells in series. The given example contains 16 layers of $n^+/n/p/p^+/n^+/n/p/p^+/n^+/n/p/p^+/n^+/n/p/p^+$. Real multi-junction solar cells (see Chapter 8) consist of even more layers. Obviously, such a layer system contains many junctions, which is characteristic of a multi-junction solar cell.

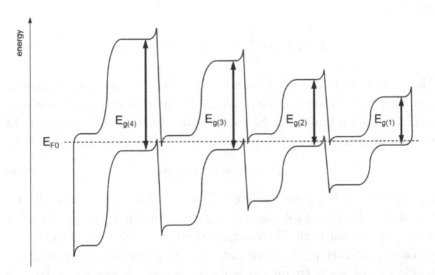

Figure 6.10. Schematic band diagram of a multi-junction solar cell with four different band gaps and its four corresponding pn-homo-junctions as charge-selective contacts and three degenerated pn-hetero-junctions as ohmic recombination contacts in thermal equilibrium.

The band gaps of multi-junction solar cells are numbered, starting with the lowest and finishing with the highest of the band gaps ($E_{g(1)}$, $E_{g(2)}$, $E_{g(3)}$, and $E_{g(4)}$ in Figure 6.10). First the sunlight passes into the ideal solar cell with the highest band gap so that all photons with energies equal or larger than $E_{g(4)}$ are absorbed. All photons with energies less than $E_{g(4)}$ pass through to the ideal solar cell with $E_{g(3)}$. All photons with energies in the interval between $E_{g(4)}$ and $E_{g(3)}$ are absorbed in the solar cell with $E_{g(3)}$. Photons with energies less than $E_{g(3)}$ pass through to the ideal solar cell with $E_{g(2)}$ where all photons with energies between $E_{g(3)}$ and $E_{g(2)}$ are absorbed. Photons with energies below $E_{g(2)}$ continue through the ideal solar cell with the lowest band gap where photons with energies in the interval between $E_{g(2)}$ and $E_{g(1)}$ are absorbed.

6.3.2 *Current-matching condition and optimum band gaps*

The spectrum of the photon flux is split into k intervals. The photons of each interval are absorbed by one separate solar cell. The lowest energy of each interval denoted by index j is assigned to a band gap $E_{g(j)}$. The short-circuit current density of the solar cell with the band gap $E_{g(j)}$ is equal to:

$$I_{SC(j)} = q \cdot \int_{E_{g(j)}}^{E_{g(j+1)}} \Phi_{sun}(E_{ph}) \cdot dE_{ph} \qquad (6.15)$$

The separate solar cells are connected in series in a multi-junction solar cell. Therefore the short-circuit current density of a multi-junction solar cell ($I_{SC(multi)}$) is limited by the lowest of the short-circuit currents of the separate solar cells ($I_{SC(i)}$).

$$I_{SC(multi)} = \min(I_{SC(i)}) \qquad (6.16)$$

Equation (6.16) is a rigid condition since the separate solar cell with the lowest $I_{SC(i)}$ limits the solar energy conversion efficiency of the complete multi-junction solar cell. Therefore, for maximum solar energy conversion efficiency, the current-matching condition of a multi-junction solar cell requires equal short-circuit current densities for all separate solar cells connected in series.

$$I_{SC(multi)} = I_{SC(i)}\big|_{i \in (1,k)} \qquad (6.17)$$

The solar energy conversion efficiency of a multi-junction solar cell can only be maximized if the current-matching condition is fulfilled. The current-matching condition is critical for attaining very high solar energy conversion efficiency in multi-junction solar cells. For ideal multi-junction solar cells the current-matching condition should be fulfilled by choosing the right values for the band gaps.

The highest I_{SC} density of a single-junction solar cell can be achieved with the lowest of the separate band gaps ($I_{SC(1)}$). For maximum $I_{SC(multi)}$, the I_{SC} density of a single-junction solar cell with the band gap of the bottom cell has to be divided into identical portions for each of the remaining solar cells

$$I_{SC(multi)} = \frac{I_{SC}^{single}(E_{g(1)})}{k} = \frac{q}{k} \cdot \int_{E_{g(1)}}^{\infty} \Phi_{sun}(E_{ph}) \cdot dE_{ph} \qquad (6.18)$$

The values of the band gaps $E_{g(j)}$ are obtained by the following condition:

$$(k + 1 - j) \cdot I_{SC(multi)}\big|_{j \in (1,k)} = q \cdot \int_{E_{g(j)}}^{\infty} \Phi_{sun}(E_{ph}) \cdot dE_{ph} \qquad (6.19)$$

The values of the forbidden band gaps in a multi-junction solar cell can be obtained from the direct application of the dependence of I_{SC} on E_g for ideal single-junction solar cells following condition (6.19). Figure 6.11 shows an example for the determination of the optimum band gaps in a multi-junction solar cell containing four ideal photovoltaic absorbers with separate band gaps ($k = 4$). A minimum band gap of $E_{g(1)} = 0.74$ eV is chosen.

The shortcircuit current density of an ideal single-junction solar cell with a band gap of 0.74 eV is 52 mA/cm^2. Therefore, with regard to Equation (6.18), $I_{SC(multi)}$ of the respective multi-junction solar cell is 13 mA/cm^2. The band gap of the solar cell with the index $j = 2$ is, with regard to Equation (6.19), equal to the band gap of a single-junction solar cell with a short-circuit current density of 39 mA/cm^2. Therefore, the value of $E_{g(2)}$ is equal to 1.1 eV. Furthermore, the values of $E_{g(3)}$, and $E_{g(4)}$ are equal to 1.47 and 1.95 eV, respectively.

Figure 6.12 shows the dependence of the band gaps $E_{g(i)}$ on $E_{g(1)}$ for tandem solar cells ($k = 2$), triple-junction solar cells ($k = 3$) and quadruple-junction solar cells ($k = 4$).

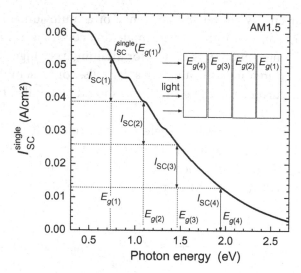

Figure 6.11. Determination of the optimum band gaps in a multi-junction solar cell with four separate solar cells and a band gap of the bottom cell of 0.74 eV.

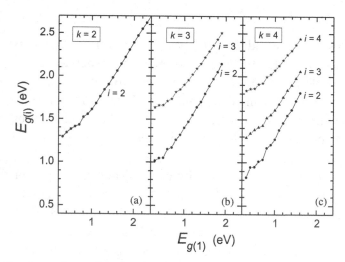

Figure 6.12. Dependence of the band gaps $E_{g(i)}$ ($i > 1$) on the band gap of the bottom solar cell ($E_{g(1)}$) for multi-junction solar cells with two (tandem solar cells, (a)), three (triple-junction solar cells (b)) and four (quadruple-junction solar cells (c)) photovoltaic absorbers under illumination at AM1.5 and optimum current-matching.

The dependencies do not follow smooth lines due to the dependence of the I_{SC} density of single-junction solar cells on the band gap. For example, the application of a band gap $E_{g(1)}$ of the bottom solar cell equal to 0.8 eV requires a band gap $E_{g(2)}$ of the top solar cell of about 1.45 eV for a tandem solar cell. The values of $E_{g(2)}$ decrease to about 1.3 eV and 1.05 eV for the respective triple- and quadruple-junction solar cells. The top solar cell of the triple-junction solar cell with $E_{g(1)}$ equal to 0.8 eV should have a band gap $E_{g(3)}$ equal to about 1.8 eV whereas the value of $E_{g(3)}$ is reduced to about 1.45 eV for the quadruple-junction solar cell. The band gap of the top solar cell of the quadruple-junction solar cell ($E_{g(4)}$) with $E_{g(1)}$ equal to 0.8 eV should be about 2.0 eV. The right combinations of optimal band gaps can be obtained in the similar way for any multi-junction solar cell.

6.3.3 Dependence of the efficiency on the number of band gaps

The open-circuit voltage of a multi-junction solar cell ($V_{OC(multi)}$) is equal to the sum of the open-circuit voltage of all separate solar cells ($V_{OC(i)}$), where k is the number of band gaps in the multi-junction solar cell (see also Figure 6.13).

$$V_{OC(multi)} = \sum_{i=1}^{k} V_{OC(i)} \qquad (6.20)$$

The ideal tunneling ohmic contacts for two neighboring separate solar cells connect the Fermi-energies of majority charge carriers with opposite sign at the same energy. The Fermi-level splitting of two neighboring separate solar cells is added by this way so that the potential drop across the external leads of a multi-junction solar cell corresponds to the sum of the Fermi-level splitting of all separate solar cells. This is shown in Figure 6.12 for the example of a multi-junction solar cell with four different band gaps under illumination at open circuit condition.

The solar energy conversion efficiency of a multi-junction solar cell (η_{multi}) is obtained from the sum of the power of the separate solar cells.

$$\eta_{multi} = \frac{q}{k} \cdot \int_{E_{g(1)}}^{\infty} \Phi_{sun}(E_{ph}) \cdot dE_{ph} \cdot \frac{\sum_{i=1}^{k}(FF_i \cdot V_{OC(i)})}{P_{sun}} \qquad (6.21)$$

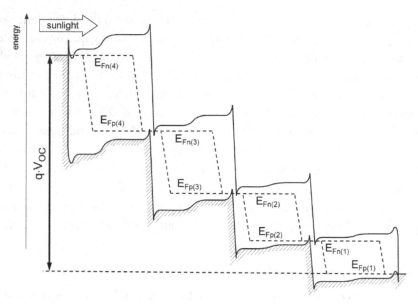

Figure 6.13. Schematic band diagram of a multi-junction solar cell with four different band gaps and corresponding four pn-homo-junctions as charge-selective contacts and three degenerated pn-hetero-junctions as ohmic recombination contacts under illumination at open-circuit condition.

Equation (6.21) shows that the solar energy conversion efficiency of multi-junction solar cells is larger than that of a single-junction solar cell due to the sum of the open-circuit voltages. In the following, η_{multi} will be analyzed as a function of band gap of the bottom cell, of the number of band gaps, and of the concentration factor.

The diode saturation current densities of the separate solar cells in a multi-junction solar cell ($I_{0(i)}$) are calculated using Equation (6.13) for each ideal solar cell with the band gap $E_{g(i)}$. The current–voltage and power–voltage characteristics are calculated for each separate ideal solar cell of the multi-junction solar cell by plugging $I_{\mathrm{SC(multi)}}$ and $I_{0(i)}$ into the diode Equation (6.11′).

Figure 6.14 shows an example for current–voltage and power–voltage characteristics of a quadruple-junction solar cell with $E_{g(1)}$, $E_{g(2)}$, $E_{g(3)}$ and $E_{g(4)}$ equal to 0.74, 1.1, 1.47 and 1.95 eV, respectively. As already mentioned, the short-circuit current density is 13 mA/cm^2 for the given quadruple-junction solar cell. The values of the open-circuit voltages are

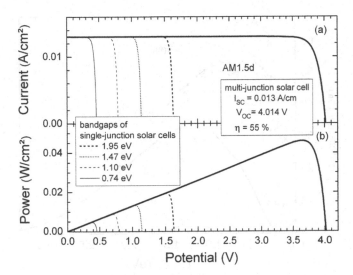

Figure 6.14. Current–voltage (a) and power–voltage (b) characteristics of an ideal quadruple-junction solar cell and the ideal single-junction solar cells with band gaps $E_{g(1)}$, $E_{g(2)}$, $E_{g(3)}$ and $E_{g(4)}$ equal to 0.74, 1.1, 1.47 and 1.95 eV (thin solid, thin dashed, dotted and thick dashed lines, respectively) within the quadruple-junction solar cell illuminated at AM1.5d.

0.47, 0.78, 1.14 and 1.63 V for $V_{OC(1)}$, $V_{OC(2)}$, $V_{OC(3)}$ and $V_{OC(4)}$, respectively. The corresponding open circuit voltage of the quadruple-junction solar cell is equal to 4.02 V. The values of the maximum power are 5, 9, 13 and 19 mW/cm^2 for the single-junction solar cells with $E_{g(1)}$, $E_{g(2)}$, $E_{g(3)}$ and $E_{g(4)}$, respectively. The total power of the given ideal quadruple-junction solar cell is therefore equal to 46 mW/cm^2. The intensity of the direct sunlight is 84.4 mW/cm^2 at AM1.5d for the given sun spectrum. Therefore the solar energy conversion efficiency is about 55% for the given quadruple-junction solar cell. The efficiency of 55% is larger than the maximum solar energy conversion efficiency in the Shockley–Queisser limit (33%).

Figure 6.15 summarizes the dependencies on $E_{g(1)}$ of the solar energy conversion efficiencies of single-, triple- and quadruple-junction solar cells and tandem solar cells. A maximum efficiency of 44% is reached for tandem solar cells with $E_{g(1)} = 1.1$ eV and $E_{g(2)} = 1.7$ eV. The increase of the efficiency is very strong for low values of $E_{g(1)}$ due to the influence of the top solar cell with $E_{g(2)}$ around 1.3–1.5 eV. Combinations of $E_{g(1)}$ in the range

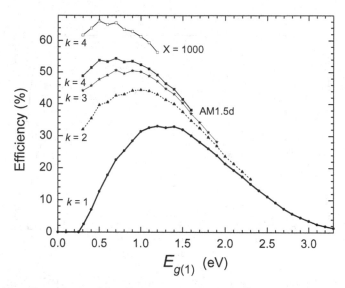

Figure 6.15. Dependence of the solar energy conversion efficiency on the band gap of the bottom solar cell ($E_{g(1)}$) of ideal single-junction solar cells ($k = 1$, circles), tandem solar cells ($k = 2$, triangles), triple-junction solar cells ($k = 3$, stars) and quadruple-junction solar cells ($k = 4$, squares) operated at room temperature and AM1.5d (filled symbols) and of quadruple-junction solar cells operated at room temperature and concentrated sunlight with concentration factor 1000 (open squares).

between 0.4 and 1.6 eV with the respective values of $E_{g(2)}$ between 1.35 and 2.0 eV are suitable for increasing the solar energy conversion efficiency of tandem solar cells above the maximum efficiency in the Shockley–Queisser limit of single-junction solar cells.

    Solar energy conversion efficiencies above 40% can be realized with ideal tandem solar cells for $E_{g(1)}$ between 0.6 and 1.3 eV ($E_{g(2)}$ between 1.37 and 1.75 eV, respectively). This shows that semiconductors with a wide range of band gaps can be applied for the realization of high-efficiency tandem solar cells. The difference between the solar energy conversion efficiency of single-junction and tandem solar cells becomes small for values of $E_{g(1)}$ above 2 eV.

    An efficiency slightly above 50% can be reached with triple-junction solar cells with the values of about 0.7, 1.15 and 1.75 eV for $E_{g(1)}$, $E_{g(2)}$ and $E_{g(3)}$, respectively. The highest efficiency of 36% at AM1.5 has been achieved for a triple-junction solar cell based on III–V semiconductors (see Chapter 8) with a similar combination of band gaps.

Solar energy conversion efficiencies above 48% can be obtained with ideal triple-junction solar cells for $E_{g(1)}$ between 0.6 and 1.1 eV corresponding to $E_{g(2)}$ between 1.15 and 1.45 eV and $E_{g(3)}$ between 1.65 and 1.95 eV, respectively.

The application of quadruple-junction solar cells presents the opportunity of increasing the solar energy conversion efficiency to values of about 54–55% for a combination of band gaps with energies of 0.7, 1.02, 1.40 and 1.92 eV for $E_{g(1)}$, $E_{g(2)}$, $E_{g(3)}$ and $E_{g(4)}$, respectively. Incidentally, the increase of the maximum efficiency is about 1.08 times for the transition from a triple- to a quadruple-junction solar cell. For purposes of comparison, the increase of the maximum efficiency is about 1.33 or 1.14 times for the transition from a single-junction solar cell to a tandem solar cell and from a tandem solar cell to a triple-junction solar cell, respectively. The relative increase of the efficiency reduces with increasing number of band gaps in multi-junction solar cells. Therefore the usefulness of multi-junction solar cells with more than 4 or 5 different band gaps has to be critically analyzed.

The solar energy conversion efficiency of ideal quadruple-junction solar cells is larger than 53% for $E_{g(1)}$ between 0.5 and 0.9 eV ($E_{g(2)}$ between 0.95 and 1.17 eV, $E_{g(3)}$ between 1.35 and 1.6 eV and $E_{g(4)}$ between 1.85 and 2.0 eV, respectively).

The solar energy conversion efficiency can be further increased by concentrating the sunlight. An example is given for the dependence of the solar energy conversion efficiency on $E_{g(1)}$ of a quadruple-junction solar cell illuminated with sunlight concentrated 1000 times. The solar energy conversion efficiency reaches 66%. This demonstrates the high potential of quadruple-junction solar cells operated at highly concentrated sunlight. The maximum efficiency is obtained for band gaps with energies of 0.5, 0.95, 1.35 and 1.84 eV for $E_{g(1)}$, $E_{g(2)}$, $E_{g(3)}$ and $E_{g(4)}$, respectively. The values of these band gaps are red shifted in comparison to the band gaps of the quadruple-junction solar cell with maximum efficiency at AM1.5. The shift to lower band gaps of a quadruple-junction solar cell with maximum efficiency at concentrated sunlight is caused by the fact that the solar energy conversion efficiency increase is stronger for lower band gaps, as mentioned before. So far, the highest solar energy conversion efficiency of 44.7% has been reached with a quadruple-junction solar cell operated at concentrated sunlight with a concentration factor of 297 (Fraunhofer, 2013).

6.4 Alternative Concepts for Increasing the Efficiency

6.4.1 *Increased quantum efficiency with higher energetic photons*

A part of thermalization losses may be avoided if the energy of one photon exceeding the energy of the doubled band gap can be converted into the excitation of two electron–hole pairs. This is possible due to impact ionization in which a free charge carrier with high kinetic energy collides with an electron in the valence band and excites it into the conduction band. The quantum efficiency of an absorption step followed by impact ionization is equal to 2 (Figure 6.16).

The quantum efficiency can exceed 1 in a solar cell (Kolodinski *et al.*, 1993) and can become even larger than 2 as reported for optical experiments on quantum dots (multi-exciton generation, see also Chapter 10) (Schaller and Klimov, 2004).

Another way to increase the quantum efficiency to values above 1 makes use of so-called down-conversion. The energy of one photon with high energy can be split into the energy of several photons with lower energy in down-conversion processes (Figure 6.17). In down-conversion, an electron is excited from a ground state with energy E_1 to an excited state E_3. The excited electron relaxes to an intermediate excited state with energy E_2 between E_1 and E_3 by a emitting a photon. Furthermore, the electron relaxes

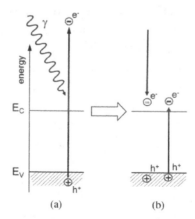

Figure 6.16. Schematic of impact ionization including the absorption of one photon with high energy (a) and the excitation of two electron–hole pairs (b).

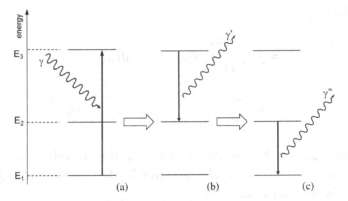

Figure 6.17. Schematic of down-conversion including the absorption of one photon with high energy (a) and the emission of two photons with lower energies ((b) and (c)).

from the state at E_2 to the ground state by emitting a second photon. The emitted photons can be absorbed by a photovoltaic absorber if the energy of the emitted photons is equal to or larger than the band gap of the photovoltaic absorber. Therefore, the quantum efficiency of a solar cell can be increased by covering the surface of a solar cell with a down-converter for photons exceeding the doubled band gap. Impact ionization and down-conversion are equivalent in this sense.

The band gap of an ideal photovoltaic absorber remains unchanged if processes of impact ionization or down-conversion are incorporated into the analysis of the maximum solar energy conversion efficiency. Therefore, only I_{SC} is changed in the diode equation of an illuminated solar cell, which can be expressed by the following equation:

$$I_{SC} = q \cdot \sum_{i=1}^{n} i \cdot \int_{i \cdot E_g}^{(i+1) \cdot E_g} \Phi_{sun}(E_{ph}) \cdot dE_{ph} \qquad (6.22')$$

Equation (6.22') describes the maximum effect of impact ionization or down-conversion. However, electrons can also be excited from states deeper in the valence band, which results in an excited electron with less energy. This behavior leads to a reduction of the influence of impact ionization on I_{SC}. Therefore, photon energies much larger than the doubled band gap are required for a significant influence of impact ionization on I_{SC}. As a more realistic assumption, the influence of impact ionization on I_{SC} will start with doubled quantum efficiency at photon energies larger than the tripled band gap.

$$I_{SC} = q \cdot \left(\int_{E_g}^{2 \cdot E_g} \Phi_{sun}(E_{ph}) \cdot dE_{ph} \right.$$

$$\left. + \sum_{i=1}^{n} i \cdot \int_{(i+1) \cdot E_g}^{(i+2) \cdot E_g} \Phi_{sun}(E_{ph}) \cdot dE_{ph} \right) \qquad (6.22'')$$

Equation (6.22′) can be rewritten in such a way that the values of the short-circuit current densities of ideal solar cells related to the Shockley–Queisser limit can be used for calculating I_{SC}

$$I_{SC} = q \cdot \sum_{i=1}^{n} \int_{i \cdot E_g}^{\infty} \Phi_{sun}(E_{ph}) \cdot dE_{ph} = \sum_{i=1}^{n} I_{SC,i \cdot E_g}^{single} \qquad (6.23)$$

Figure 6.18 compares the dependencies of the solar energy conversion efficiency on the band gap of an ideal single-junction solar cell for the Shockley–Queisser limit and for considering impact ionization.

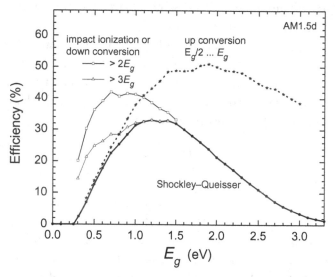

Figure 6.18. Comparison of the Shockley–Queisser limit (thick solid line) with the dependence of the efficiency on the band gap of an ideal solar cell taking into account impact ionization or down-conversion beginning at $2 \cdot E_g$ (squares) or $3 \cdot E_g$ (triangles) and taking into account up-conversion for photons with energies between $E_g/2$ and E_g (stars).

Maximum impact ionization or optimum down-conversion may play a significant role for band gaps below 1.5 eV since the photon flux is already very low at photon energies above 3 eV. For maximum impact ionization or optimum down-conversion, the solar energy conversion efficiency reaches a value of about 42% at a band gap of 0.7 eV and values above 39% for band gaps between 0.6 and 1.2 eV. These values are higher than the maximum efficiency in the Shockley–Queisser limit and demonstrate the potential for impact ionization and/or down-conversion. However, one has to take into account that the solar energy conversion efficiency will not exceed the maximum efficiency in the Shockley–Queisser limit if considering the more realistic assumption expressed by Equation (6.22″). Therefore, it seems difficult to improve significantly the solar energy conversion efficiency if assuming even moderate losses on the maximum influence of impact ionization or down-conversion. The experimental task of improving high-efficiency solar cells with down-conversion is challenging.

6.4.2 Up-conversion of lower energetic photons

The use of photons with energies below the band gap of a photovoltaic absorber in a solar cell has a large potential for the increase of the solar energy conversion efficiency, and especially so for semiconductors with wide band gaps. For this purpose, the energy of two photons with energy below the band gap should be converted into the energy of one photon with energy equal to or larger than the band gap of the semiconductor. Processes in which the energy of two photons is converted into a higher energy of one photon are called up-conversion. The quantum efficiency is 0.5 or less for up-conversion.

Figure 6.19 shows an example of a transition scheme for up-conversion in a system with six energy levels. An electron is excited from its ground state E_1 to an excited state E_2 by a photon with low energy. The excited electron relaxes very rapidly into a state E_3 at a slightly reduced energy. The excited electron is metastable in state E_3, i.e. further relaxation to lower energies takes a much longer time compared to the time it takes for the second excitation step to the energy E_4, again caused by another photon with low energy. The electron relaxes from state E_4 to state E_5 at a slightly reduced energy. Relaxation of the excited electron from state

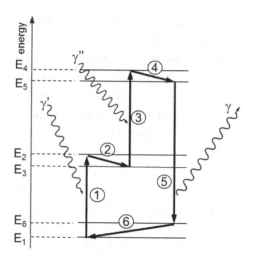

Figure 6.19. Transition scheme for up-conversion of photon energies including elementary steps of excitation of an electron from a ground state E_1 to a state E_2 (1), fast relaxation of the electron from state E_2 to state E_3 from where recombination to states at lower energy is strongly reduced (2), subsequent excitation from state E_3 to state E_4 (3), fast relaxation of the electron from state E_4 to state E_5 from where recombination to states E_2 and E_3 is strongly reduced (4), radiative relaxation from state E_5 to a state E_6 including photon emission (5) and relaxation from state E_6 back to the ground state E_1 (6).

E_5 to the states at energies E_2 and E_3 is much slower than relaxation to the state E_6. Therefore, the electron relaxes from state E_5 at high energy to state E_6 at low energy under the emission of a photon with high energy. In the last step, the electron relaxes from the state at energy E_6 to the ground state at energy E_1 from which a next up-conversion cycle can start.

Up-conversion demands sophisticated optical or molecular systems with reliable transitions. It seems that up-conversion is practically impossible for a broad spectrum of incoming photons. The increase of the waiting time of an excited electron before the second excitation occurs is challenging since relaxation back to the ground state must be efficiently suppressed during this waiting time. The waiting time can be reduced under concentrated sunlight.

Up-converting optical elements can be incorporated into solar cells, for example, between the back surface of the solar cell and a back reflector without changing the photovoltaic absorber. An optimum up-converter will be able to convert all photons with energy between the half of the band gap of the ideal photovoltaic absorber and its band gap to photons with energies above the band gap. Therefore the short-circuit current density can be calculated in the following way for a solar cell with an optimum up-converter

$$I_{SC} = \frac{q}{2} \cdot \int_{E_g/2}^{E_g} \Phi_{sun}(E_{ph}) \cdot dE_{ph} + q \cdot \int_{E_g}^{\infty} \Phi_{sun}(E_{ph}) \cdot dE_{ph} \quad (6.24)$$

The dependence of the solar energy conversion efficiency on the band gap of a single-junction solar cell with optimal up-conversion is plotted in Figure 6.18. The relative increase of the solar energy conversion efficiency is tremendous for solar cells with large band gaps. For example, the efficiency increases from about 3.5% for a solar cell with a band gap of 3 eV in the Shockley–Queisser limit to about 39% for a solar cell with the same band gap but with additional optimum up-conversion.

The maximum solar energy conversion efficiency of about 50% may be reached for a single-junction solar cell with up-conversion and a band gap of 1.9 eV. The solar energy conversion efficiency is above 48% for solar cells with band gaps between 1.4 and 2.2 eV and optimum up-converters. However, search for reliable up-converters is still ongoing.

Electrolytic water splitting requires potential energies of more than 1.23 eV (Walter *et al.*, 2010). Potentials above 1.23 V can be reached with single-junction solar cells with high band gap. Therefore, the successful development of suitable up-converters may become important, especially for the production of solar fuels by photocatalytic water splitting.

6.5 Summary

All losses that can be avoided in ideal solar cells have to be minimized to zero for achieving the maximum solar energy conversion efficiency. These losses include (i) optical losses caused by reflection and transmission, (ii) resistive losses caused by series and shunt resistances and (iii) recombination losses caused by defects in the bulk und at the surfaces of the photovoltaic absorber. Furthermore it is assumed that radiative recombination is much stronger than Auger recombination so that Auger recombination can be neglected for the analysis of the maximum efficiency. As a consequence, the maximum efficiency depends only on the band gap (E_g) of a photovoltaic absorber, on its temperature and on the spectrum and intensity of sunlight.

A photovoltaic absorber absorbs the part of the blackbody radiation with photon energies equal to or larger than E_g resulting in thermal generation of free electrons and holes. The thermal generation rate ($G_0(E_g, T_0)$) is balanced by the radiative recombination rate so that the rate constant can be derived as a function of the temperature and E_g.

Under illumination, the radiative recombination rate is proportional to the product of $G_0(E_g, T_0)$ and the exponential of the Fermi-level splitting ($E_{Fn} - E_{Fp}$) divided by $k_B \cdot T_0$. In an ideal solar cell, the Fermi-level splitting is equal to the potential difference at the external leads multiplied by the elementary charge ($q \cdot U$) due to ideal charge-selective and ohmic contacts. The current density of an illuminated solar cell follows from the radiative recombination, thermal generation and photo-generation rates. The resulting diode saturation current density is given by $q \cdot G_0(E_g, T_0)$. The calculation of the diode saturation current density from the thermal generation rate is key for the analysis of the maximum solar energy conversion efficiency of solar cells.

At a given temperature and for a given sun spectrum, the maximum efficiency depends only on E_g. The Shockley–Queisser limit describes the dependence of the maximum efficiency on E_g under illumination at AM1.5 and at a temperature of 25°C. The maximum efficiency in the Shockley–Queisser limit is about 32% for solar cells with E_g between 1.1 and 1.5 eV. A comparison of the efficiency of real solar cells with the efficiency in the Shockley–Queisser limit shows the potential for solar energy conversion of a particular PV absorber and of advanced technologies for producing solar cells with minimum losses. The lowest deviation between the Shockley–Queisser limit and the efficiency of record solar cells is obtained for a GaAs absorber.

Losses are dominated by thermalization in the Shockley–Queisser limit. A (realistic) further increase of the maximum efficiency of solar cells is only possible by reducing thermalization losses. For this purpose (i) spectral splitting of the sun spectrum and (ii) separate conversion with an optimum E_g for each part of the sun spectrum are required.

Solar energy conversion efficiencies above the Shockley–Queisser limit can be realized with stacked multi-junction solar cells where the separate solar cells are connected in series. The bottom cell has the lowest band gap ($E_{g(1)}$). The top cell has the highest band gap ($E_{g(k)}$, k is the number of band gaps in the multi-junction solar cell). Each subsequent solar cell has a band gap ($E_{g(i)}$) lower than the band gap of the solar cell in front $E_{g(i+1)}$ but larger than the band gap of the solar cell that follows ($E_{g(i-1)}$). Therefore, each solar cell with $E_{g(i)}$ absorbs photons in the interval between $E_{g(i+1)}$ and $E_{g(i)}$.

In a multi-junction solar cell, the short-circuit current density (I_{SC}) has to be the same in each separate solar cell (current-matching condition) since I_{SC} of the multi-junction solar cell is limited by the solar cell with the lowest I_{SC}. The optimum combinations of band gaps in multi-junction solar cells, respectively, can be easily derived from the current-matching condition. The optimum band gaps of the bottom and top cells are 1.1 and 1.7 eV, respectively, in a tandem solar cell.

The maximum solar energy conversion efficiency is about 44%, 50% and 55% for tandem solar cells, triple-junction solar cells and quadruple-junction solar cells, respectively, operated at room temperature and illuminated at AM1.5. The maximum efficiency can be further increased under illumination with concentrated sunlight, for example, up to 66% for a quadruple-junction solar cell illuminated with sunlight concentrated 1000 times. A solar energy conversion efficiency as high as 44.7% has been achieved with a quadruple-junction solar cell illuminated with concentrated sunlight (Green *et al.*, 2015).

The maximum efficiency of solar cells may also be increased by increasing the quantum efficiency of photogeneration. For this purpose, impact ionization can be applied. The number of photons absorbed by the solar cell can be increased by introduction of optical down-converters for photons with energies above the doubled band gap of the photovoltaic absorber. Impact ionization and down-conversion are equivalent. The implementation of optical up-converters at the back side of solar cells provides a further alternative with which to overcome the Shockley–Queisser limit. However, in reality increasing the solar energy conversion efficiency with concepts of impact ionization, down-conversion and up-conversion remains challenging.

6.6 Tasks

The tasks T6.1–T6.4 are well-suited for working in groups. Figures T6.1 and 2.4. are used as nomograms for ascertaining values of I_{SC} and fill factors (FF) by interpolation without the need for integrating over the sun spectrum or for simulating current–voltage characteristics.

T6.1: Shockley–Queisser limit

Ascertain the Shockley–Queisser limit for solar cells with ideal absorbers with band gaps of (a) 0.3, (b) 1.3, (c) 2.0 and (d) 2.7 eV by using Figures T6.1 and 2.4.

T6.2: Maximum efficiency of tandem solar cells

Ascertain the efficiency of ideal tandem solar cells with a band gap of the bottom cell of (a) 0.5, (b) 1.0 and (c) 1.5 eV operated at 50°C by using Figures T6.1 and 2.4.

T6.3: Band gaps of absorbers in multi-junction solar cells

Ascertain the optimum band gaps of the absorbers in a multi-junction solar cell with five different absorbers if the band gap of the bottom cell is (a) 0.5 or (b) 1.0 eV.

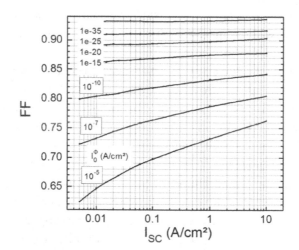

Figure T6.1.   Fill factor as a function of I_{SC} and I_0.

T6.4: Concentrator multi-junction solar cells

Calculate the maximum solar energy conversion efficiency of an ideal triple-junction solar cell with a band gap of the bottom cell of 0.7 eV under illumination with concentrated sunlight for concentration factors of 20, 200, 500 and 1000 by using Figures T6.1 and 2.4.

T6.5: Maximum concentration factor of sunlight

Ascertain the maximum concentration factor of sunlight by using the balance between the power of sunlight and the power of thermal radiation.

T6.6: Thermodynamic limit of solar energy conversion

Ascertain the maximum solar energy conversion efficiency in the thermodynamic limit. Heating of an absorber or medium from which work is obtained via a Carnot process is taken into account.

T6.7: Critical remark on solar cells with metal intermediate band

Some people believe that solar energy conversion efficiencies above the Shockley–Queisser limit are possible for solar cells with an absorber which has a so-called metal intermediate band between the valence and conduction band edges. Demonstrate that the efficiency cannot exceed the Shockley–Queisser limit in that case by taking into account ideal absorption.

Part II

Materials Specific Concepts

7

Solar Cells Based on Crystalline Si

Crystalline silicon (c-Si) solar cells are the backbone of large-scale photovoltaic (PV) energy production. Over more than half a century of technological development, the solar energy conversion efficiency of c-Si solar cells has been increased from 6% (Chapin *et al.*, 1954) to 25.6% (Masuko *et al.*, 2014) and the production costs have diminished drastically. This chapter starts with principle design rules of classical c-Si solar cells based on 200–400 μm thick c-Si wafers and consisting of a thick p-type doped base, a thin n-type doped emitter, a front contact finger grid, a back contact and an antireflection coating. Architectures for reducing shading losses by the front contact and for reducing resistive losses by the front and back contacts are discussed briefly. The role of segregation at the liquid–solid interface is illuminated for the fast growth of homogeneously doped and extremely pure c-Si crystals. Wafering by means of wire sawing and by kerfless transfer is mentioned. Processes employed for the emitter formation, such as diffusion of phosphorus in c-Si, laser assisted doping and ion implantation are described. The formation of well-passivated charge-selective hetero-junctions between c-Si and undoped/doped amorphous silicon double layers is explained. Recombination losses at c-Si surfaces are minimized by cleaning, structuring and passivation treatments including etching, oxidation, hydrogen passivation, deposition of amorphous silicon nitride or aluminum oxide and formation of a back surface field. Concepts of high-efficiency c-Si solar cells are compared at the end of this chapter.

7.1 Principle Architecture of c-Si Solar Cells and Resistive Losses

7.1.1 Base and emitter of conventional c-Si solar cells

The thickness of a photovoltaic (PV) absorber should be about three times larger than its absorption length (see Chapter 2). Without sophisticated measures of light trapping, the thickness of a c-Si absorber has to be of the order of 200–400 μm with regard to the absorption spectrum of c-Si (see Figures 2.8 and 2.11). A flat silicon crystal with a thickness of the order of 200–400 μm is stiff and mechanically stable. Such self-supporting silicon crystals are known as wafers. The stiffness of c-Si wafers defines their handling during the production of c-Si solar cells (wafer-based technology). Wafer-based technologies have strong advantages such as high throughput, lightweight handling and excellent binning into narrow intervals for the separation of different power classes into modules.

The thick light-absorbing layer of a c-Si solar cell is called the base. To achieve a high solar energy conversion efficiency (η), the lifetime of minority charge carriers in the base of a c-Si solar cell has to be significantly larger than the minimum lifetime (about 100 μs for c-Si; see Chapter 3) so that the diffusion length of minority charge carriers exceeds the absorption length by about ten times (see also Figure 4.11).

The lifetime of minority charge carriers is limited by Auger recombination in c-Si absorbers (Figure 3.14). With regard to the definition of the diffusion length (Equation (3.5)), of the lifetime (Equation (3.2)) and with reference to the dependence of the Auger recombination lifetime on the density of free charge carriers in doped c-Si (Equation (3.15)), the conditions for the densities of majority free electrons or holes is given by the following equations:

$$n_0 \leq \sqrt{\frac{D_p}{100 \cdot C_A \cdot d_{abs}^2}} \qquad (7.1')$$

$$p_0 \leq \sqrt{\frac{D_n}{100 \cdot C_A \cdot d_{abs}^2}} \qquad (7.1'')$$

The diffusion coefficient of electrons (D_n) is 26 cm^2/s in moderately p-type doped c-Si, whereas the diffusion coefficient of holes (D_p) in n-type doped c-Si is about 3.5 times less than D_n (Jacoboni *et al.*, 1977). The Auger recombination coefficient is about $2 \cdot 10^{-30}$ cm^6/s for c-Si (Yablonovitch and Gmitter, 1986a).

The densities of donors and acceptors of the *n*-type and *p*-type doped regions, respectively, should be as high as possible in order to maximize the diffusion potential of a *pn*-junction (Equation 4.12) since the diffusion potential is the upper limit of the open-circuit voltage (V_{OC}; see Chapter 4).

The c-Si wafer defines the doping of the base of c-Si solar cells. A *p*-type doped c-Si base has two significant advantages in comparison to an *n*-type doped c-Si base. First, a *p*-type doped c-Si absorber can be higher doped than an *n*-type doped c-Si absorber with the same d_{abs} (thickness of the absorber layer) due to the higher diffusion coefficient of electrons, i.e. a higher diffusion potential can be reached with a *p*-type doped base in such a case. Second, and this is the most important technological reason, silicon crystals can be very homogeneously *p*-type doped because the segregation coefficient of boron is close to 1 over a wide range of growth parameters (as discussed in this chapter). Therefore, the base of classical c-Si solar cells is *p*-type doped with a density of boron atoms of about $2-4 \cdot 10^{16}$ cm^{-3}.

The resistance of the base can be calculated if the density and the mobility of free holes (μ_p) are known. The conductivity of *p*-type doped c-Si is

$$\sigma_p = q \cdot p_0 \cdot \mu_p \qquad (7.2)$$

With a typical mobility of holes of about 270 cm^2/(Vs) for moderately *p*-type doped c-Si (Jacoboni *et al.*, 1977), the resistance of the base of a c-Si solar cell (R_{base}^{c-Si}) is

$$R_{base}^{c-Si} = \frac{d_{abs}}{q \cdot p_0 \cdot \mu_p} \approx 0.02 \; \Omega\text{cm}^2 \qquad (7.3)$$

For purposes of comparison, the tolerable series resistance (Equation (1.29′)) is about 0.17 Ωcm^2 for a c-Si solar cell with V_{OC} and I_{SC} (short-circuit current density) equal to 0.709 V and 40.7 mA/cm^2, respectively, at air mass (AM)1.5 (Zhao *et al.*, 1995). Therefore, the resistance of the base does not limit η of a c-Si solar cell illuminated at AM1.5. However, limitation of η of c-Si solar cells by the resistance of the base becomes important under illumination at concentrated sunlight when R_{base}^{c-Si} becomes larger than the tolerable series resistance.

The n-type doped layer of a conventional c-Si solar cell is called the emitter. The density of donors in the emitter should be as high as possible for maximizing the diffusion potential of the pn-junction. It is useful to increase the density of donors in the emitter to a value close to the effective density of states at the conduction band edge (N_C about $3 \cdot 10^{19}$ cm^{-3} for c-Si (Sze, 1981)). For a maximum density of donors, the maximum useful thickness of the emitter (d_{em}) can be obtained.

$$d_{em} \leq \sqrt{\frac{D_p}{100 \cdot C_A \cdot n_0^2}} \tag{7.4}$$

Therefore, the thickness of the n-type doped emitter should be of the order of 100 nm in c-Si solar cells. Usually d_{em} is a factor of two to four times larger than 100 nm for technological and/or design reasons of the c-Si solar cell. An n-type doped base and a p-type doped emitter can be considered in a similar way in general (see task T7.1 at the end of this chapter).

The diffusion potential of classical c-Si solar cells ranges between 0.90 and 0.945 V. An additional increase of the diffusion potential is possible by increasing the density of acceptors in the base. For this purpose, the thickness of the base has to be reduced (Equation (7.1″)). In such a case, the optical path of light has to be increased by introducing light trapping (see Chapter 2).

Incidentally, for many transition metal impurities, the capture cross-section of electrons (σ_e) is much larger than then the capture cross-section of holes (σ_h), for example σ_e and σ_h are $1.04 \cdot 10^{-14}$ and $8.3 \cdot 10^{-16}$ cm^2 (see, for example (Davies *et al.*, 1980). Therefore, in the case of identical densities of transition metal impurities, the lifetime of photogenerated minority charge carriers is much longer in n-type than in p-type silicon. For this reason, world record c-Si solar cells are produced on n-type doped wafers (Masuko *et al.*, 2014).

7.1.2 *Front-contact finger grid and emitter resistance*

A c-Si solar cell is illuminated from the emitter side for absorbing photons with a low absorption length as close as possible to the pn-junction. The area of the metal front-contact should be minimized so that sunlight can penetrate to the emitter and base. The front-contact consists of thin grid-like metallic fingers collecting electrons from the emitter. The single grid fingers are connected via bus bars with an external lead.

The high reflectivity of c-Si is reduced with an antireflection coating layer (see Chapter 2). Figure 7.1 depicts a schematic of the simplest c-Si solar cell.

The series resistance of the emitter ($R_{em}^{c\text{-}Si}$) is defined by d_{em}, by the distance between grid fingers (a) and by the specific resistance of the n-type doped emitter (ρ_n). Figure 7.2 shows the geometry of the emitter and two neighboring grid fingers under photocurrent generation.

The photocurrent flows homogeneously into the emitter and is collected by equal parts at the neighboring grid fingers (length of a grid finger denoted by L_f). The emitter current (I) is 0 in between the neighboring grid fingers and increases linearly towards the grid fingers to a value equal to the photocurrent density multiplied by half of the area between the neighboring grid fingers. The current density at the maximum power point (I_{mp}) is used for analysis.

$$I(x) = I_{mp} \cdot L_f \cdot x \qquad (7.5)$$

The power loss in a small slice at position x in the emitter ($dP_s(x)$) is given by the product of the squared current at x and the resistance of the slice (dR).

$$dP_s(x) = I(x)^2 \cdot dR \qquad (7.6)$$

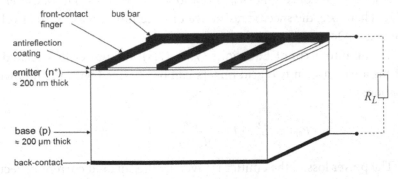

Figure 7.1. Schematic of the simplest c-Si solar cell.

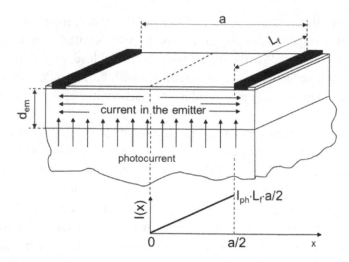

Figure 7.2.   Photocurrent and current in the emitter between two grid fingers.

The resistance of the slice depends on ρ_n, d_m, L_f and the thickness of the slice (dx).

$$dR = \frac{\rho_n}{d_{em} \cdot L_f} \cdot dx \qquad (7.7)$$

The specific resistance of the emitter is equal to the following expression:

$$\rho_n = \frac{1}{q \cdot n_0 \cdot \mu_{n+}} \qquad (7.8)$$

where q is the elementary charge ($1.6 \cdot 10^{-19}$ As). The electron mobility in highly n-type doped c-Si (μ_{n+}) is about 100 cm^2/(Vs) (Jacoboni *et al.*, 1977). Therefore, the specific resistance of the emitter in a c-Si solar cell is about 0.5 mΩcm.

Taking into account Equations (7.5)–(7.8), the power loss related to half the area between two grid fingers can be obtained with the following equation:

$$P_{em} = I_{mp}^2 \cdot L_f^2 \cdot \frac{\rho_n}{d_m \cdot L_f} \cdot \int_0^{a/2} x^2 \, dx \qquad (7.9)$$

The power loss in the emitter is given by the squared current collected over half the distance between two grid fingers and the emitter resistance

$(R_{em}^{c\text{-Si}})$ normalized to the considered generation area.

$$P_{em} = \left(I_{mp} \cdot L_f \cdot \frac{a}{2}\right)^2 \frac{\cdot R_{em}^{c\text{-Si}}}{\frac{a}{2} \cdot L_f} \tag{7.10}$$

The combination of Equations (7.9) and (7.10) leads to the following expression for $R_{em}^{c\text{-Si}}$:

$$R_{em}^{c\text{-Si}} = \frac{\rho_n \cdot a^2}{12 \cdot d_{em}} \tag{7.11}$$

The distance between grid fingers is usually of the order of 0.2 cm and the thickness of the emitter is about 200 nm for c-Si solar cells. This results in an emitter resistance of about 0.05 Ωcm^2. Therefore, the resistance of the emitter is significantly smaller than the tolerable series resistance of a c-Si solar cell illuminated at air mass (AM)1.5.

The sum of the base and emitter resistances is about 0.07 Ωcm^2. Therefore the sum of the resistances of the ohmic emitter and base contacts, the grid fingers and the bus bar should be equal to or less than 0.1 Ωcm^2 for c-Si solar cells with high solar energy conversion efficiency (η).

7.1.3 *Minimization of shading caused by the front-contact finger grid*

The metal finger grid shades a part of the absorber from the irradiation of sunlight. To minimize shading losses, the area of the front-contact finger grid should be as small as possible. In large-scale production of solar cells, the front-contact finger grid is usually screen-printed from a paste containing silver particles. The width of grid fingers and the distance between the grid fingers are about 100 μm and 2 mm, respectively. The bus bars occupy a similar area as the grid fingers. The corresponding shading losses reduce the active area of a simple c-Si solar cell by about 8–9%.

Shading losses can be only reduced if the area occupied by the front-contact finger grid is decreased (Figure 7.3). This can be done, for example, by increasing the distance between grid fingers and increasing the thickness of the emitter in order to avoid increasing resistive losses. However, recombination losses would increase in such a case since the quantum efficiency decreases for photons absorbed close to the emitter surface if the thickness of the emitter increases. Therefore, a significant reduction of

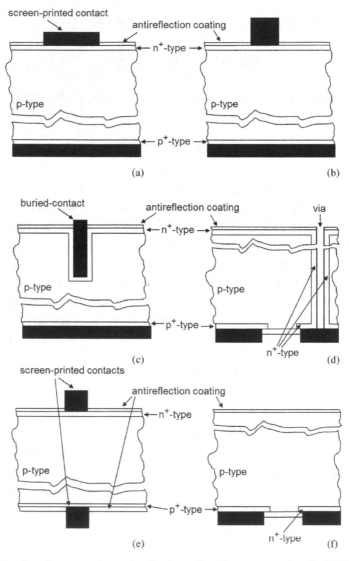

Figure 7.3. Local cross-sections of c-Si solar cells with a screen-printed grid finger (a), a double screen-printed grid finger (b), a buried–front-contact grid finger (c), a back contacted wrap-through-emitter (d), two grid fingers screen printed on the front- and back-contacts (bifacial solar cell) (e) and with charge-selective and electron and hole collecting contacts at the back surface (f). The solar cells are illuminated from the top. Incident sunlight and sunlight reflected from the ground (albedo) is absorbed in the bifacial solar cell. Light scattering surface structures such as texture (Chapter 2) are not considered.

shading losses is practically impossible without changing the architecture of front-contacts.

The area occupied by the front-contact finger grid can be reduced by increasing the thickness of the grid fingers. The ratio between the height and the width is also called the aspect ratio. The resistance of a grid finger and hence its cross-section should be kept constant when increasing the aspect ratio. The aspect ratio is limited for screen-printing. The aspect ratio can be increased by screen-printing the finger grid twice one on top of the other (see, for example, Dullweber *et al.*, 2012), as shown in Figure 7.3(b).

The aspect ratio can be strongly increased if filling trenches or grooves in the emitter with a metal (buried-contact approach (Wenham *et al.*, 1994), Figure 7.3(c)). Trenches that are about ten times finer than grid fingers produced by screen-printing can be formed, for example, by laser scribing and subsequent etching (Colville, 2009). Shading losses can be reduced to 3% or less by the buried-contact approach.

In addition to reduction of shading losses, the buried-contact approach has more advantages. First, the emitter resistance can be significantly reduced for buried-contacts due to a finer grid in comparison to printed contacts. For a finer grid, d_{em} can be reduced as well which leads to a reduction of recombination losses in the emitter. Second, buried-contacts can be combined with selective doping along the groove walls of buried-contacts so that recombination losses in the emitter are further reduced. Third, grooves can be coated with diffusion barriers for copper so that silver, which is usually part of pastes for screen printing, can be replaced by copper. The replacement of silver by copper is important since the limited availability of silver in the earth's crust limits the worldwide implementation of photovoltaic energy production.

The deposition of ohmic front-contacts at the back side of the solar cell is the most radical step for reduction of shading losses. This is demonstrated for the so-called emitter-wrap-through concept and for the back-contacted architectures in Figures 7.3(d) and 7.3(f), respectively.

In the emitter-wrap-through concept, the emitter is extended from the front side to the back side of the solar cell by using via holes. The base and the emitter contacts have to be structured such that the emitter and the base can be contacted separately with an interdigitated finger grid at the back side of the c-Si solar cell (see, for example, Ulzhöfer *et al.*, 2008). Via holes can be drilled with strong laser pulses into silicon wafers (rate of more than 10^4 holes/s (Patwa *et al.*, 2013)). The area occupied by via holes is much smaller than the area occupied by buried-contacts. In addition, charge separation occurs on the front and back sides of a solar cell with the emitter-wrap-through concept, i.e. the diffusion length of the c-Si wafer can be reduced so that the base can be higher doped and therefore the diffusion potential can be increased.

In the concept of back-contacted solar cells, the emitter is eliminated from the front side of the solar cell and a *pn*-junction contact with an interdigitated finger grid is deposited at the back side of the solar cell. The elimination of the emitter from the front side has the additional advantages that the emitter resistance can be strongly reduced and that the quantum efficiency can be increased for photons with higher energies. However, the diffusion length of the base has to be increased by a factor of two in comparison to a conventional c-Si solar cell and excellent surface passivation of the front side of the back-contacted solar cell is required.

The concept of electron and hole collecting contacts at the back side of the solar cell can be realized well by laser-assisted technologies (see, for example, Huljic *et al.*, 2006) and has a strong potential for reaching very high η. The structuring of back contacts on large areas demands the application of powerful lasers being used for laser ablation of material, i.e. for precise removal of silicon and of protective coatings (see, for example, Engelhardt *et al.*, 2007).

Bifacial solar cells (Figure 7.3(e)) have a finger grid for both the front and back contacts (see, for example, Bordina *et al.*, 1975; Duran *et al.*, 2010). Therefore, bifacial solar cells absorb the incident light and the light back-reflected from the ground (albedo). Examples for grounds with high albedo are snow fields and concrete.

7.1.4 *Resistances at emitter and base contacts*

The sum of the contact resistances of the emitter and of the base should be of the order of 0.05 Ωcm^2 or less. Resistive losses through contacts increase with decreasing contact area of the front contact (A_{fc}) if the current remains nearly unchanged. In a first approximation, the series resistance of a c-Si solar cell caused by the front contact ($R_{fc}^{c\text{-Si}}$) is related to the tunneling resistance (see Equation 5.30) of electrons and scales with the ratio between

A_{fc} and the area of the solar cell (A_{cell}).

$$R_{fc}^{c\text{-}Si} \approx R_{c,tunn,e} \cdot \frac{A_{fc}}{A_{cell}} \qquad (7.12)$$

Front- and back-contacts are screen printed on c-Si solar cells in conventional technology. The screen-printing paste contains metallic particles, a frit and a binder giving the viscosity of the paste needed for printing. The contacts are fired after the screen-printing step so that a conducting metal layer is formed and the organic binder is removed. The formation of the tunneling junction at ohmic contacts is a rather critical technological step in which additional barriers arising between the grid finger and the emitter need to be avoided. For this reason, the screen-printing paste contains additional chemical components such as a frit, which enables the penetration of silver onto the c-Si surface by adding a small amount of lead (2–4%, Schubert *et al.*, 2006). Resistances of tunneling contacts can be of the order of $10^{-3} \cdot 10^{-7}$ Ωcm^2 depending on doping of c-Si and the barrier height between the metal and c-Si (see Chapter 5).

Figure 7.4 shows a typical idealized diffusion profile of phosphorus atoms across an emitter between the ohmic front contact and the base.

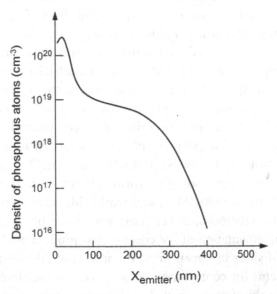

Figure 7.4. Typical doping profile of phosphorus atoms across the *n*-type doped emitter in a c-Si solar cell. Zero corresponds to the ohmic contact, 10% of compensation with boron in the base is reached at about 350 nm for the given example.

The density of phosphorus atoms is above 10^{20} cm^{-3} at the ohmic metal contact and close to the solubility limit. The tunneling resistance of the ohmic contact is very low for a very high density of phosphorus atoms (see Chapter 5). The density of phosphorus atoms underneath the highly doped contact layer is about 10^{19} cm^{-3}. Full compensation with boron atoms in the p-type doped base is reached at about 400 nm. The area of the ohmic contact at the emitter is about one to two orders of magnitude lower than area for screen-printed or buried-contact solar cell, respectively. The reduced emitter contact area is not critical for the series resistance of c-Si solar cells.

The ohmic back contact with the p-type doped base can cover the whole contact area. Aluminum atoms are acceptors in c-Si. Therefore, aluminum is used for the back contact since it easily forms a metal/p^+/p ohmic contact (see Chapter 5) due to diffusion of aluminum into the near contact region of the p-type doped base of a conventional c-Si solar cell.

7.1.5 Connection of c-Si solar cells in PV modules and role of silver

Separate c-Si solar cells are connected in series by contacting the bus bars of the front contact of a solar cell with the bus bars of the back contact of the following solar cell. For this purpose, the bus bars contain silver which can be easily soldered (in comparison with aluminum). Soldering bars containing silver are deposited on aluminum back contacts.

Silver is widely used in c-Si solar cells since it has the highest conductivity of the metals. It solders also well, and its particles form well-connected layers during the firing of screen-printed contacts. The amount of silver needed for 1 W_p of a c-Si solar cell is of the order of 0.1 g/W_p (Herron and Podewils, 2011) (also see task T7.2 at the end of this chapter). For comparison, the worldwide silver production is of the order of $2 \cdot 10^4$ t/a (Brooks, 2011) and worldwide silver reserves are given as about $5 \cdot 10^5$ t (Brooks, 2011). This would not be sufficient for the worldwide implementation of PV energy mass production. The complete replacement of silver in screen-printed contacts of c-Si solar cells and in soldering systems for connecting c-Si solar cells is therefore strategically important. For this purpose, plated copper contacts in combination with diffusion barriers based on nickel and silicon nitride can be applied for

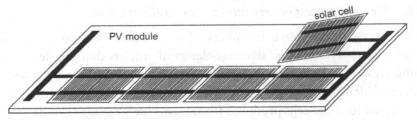

Figure 7.5. Schematic of series and parallel connection of c-Si solar cells in a PV module.

the production of solar cells and PV modules with high long-term stability (Kraft *et al.*, 2015).

Solar cells connected in series are called strings of solar cells. Several strings of solar cells are connected in parallel in PV modules. Figure 7.5 shows a schematic for connecting c-Si solar cells in series and parallel in a PV module. Strings of solar cells are produced with so-called stringers using inductive energy transfer for heating and soldering. The relatively high number of mechanical handling steps needed for the conventional soldering of each solar cell is a disadvantage for the conventional production of PV modules.

Materials can be locally heated with precise lasers. Metal layers absorbing laser light with photons of a certain wavelength can be combined with substrates and laminates, which are transparent to those wavelengths. Laser pulses penetrate the laminate and are absorbed by tinned copper ribbons (Gast *et al.*, 2008). Soldering with laser pulses reduces the number of handling steps in the production of PV modules.

7.2 Homogeneously Doped c-Si Wafers with Low Density of Defects

A c-Si wafer is a slice of a crystal which properties which decisively influence the performance of the solar cell that will be built on the wafer. Homogeneously doped wafers with a very low density of defects in the bulk are required for a maximum of photocurrent generation and a minimum diode saturation current density. Only crystallization via solid–liquid interfaces enables the growth of large silicon crystals within reasonable time frame. Wafers are usually cut by wire sawing from large c-Si crystals.

7.2.1 *Siemens process of very pure polycrystalline silicon*

Silicon crystals are grown from very pure so-called electronic grade or solar grade polycrystalline silicon. Solar grade means that the density of impurity atoms remaining in silicon crystals is reduced to a value at which a defined lifetime of minority charge carriers is guaranteed. A certain η may be reached for solar cells produced from a given solar grade crystal.

Distillation of trichlorsilane ($SiHCl_3$) is key for the production of polycrystalline silicon with highest purity. Figure 7.6 shows a schematic of the very robust Siemens process (developed in the 1950s; see, for example, Gutsche, 1958) consisting of chemical vapor deposition (CVD) of silicon from H_2 and $SiHCl_3$ on hot silicon rods (Bischoff, 1954; Schweickert *et al.*, 1956).

The silicon rod is heated resistively to a temperature of 1100°C. Trichlorsilane is reduced under hydrogen atmosphere (the simplified overall reaction is given by Equation (7.13)) at the surface of a silicon rod.

$$SiHCl_3 + H_2 \rightarrow Si + 3HCl \tag{7.13}$$

A disadvantage of the Siemens process is the high amount of energy needed for production of the very pure polysilicon. The amount of energy and

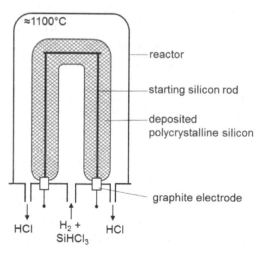

Figure 7.6. Schematic of the Siemens process of silicon deposition from trichlorsilane and hydrogen on hot silicon rods. The vast majority of pure silicon for c-Si solar cells is produced using the Siemens process.

the deposition time required for the Siemens process can be reduced if decomposing SiHCl$_3$ or silane (SiH$_4$) at a lower temperature, for example, in a continuously working fluidized bed reactor (Filtvedt *et al.*, 2012). In a fluidized bed reactor, silicon particles flow in a fluidization gas and grow permanently in the reaction gas. The heaviest particles fall down and can be extracted at the bottom of the reactor vessel.

7.2.2 *Methods of c-Si crystal growth with liquid–solid interfaces*

The very pure polycrystalline silicon is then further processed into large crystals for c-Si solar cells. Three types of crystallization are used most often in c-Si photovoltaics. In the first method, a growing crystal is pulled out from a silicon melt. This method is also known as Czochralski growth (Czochralski, 1918; see Figure 7.7(a)).

In the second method, a silicon rod is fixed and a narrow molten zone is floated along the silicon rod from one end to the other end during controlled crystallization. This method is called float zone (FZ) growth (see Figure 7.7(b)).

Czochralski growth and FZ growth of c-Si crystals start at a seed crystal defining the orientation of the growing crystal. Silicon single crystals grown

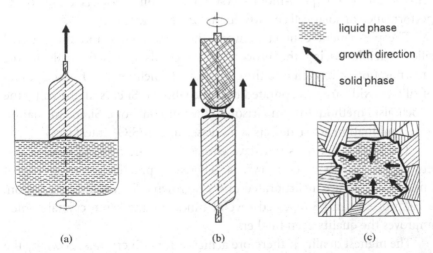

Figure 7.7. Cross-sections for directions of crystallization during Czochralski (a) and float zone (b) growth of c-Si single crystals and during block casting of multi-crystalline c-Si blocks (c). The arrows mark the pulling ((a) and (b)) and the growth directions (c).

by the Czochralski and FZ techniques have a cylindrical shape and can be as large as 40 cm in diameter and 2 m in length (Zulehner, 2000).

In the third method, troughs are filled with molten silicon and crystallization starts under well-controlled cooling. This method is known as block casting (see Figure 7.7(c)). Block casting results in large prismatic crystals of polycrystalline or multi-crystalline silicon (mc-Si).

Each of these three methods has advantages and disadvantages. From the point of view of production costs, block casting is the cheapest method and FZ growth of c-Si crystals is the most expensive. However, in contrast to Czochralski and FZ growth, the growth direction and velocity cannot be so well controlled in a block-casting process. Grain boundaries between crystals of different size and orientation contain a large number of unsaturated silicon bonds (so-called dangling bonds) which are recombination active defects. Therefore, η that can be reached with mc-Si solar cells produced by block casting is lower than that for c-Si solar cells produced by Czochralski or FZ growth.

As an alternative to conventional block casting, directed solidification on large mono-crystalline silicon seed plates is possible under precise temperature control and large quasi-mono-crystalline silicon ingots can be obtained (see, for example, Toor, 2012). The performance of solar cells produced on quasi-mono-crystalline silicon ingots is close to the performance of solar cells produced on Czochralski grown c-Si.

The permanent contact between the silicon melt and the internal silica crucible during the process of Czochralski growth leads to the incorporation of oxygen into the c-Si crystal (Zulehner, 2000). Precipitates of silicon oxide are incorporated over the whole c-Si crystal grown by the Czochralski method. In comparison to the pure bulk of c-Si, the formation of recombination active defects is increased at c-Si/SiO$_2$ interfaces.

The silicon rod does not have contact with the moving heating coil during FZ growth of c-Si crystals. Therefore practically no additional impurity atoms are incorporated into FZ-grown c-Si crystals. In addition, a FZ process can be repeated several times for the same crystal, which improves the quality even further.

The highest quality is therefore achieved for c-Si crystals grown by the FZ method. Float zone-grown c-Si crystals are necessary for getting very high η of c-Si solar cells. Incidentally, the world record for c-Si solar cells was kept over decades of years by c-Si solar cells prepared on FZ-grown silicon (Zhao *et al.*, 1998).

Large silicon crystals are grown at liquid–solid interfaces. New crystal unit cells are formed at the crystal surface during the crystal growth. Impurity atoms present during the formation of new crystal unit cells are able to escape back into the liquid by diffusion from the region where new crystal unit cells are formed (see Figure 7.8) so that silicon atoms can occupy vacant lattice positions.

The escape distance of impurity atoms from the growing crystal back into the liquid is of the order of 1–2 lattice constants (a_{lc}). The time that is needed for the escape of impurity atoms from the growing crystal surface back into the liquid is given by the squared escape distance divided by the diffusion coefficient of the impurity atom in the crystal at a temperature close to the melting point (D_{ic}). Then, the upper limit of an optimum velocity of crystal growth (v_{cg}) can be estimated:

$$v_{cg} = \frac{D_{ic}}{a_{lc}} \tag{7.14}$$

The lattice constant of c-Si is 0.543102 nm (Massa *et al.*, 2009). The D_{ic} of impurity atoms in the crystal near the melting point vary over a wide range (Figure 7.9). The lower values of D_{ic} are important for the crystal growth and are of the order of 10^{-10} cm^2/s (Tang *et al.*, 2009). Therefore the growth rate of c-Si crystals should be of the order of, or below, 0.2 mm/s.

A usual growth rate of c-Si crystals is about 0.1 mm/s. This means that the growth of a 2 m long silicon crystal takes more than three hours. The distance over which c-Si crystals are grown is shorter by about one

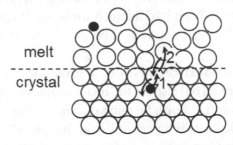

Figure 7.8. Schematic of an interface between a crystal and a melt with an escape sequence of an impurity atom from the crystal surface into the liquid.

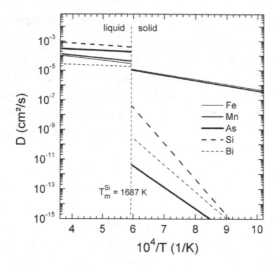

Figure 7.9. Dependence of the diffusion coefficients of impurity atoms in liquid and solid silicon and of the self-diffusion coefficient of silicon on the reverse temperature (data obtained by using equations given by Tang *et al.*, 2009).

order of magnitude for block casting than for Czochralski or FZ growth of silicon crystals. The resulting much shorter growth time is one reason for reduced production costs of block casted crystalline silicon solar cells. As remark, about half of the large silicon crystals are lost during wire sawing (kerf loss).

7.2.3 Homogeneous doping of large silicon crystals

Dopants are impurities which are intentionally incorporated into the growing c-Si crystals in order to achieve a well-defined *n*-type or *p*-type conductivity. Silicon crystals are doped during the growth process by adding highly doped silicon into the melt for block casting or Czochralski growth, or by adding gases such as diborane to the ambient gas during FZ growth.

The solid–liquid interface plays a decisive role not only for the quality of the growing c-Si crystal but also for the incorporation of both dopants, and unwanted impurities, i.e. atoms forming recombination active defects. The solubility of the elements is higher in liquid than in solid silicon resulting in an increase of the concentration of impurity atoms in the liquid at the solid–liquid interface during crystallization. The enrichment of impurity atoms at the freezing solid–liquid interface is called segregation. The incorporation of impurity atoms into a growing c-Si crystal is controlled by segregation.

In the following, the dependence of the concentration of a given impurity will be obtained as a function of the volume of the growing crystal. The initial volume of the melt (V_0) and an initial number of impurity atoms in the melt (N_{i0}) are given. The initial concentration of impurity atoms in the melt is denoted by C_{i0}. The volume of the crystal (V_s) increases and the volume of the melt (V_m) decreases during crystal growth. The concentration of impurity atoms in the melt (C_m) is given by the number of impurity atoms in the melt (N_{im}) and the difference between the initial volume and the volume of the growing crystal.

$$C_m = \frac{N_{im}}{V_0 - V_s} \tag{7.15}$$

The decrease of the number of impurity atoms in the melt (dN_{im}) is proportional to the concentration of impurity atoms in the melt and the increase of the volume of the growing crystal (dV_s). The proportionality factor is the segregation coefficient (k_s).

$$dN_{im} = -k_s \cdot C_m \cdot dV_s \tag{7.16}$$

The segregation coefficient is specific for different impurities and depends on diffusion of the impurity in the melt, on temperature, on the lattice side where a given impurity can be incorporated into a growing crystal, on the velocity of crystal growth and other factors such as crystal orientation and morphology of the solid–liquid interface.

The concentration of impurity atoms in the melt can be replaced by Equation (7.15), and the resulting equation can then be transformed:

$$\int_{N_{i0}}^{N_{im}} \frac{dN_{im}}{N_{im}} = -k_s \cdot \int_0^{V_s} \frac{dV_s}{V_0 - V_s} \tag{7.17}$$

Equation (7.17) has the following solution:

$$N_{im} = N_{i0} \cdot \left(1 - \frac{V_s}{V_0}\right)^{k_s} \tag{7.18}$$

The number of impurity atoms in the melt remains unchanged if the segregation coefficient is equal to 0, i.e. impurity atoms are not incorporated into the growing crystal. This is desired for unwanted impurities atoms, which would form recombination active defects in the c-Si crystal.

The concentration of impurity atoms in the growing crystal (C_{is}) corresponds to the decrease of the number of impurity atoms in the melt divided by the change of the volume of the growing crystal.

$$C_{is} = -\frac{dN_{im}}{dV_s} \tag{7.19}$$

Equation (7.19) can be solved with ease by using Equation (7.18). The following dependence of the concentration of impurity atoms in a growing crystal on the segregation coefficient is obtained:

$$\frac{C_{is}}{C_{i0}} = k_s \cdot \left(1 - \frac{V_s}{V_0}\right)^{k_s-1} \tag{7.20}$$

If k_s is equal to 1 then the concentrations of the impurity are equal in the melt and in the growing crystal. A segregation coefficient close to 1 is desired for dopants in silicon crystals. The advantage of a dopant atom is that it replaces a silicon atom at the correct position in the c-Si lattice. Therefore, the segregation coefficient of dopant atoms can be close to 1. This is the case for boron (Hall, 1953).

The segregation coefficient of boron (B) is close to 1 over a wide range of low growth rates of silicon crystals ($k_s = 0.8$ near the melting point (Trumbore, 1960)). For purposes of comparison, the segregation coefficients of phosphorus (P) and aluminum (Al) are 0.35 and 0.002, respectively, near the melting point of silicon (Trumbore, 1960). Therefore, silicon crystals can be doped with boron homogeneously much easier than with the other dopants. This is one important reason why boron-doped silicon crystals are preferred for solar cells.

A homogeneous doping of silicon crystals is important for a narrow distribution of the lifetime and of the diffusion potential for solar cells produced on wafers cut from the grown crystal. The concentrations of impurity atoms in a silicon crystal normalized to C_{i0} is shown in Figure 7.10 as a function of the volume fraction of the growing crystal for segregation coefficients of 0.99, 0.8 (B), 0.35 (P) and 0.02 (Al).

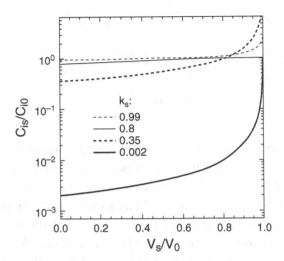

Figure 7.10. Density of impurity atoms in the crystal normalized to the density of impurity atoms in the liquid before starting crystal growth as a function of the volume fraction of the growing crystal.

The concentration of impurity atoms generally increases with increasing volume fraction of the growing crystal due to the growing concentration of impurity atoms in the liquid. for a k_s of 0.99, the concentration ratio increases from 0.99 at the beginning of crystallization to 0.997 and 1.014 after crystallization of 50% and 90%, respectively, of the initial liquid volume. For k_s equal to 0.8, 0.35 and 0.002, the concentration ratios increase by 15%, 57% and 100%, respectively, after crystallization of only half of the volume. The concentration ratios strongly increase after crystallization of 90% of the initial liquid volume and, consequently, these parts of the silicon crystal are not used for producing solar cells.

Growing crystals are rotating during the crystal growth for reaching high lateral homogeneity. Segregation can be limited by the access of impurity atoms to the surface of the growing crystal. This means that segregation depends on diffusion and on the normal flow velocity, i.e. on laminar flow near the surface due to rotation of the crystal in the liquid and on the pulling speed of the crystal (v_i, the growth rate).

The segregation coefficient becomes larger than the segregation coefficient in equilibrium (k_{s0}) with increasing growth rate. The following model for the dependence of k_s on the growth rate and on the rotation frequency (ω) was obtained (Burton *et al.*, 1953):

$$k_s = \frac{k_{s0}}{k_{s0} + (1 - k_{s0}) \cdot e^{-\Delta_i}} \qquad (7.21')$$

$$\Delta_i = \frac{1.6 \cdot v_i \cdot \eta_{\mathrm{vis}}^{1/6}}{D^{2/3} \cdot \omega^{1/2}} \qquad (7.21'')$$

The diffusion coefficient of impurities and the viscosity of molten silicon (η_{vis}) are of the order of 10^{-4} cm^2/s (see Figure 7.9) and $2.4 \cdot 10^{-3}$ cm^2/s (Assael *et al.*, 2012), respectively. Figure 7.11 shows the dependence of k_s on v_i for B, P and Al as dopants and for Fe as an unwanted impurity in c-Si at two values of ω. For doping with B, the segregation coefficient is practically equal to 1 for growth rates larger than 0.1 mm/s at ω of $1\,\mathrm{s}^{-1}$. For P and Al, k_s becomes close to 1 at v_i of about 0.2 and 0.4 mm/s, respectively, at ω of $1\,\mathrm{s}^{-1}$. For higher ω, the increase of k_s towards 1 shifts to higher v_i. For example, k_s of Al is about 0.1 at v_i and ω equal to 0.4 mm/s and $10\,\mathrm{s}^{-1}$, respectively.

The dependence of k_s on the growth rate is key for very homogeneous doping of large silicon crystals within the PV industry. For this purpose,

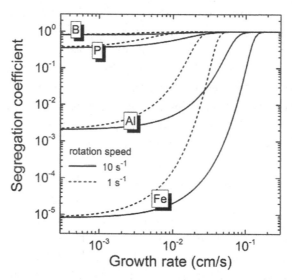

Figure 7.11. Dependencies of the segregation coefficient of B, P, Al and Fe on the growth rate for rotation speeds of 10 and $1\,\mathrm{s}^{-1}$ (solid and dashed lines, respectively).

the growth rate and the temperature are controlled very precisely during the entire growth process of large silicon crystals.

7.2.4 Refinement of silicon crystals by segregation

The values of segregation coefficients usually range between 0 and 1. In some cases, a k_s practically equal to 1 or even a little larger than 1 is possible, for example, for incorporation of oxygen (O) into silicon crystals during FZ growth in oxygen atmosphere (k_s slightly lower than 1 (Carlberg, 1986) or equal to 1.25 ± 0.17 (Yatsurugi *et al.*, 1973)).

Unwanted impurities are atoms which cannot usually replace silicon atoms in their host lattice positions. For this reason, their solubility limit in c-Si and therefore k_s are much lower than for dopants (refining effect, see Zulehner, 2000). For example, the k_s of iron is only $8 \cdot 10^{-6}$, i.e. less than one iron atom from 10^5 iron atoms in the melt is incorporated into the growing crystal (Trumbore, 1960). This fact makes it possible to grow doped silicon crystals with Shockley–Read–Hall (SRH) recombination lifetimes longer than the Auger recombination lifetime (see Chapter 3).

The k_s of impurity atoms decreases with decreasing growth rate (see k_s for Fe in Figure 7.11). This effect gives the opportunity of further reducing the concentration of unwanted impurities at very low growth rates.

The concentration of unwanted impurities with low k_s can be reduced drastically by repetitive melting and crystallization during FZ processing of the same crystal. Unwanted impurities are collected in the melt and transported in the floating zone to the end of the long silicon crystal. Silicon crystals of the absolute highest purity are produced by repetitive FZ processing (Pfann, 1957).

In silicon crystals, segregation of impurity atoms, which cannot replace silicon atoms in their host lattice positions, is strongly enhanced in regions with imperfections such as precipitates or grain boundaries. The directed segregation of unwanted impurities at precipitates is also known as gettering (see Chapter 3). One has to take into account that defects at the interface between the silicon host crystal and a precipitate can cause electronic states in the band gap of c-Si. The additional incorporation of so-called getterers into silicon crystals is useful if the number of gettered atoms forming recombination active defects is significantly larger than the number of defects introduced by the getterer.

Segregation or solidification refining can be applied for alternative production technologies of solar grade polycrystalline silicon directly from metallurgical silicon. The required concentration of most impurities, except for O, P, B and C, can be achieved by applying solidification refining twice (Morita *et al.*, 2003). As a practical example, metallurgical silicon with a purity of 98% can be cleaned by solidification from an Al-Si melt to purity better than 99.9999% (Sollmann, 2009). In this process, metallurgical silicon is dissolved in an aluminum melt. Silicon crystallizes in the form of flakes during cooling of the Al–Si melt. This is the basic refinement step since most of the impurities remain in the melt due to low segregation coefficients. After draining the liquid aluminum, the silicon flakes are washed in an acid to get rid of the thin Al layer coating the flakes. The temperature of the silicon flakes is increased to the melting temperature of c-Si and a rest of the Al segregates at the surface. The concentration of unwanted impurities can be further reduced during subsequent Czochralski growth of large silicon crystals due to gettering at oxide precipitates. However, the concentration of unwanted impurities remains relatively high, and so the η of c-Si solar cells produced with solar grade silicon from an Al–Si process is significantly lower than for c-Si solar cells fabricated with silicon from the Siemens process. Incidentally, aluminum drained after crystallization of silicon flakes contains about 10–12% of Si, which makes the aluminum hard. Hard aluminum is widely used for lightweight constructions, i.e. solar grade silicon can be obtained as a by-product in the production of hard aluminum.

7.2.5 *Wafering of c-Si crystals*

Practically all c-Si wafers are cut by wire sawing from large silicon crystals or ingots (see Figure 7.12(a)). A long steel wire moves rapidly through the ingot as a slurry is continuously added to the kerf. Diamond or silicon carbide crystallites in the slurry cause the abrasion between the wire and the silicon. The slurry can be omitted and the speed of wire sawing can be increased by several times, for example, to 0.8 mm/min, if using steel wires coated with diamond crystallites (Watanabe *et al.*, 2010). Hundreds wafers are cut from a silicon crystal by wire sawing within one run. The wafers are planar after wire sawing. A defect layer with a thickness of about 10 μm at

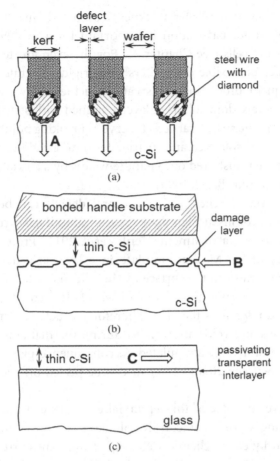

Figure 7.12. Principles of wafering for solar cells based on c-Si: wire sawing of many wafers for the same time (a), transfer of thin c-Si onto a handle substrate by cutting across a damage layer (b) and liquid-phase crystallization of thin c-Si on glass that will be cut into wafers (c). The arrows A, B and C denote the directions of wire sawing, cutting and crystallization, respectively.

the silicon wafer has to be removed by polishing and wet chemical etching after wire sawing.

About half of a silicon crystal is lost during wire sawing due to abrasion. This is a major disadvantage of wire sawing. A further disadvantage of wire sawing is the limitation to a certain thickness of c-Si wafers since the wafers should remain stiff. For example, silicon wafers with a thickness of about 50 μm are not stiff enough for self-supporting.

The reduction of the wafer thickness combined with light trapping and advanced surface passivation is important for a further increase of the η of c-Si solar cells (see Chapter 4). For the even greater increase of the η, a transfer of flexible c-Si wafers onto rigid substrates is required during subsequent processing. This can be achieved by cutting wafers from silicon crystals along a buried layer enriched with structural defects (Figure 7.12(b)). The structural defects serve as breaking points for the lift-off of a flexible c-Si solar cell and its transfer onto a rigid substrate. Thin c-Si wafers can be transferred onto handle wafers by a so-called smart cut and further processed (Bruel, 1995).

A buried layer enriched with structural defects can be created by implantation of protons at high dose (beam-induced solar cell wafering) combined with a thermal treatment (Henley *et al.*, 2011). Protons implanted with an energy of 2–4 MeV accumulate in a layer at a distance of the order of 50–100 μm from the surface of the silicon crystal. The solubility of hydrogen in c-Si is only of the order of $10^{15}\,\mathrm{cm}^{-3}$ (see Van de Walle, 1994 and references therein). Therefore, structural defects such as dislocation loops and voids are formed during thermal annealing in the regions where protons are accumulated. As consequence, planar wafers with thicknesses of between 20 and 150 μm can be peeled off (beam-induced cleaving).

Kerfless layer transfer of thin epitaxial c-Si foils can also be realized by implementing a porous silicon double layer. A porous silicon double layer is formed by electrochemical etching at different current densities. The porous silicon top layer is reorganized during sintering in hydrogen atmosphere and the epitaxial c-Si layer can be deposited at 1100°C in $\mathrm{SiHCl_3/H_2}$ atmosphere (Haase *et al.*, 2012). The deposition rate of silicon epitaxial layers from $\mathrm{SiHCl_3/H_2}$ is between 0.4 and 3 μm/min (Hammond, 2001). This technology can be combined well with back-contacted solar cells (Haase *et al.*, 2012) and is suitable for detachment of large c-Si foils with lifetimes much longer than the minimum lifetime (100 μs, see Chapter 3) (Radhakrishnan *et al.*, 2015) so that only 40 μm thin solar cells (area 243 cm^2) with η above 20% can be produced (Kapur *et al.*, 2013).

The roughness and the thickness variation of wafers cut by beam-induced solar cell wafering are less by about one order of magnitude than for wafers produced by wire sawing. Beam-induced solar cell wafering provides a tool for possible production of c-Si solar cells with a very high η at low cost.

Peeling of 25 μm thin c-Si layers is possible on large areas by exfoliation, which makes use of stress induced between a c-Si crystal and a metal layer during cooling (Rao *et al.*, 2012).

Liquid-phase crystallization of c-Si on glass (Haschke *et al.*, 2016) is a promising alternative to form substrates with thin c-Si foils (Figure 7.12(c)). In this technique, a melting zone of silicon is scanned over the coated substrate with electron or laser beams. The interface layer between the glass and the silicon is critical due to the high gradient of temperature at the interface during melting. The interface between the glass and the c-Si can be well controlled with a stack of silicon oxide/silicon nitride/silicon oxide (Haschke *et al.*, 2016). Thin c-Si films on glass can be processed in a wafer-based technology similarly to back-contacted solar cells. From the point of view of wafering, the major advantages of liquid-phase crystallization of c-Si on glass are the absence of material losses due to abrasion and the abolition of polishing.

7.3 Formation of the Emitter

7.3.1 *Diffusion from inexhaustible and exhaustible sources*

In conventional c-Si solar cells, the *n*-type doped emitter is formed by diffusion of P from an inexhaustible and/or from an exhaustible source into the *p*-type doped base while the density of donors in the emitter is much higher than the density of acceptors in the base (overcompensation of acceptors). A quasi-unlimited source for diffusion of *P* is typically created with phosphorus oxychloride (POCl$_3$) and some oxygen that form a phosphorus–silica glass on top of the silicon wafer.

The propagation in time (t) and space (x in one dimension) of the density of a diffusing impurity ($N_i(x, t)$) with a diffusion coefficient D_i can be calculated by solving the one-dimensional diffusion equation.

$$\frac{\partial N_i(x \cdot t)}{\partial t} = D_i \cdot \frac{\partial^2 N_i(x, t)}{\partial x^2} \tag{7.22}$$

An inexhaustible phosphorus source permanently provides *P* atoms at the solubility limit in c-Si. Therefore, the density of impurity atoms at the surface is constant to N_{is} (saturation density of diffusing species) during the whole diffusion process

$$N_i(0, t) = N_{is} \qquad (7.23)$$

N_{is} increases with increasing temperature. For the diffusion of P, for example, N_{is} is equal to $3 \cdot 10^{20}$ or $5.5 \cdot 10^{20}$ cm^{-3} at 800°C or 900°C, respectively (Solmi *et al.*, 1996). Incidentally, at temperatures above 750°C the saturation density of mobile P atoms becomes larger than the density of activated P atoms (Solmi *et al.*, 1996).

The density of impurities beyond the emitter surface is 0 at the start of diffusion. Then the solution of Equation (7.22) is equal to the product of N_{is} and the complementary error function (erfc), the values of which are tabulated.

$$\frac{N_i(x, t)}{N_{is}} = \text{erfc} \left(\frac{x}{2 \cdot \sqrt{D_i \cdot t}} \right) \qquad (7.24)$$

Figure 7.13 shows the dependence of the normalized density of diffusing impurity atoms as a function of the erfc. The depth at which

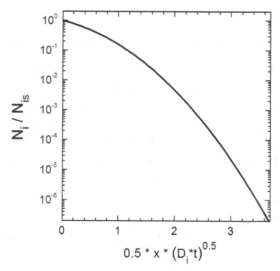

Figure 7.13. Dependence of the normalized density of diffusing impurity atoms from an inexhaustible source as a function of the complementary error function of the quotient of the distance and the doubled square root of the product of the diffusion coefficient of the impurity and time.

the density of P atoms reaches the density of acceptors in the base can be considered to be the thickness of the emitter (H_n). Therefore, the solution of the complementary error function can be found from the ratio between the density of acceptors in the base and N_{is}. The penetration depth of phosphorus is reached for a solution of the complementary error function between 2.6 and 3.0. So H_n can be estimated from D_i and t

$$H_n \approx 5.6 \cdot \sqrt{D_i \cdot t} \tag{7.25}$$

Diffusion from an inexhaustible source of P is performed at a temperature of between 800°C and 900°C. This first diffusion step defines the total amount of P in the emitter (Q_{is}, given as a real density. The inexhaustible source is removed by etching the phosphorus silica glass in hydrofluoric acid.

The optimal density of free electrons in the emitter is much lower than the saturation density of P atoms in c-Si. Therefore, a second diffusion step from an exhaustible source is applied for forming the optimum doping profile of the emitter. For diffusion of phosphorus, the second diffusion step is usually performed at temperatures of about 1000°C. Diffusion times of thousands of seconds are needed to form the right doping profile in the emitter.

In the simplest case, a delta function can be assumed as the initial distribution function before starting diffusion. The solution of the diffusion equation with an exhaustible source is then given by an exponential function. Equation (7.26) gives an idea of the development of a doping profile during diffusion:

$$N_i(x, t) = \frac{Q_{is}}{\sqrt{\pi \cdot D_i \cdot t}} \cdot \exp\left(-\frac{x^2}{4 \cdot D_i \cdot t}\right) \tag{7.26}$$

As an example, Figure 7.14 shows diffusion profiles for an initial density of dopants of 10^{14} cm^{-2} and a D_i for P atoms of $3 \cdot 10^{-15}$ cm^2/s (Fahey *et al.*, 1989) after diffusion times of 20, 200 and 2000 s.

The D_i of P atoms in c-Si has to be known for the precise control and calculation of diffused doping profiles across the emitter. The diffusion coefficient of impurity atoms is thermally activated and depends on the

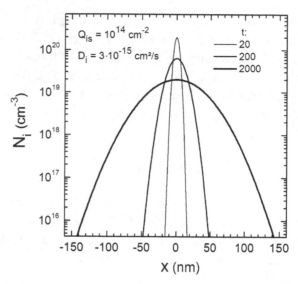

Figure 7.14. Profile of the density of impurity atoms diffusing from an exhaustible source with an areal density of 10^{14} cm^{-2} at position 0 after different diffusion times for a diffusion coefficient of $3 \cdot 10^{-15}$ cm^2/s.

diffusion mechanism (activation energy E_{Ai}):

$$D_i = D_{i0} \cdot \exp\left(-\frac{E_{Ai}}{k_B \cdot T}\right) \quad\quad (7.27)$$

Dopants change their lattice position during diffusion. The change of the lattice position of an impurity atom corresponds to a local chemical reaction, which can involve different reaction partners such as point defects. Vacancies, i.e. atoms missing in their lattice position, or interstitials, i.e. atoms in a place between lattice positions, are prominent point defects.

Diffusion of phosphorus in c-Si can be provided by different elementary steps such as kick-out reactions with self-interstitials, via vacancies or other point defects whereas the charge state of the defect has to be considered as well (see, for example, Fahey *et al.*, 1989). Figure 7.15 illustrates diffusion via interstitials and via double negatively charged vacancies.

Diffusion of phosphorus is dominated at low densities of phosphorus atoms by kick-out reactions of interstitial phosphorus atoms (P_i) with silicon atoms at lattice positions (Si_{Si}). This process is in equilibrium with

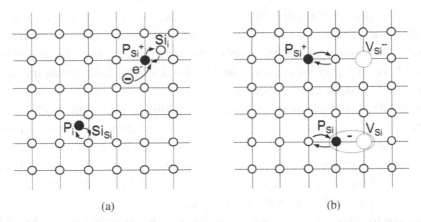

(a)   (b)

Figure 7.15. Illustration of diffusion of phosphorus via interstitials (a) and via double charged vacancies (b).

kick-out reactions of self-interstitials (Si_i) with ionized phosphorus atoms at lattice positions (P_{Si}^+) and free electrons (Bentzen *et al.*, 2006)

$$Si_{Si} + P_i \leftrightarrow Si_i + P_{Si}^+ + e^- \tag{7.28}$$

The density of double negatively charged vacancies in c-Si increases with increasing density of phosphorus atoms so that diffusion via vacancies becomes dominating at very high densities of phosphorus atoms (Bentzen *et al.*, 2006):

$$V_{Si}^- + P_{Si}^+ \leftrightarrow (V_{Si} - P_{Si})^- \tag{7.29}$$

The dominating diffusion mechanism can change depending on the density of diffusing species, on the charge of diffusing species, on the presence of other charged or compensating species in c-Si, and on temperature. Furthermore, the superposition of different diffusion processes has to be taken into account.

The experimental values of the diffusion coefficient of phosphorus in c-Si are about 10^{-16}, 10^{-15} and 10^{-12} cm^2/s at 800°C, 900°C and 1200°C, respectively (Fahey *et al.*, 1989). The diffusion coefficient of copper in c-Si is, for example, $5 \cdot 10^{-5}$ cm^2/s at 900°C (Istratov *et al.*, 1998), which is about ten orders of magnitude larger than the diffusion coefficient of phosphorus in c-Si. Therefore, copper contamination must be avoided during thermal processing of c-Si solar cells.

Diffusion of the phosphorus emitter occurs around the whole c-Si wafer. Therefore, the *pn*-junction has to be disconnected between the front and back contacts of c-Si solar cells. This is realized by the edge isolation which can be performed, for example, by a strong laser damaging the *pn*-junction at the edges of the wafer.

Aside from the formation of the emitter, during phosphorus diffusion, the phosphorus–silica glass acts as a sink for impurity atoms which can diffuse from the bulk silicon towards the surface. This so-called phosphorus-gettering leads to a reduction of transition metals especially in mc-Si and therefore to a strong increase of the lifetime of minority charge carriers in the mc-Si up to 200 μs (Cuevas *et al.*, 1997).

The high content of oxygen in silicon crystals grown by the Czochralski technique leads to the formation of a recombination-active and metastable B–O complex (B$_i$O$_i$ pair defect, here, the index *i* denotes interstitial) in boron doped c-Si under illumination of the solar cell (Glunz *et al.*, 2001). The formation of B–O complexes can be avoided by getting rid of oxygen in c-Si or by changing to phosphorus (*n*-type) or gallium (*p*-type) doping of silicon crystals grown by the Czochralski technique (Glunz *et al.*, 2001). The present world record c-Si solar cell is based on *n*-type doped c-Si grown by the Czochralski technique (Masuko *et al.*, 2014).

7.3.2 *Laser-assisted doping and ion implantation of the emitter*

With a thickness of 200–400 nm or even less, the emitter is thinner than the silicon wafer by about 1000 times. It saves a lot of energy if only the region of the emitter is heated during the formation of the emitter instead of heating the whole silicon wafer as in a conventional diffusion process. This is possible by inducing melting with laser pulses with a duration time of nanoseconds.

Heat can be locally transferred to silicon wafers during pulsed laser irradiation. Heat transfer depends on heat conductivity, heat capacitance and density of a given medium as well as on phase transitions. The values of the heat conductivity ($k(T)$), heat capacitance ($C(T)$) and density ($\rho(T)$) depend on temperature (T). For a given energy source (Q_{source}), the following equation of heat transfer can be written:

$$\frac{\partial}{\partial t}[C(T) \cdot \rho(T) \cdot T(x,t)] = \frac{\partial}{\partial x}\left[k(T) \cdot \frac{\partial T(x,t)}{\partial x}\right] + Q_{source}(x,t) \quad (7.30)$$

For a given initial temperature distribution, an equation for the temperature similar to the diffusion equation can be obtained, and a thermal diffusivity corresponding to a thermal diffusion coefficient (D_{th}) can be defined.

$$D_{th}(T) = \frac{k(T)}{C(T) \cdot \rho(T)} \qquad (7.31)$$

At 400 K, D_{th} of c-Si is of the order of 0.5 cm^2/s (k, C and ρ of the order of 1 W/(cm · K), 0.8 W · s/(g · K) and 2.33 g/cm^3, respectively (Shanks *et al.*, 1963). The thermal diffusivity decreases to about 0.1 cm^2/s near to the melting temperature of c-Si (Shanks *et al.*, 1963).

A thermal diffusion length can be estimated for the time of interaction between the laser pulse and the silicon wafer (Δt):

$$L_{th}(T) = \sqrt{D_{th}(T) \cdot \Delta t} \qquad (7.32)$$

The dependence of L_{th} on Δt is plotted in Figure 7.16 for constant D_{th} of 0.5 and 0.1 cm^2/s. For example, heat is transferred over about 100–200 nm within one nanosecond or over about 5–10 μm within one

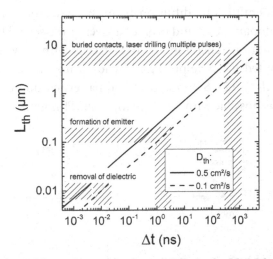

Figure 7.16. Estimated dependence of the thermal diffusion length on the interaction time between a laser pulse and silicon for constant thermal diffusion coefficients of 0.5 (moderate heating) and 0.1 (heating near to melting) cm^2/s. The dashed areas mark the typical ranges for the formation of the emitter, of the removal of the dielectric and of the formation of buried contacts and via holes.

microsecond. Figure 7.16 gives an idea about the duration time of laser pulses necessary for different processing steps. For example, local melting of a c-Si surface layer with a thickness of the order of 100 nm is possible with nanosecond laser pulses strongly absorbed in c-Si with energies of the order of 0.5 J/cm^2.

The required density of phosphorus atoms for the emitter on a p-type doped base can be adjusted by diffusion of phosphorus from a precursor layer into a molten surface layer within a time of about 100 ns (laser-assisted doping, (Eisele *et al.*, 2009). This is schematically shown in Figure 7.17(a).

Dopants for the emitter can also be provided by ion implantation. Ion implantation means that dopants are ionized and accelerated in the high electric field of an accelerator and directed onto a substrate. The penetration depth of ionized dopants into c-Si is adjusted by their kinetic energy, i.e. by the acceleration voltage. Implanted dopants are activated by recrystallization after laser-induced melting. Ion implantation is applied, for example, also for the formation of a p-type doped emitter on an n-type doped base (see, for example, Pawlak *et al.*, 2012). This is shown schematically in Figure 7.17(b).

Ion implantation has strong advantages in comparison to the formation of phosphorus emitters by diffusion from inexhaustible and exhaustible sources. First, dopant doses and profiles are very precise and homogeneous for ion implantation. Therefore, c-Si solar cells with very high and very narrow distributed η can be produced by ion implantation of the emitter (Hieslmair *et al.*, 2012). Second, local and patterned doping with donors and acceptors is possible with ion implantation. This gives the opportunity

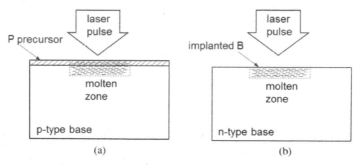

Figure 7.17.   Laser doping of an n^+-doped emitter on a p-type doped base (a) and laser-induced activation of an implanted p-type doped emitter on an n-type doped base (b).

of applying high-efficiency concepts of c-Si solar cells (buried-contact solar cells, back-contacted solar cells) to mass production for which, for example, patterning by lithography would be very expensive. Third, the number of processing steps and the amount of processing energy are reduced for ion implantation. For example, the step of edge isolation is not needed after ion implantation.

7.3.3 *Amorphous silicon emitter*

Amorphous silicon (a-Si:H, see also Chapter 9) is a disordered semiconductor that can be doped and which has an optical band gap between 1.45 and 1.85 eV depending on disorder (Cody *et al.*, 1981). A doped a-Si:H/c-Si hetero-junction can be well applied for charge-selective contacts in c-Si solar cells.

At a-Si:H/c-Si hetero-junctions, the valence band offset (ΔE_V), i.e. the differences between the valence band edges of a-Si:H and c-Si (see also Figure 7.18), is about 0.46 eV (Schmidt *et al.*, 2007). The conduction band offset (ΔE_C) is about 0.14 eV, with respect to the band gaps of c-Si and a-Si:H and ΔE_V.

The density of deep defect states has to be as low as possible at a charge-selective contact (Chapter 4). In contrast to doped a-Si:H (see also

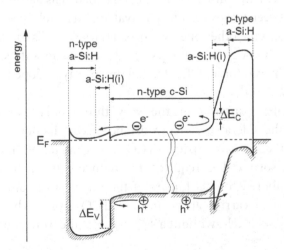

Figure 7.18. Schematic band diagram across a HIT solar cell with *n*-type a-Si:H/ a-Si:H(i)/*n*-type c-Si/a-Si:H(i)/*p*-type a-Si:H in thermal equilibrium. Defect states in the band gap of a-Si:H are omitted for clarity. One of the ohmic contact layers is transparent.

Chapter 9), the density of deep defect states in undoped a-Si:H can be lower than 10^{15} eV^{-1}cm^{-3} (Tiedje *et al.*, 1981b). Furthermore, dangling bonds at the c-Si surface are passivated very well with hydrogen, which is present in a-Si:H. If taking into account tunneling of electrons and holes between c-Si and a-Si:H(i), the effective density of surface defects (N_{st}) is as low as 10^8 cm^{-2} at the a-Si:H(i)/c-Si interface. Such a low N_{st} results in a very long surface recombination lifetime of about 5 ms for a 100 μm thick c-Si wafer (see Chapter 3). For this reason a very thin layer of undoped a-Si:H (a-Si:H(i), thickness several nm) is placed between the doped c-Si and doped a-Si:H layers (Tanaka *et al.*, 1992). With regard to the decisive charge-selective and passivating component, c-Si solar cells with a-Si:H contacts are known as hetero-junction with intrinsic thin layer (HIT) solar cells.

Figure 7.18 shows a schematic band diagram across a HIT solar cell with a layer structure of *n*-type a-Si:H/a-Si:H(i)/*n*-type c-Si/a-Si:H(i)/*p*-type a-Si:H. The doped a-Si:H layers are about 10 nm thick. Photogenerated electrons are reflected at the barrier of the charge-selective *n*-type c-Si/a-Si:H(i)/*p*-type a-Si:H contact. Photogenerated holes can penetrate the thin barrier by tunneling via defect states (see Section 5.3.5). On the other side, photogenerated holes are reflected at the electron extracting contact due to the large valence band offset. Therefore, HIT solar cells have a c-Si absorber embedded between two excellently passivated charge-selective contacts, one for electrons and the other one for holes. HIT solar cells can be fabricated by the same technology at *n*- and *p*-type doped silicon wafers. The highest η has been achieved for a HIT solar cell on an *n*-type doped silicon wafer (Masuko *et al.*, 2014).

The specific resistance of the very thin a-Si:H emitter is much higher than the tolerable series resistance (see Chapter 1). Therefore, a transparent conductive oxide (TCO, see Chapter 9) layer covering the complete solar cell is required for reducing the emitter resistance between neighbored contact fingers of the finger grid. Some optical loss is caused by free carrier absorption in TCO layers. Therefore, back-contacted HIT solar cells without a TCO layer at the front contact have the highest η.

The η of HIT solar cells can be very high despite the fact that only low-temperature processes (such as deposition of a-Si:H at 200°C by plasma enhanced chemical vapor deposition, see Chapter 9) are applied for the formation of the charge-selective and passivating contacts. Therefore, less energy is demanded for the formation of a-Si:H/c-Si charge-selective contacts than for a diffused *pn*-junction.

7.4 Passivation and Structuring of c-Si Surfaces

7.4.1 *Cleaning and structuring of silicon surfaces by etching*

Silicon surfaces have to be cleaned by etching to get rid of structural defects and contaminating impurities before depositing passivation, antireflection or contact layers. Isotropic etching of surfaces is also called chemical polishing. Many cleaning procedures of c-Si surfaces are based on oxidation of surface atoms or molecules followed by dissolution of oxidized species. Volatile oxidized species of surface contaminations can leave the solution and enter the surrounding atmosphere. An example of an oxidizing species is hydrogen peroxide (H_2O_2).

Structural defects introduced during wafering of silicon crystals are removed by mechanical and chemical polishing. Chemical polishing of silicon wafers is performed in solutions containing hydrofluoric acid (HF), nitric acid (HNO_3), H_2O and, for example, acetic acid ($HC_2H_3O_2$) (Robbins and Schwartz, 1959, 1960). Silicon dioxide (SiO_2) is etched in aqueous HF solutions (see, for example, Knotter, 2000). Etch rates are controlled by temperature and concentrations of components in etch solutions.

The so-called RCA clean (RCA: Radio Corporation of America; Kern and Puotinen, 1970; Kern, 1984) is an industrial standard process for cleaning of c-Si surfaces. In a precleaning step, the c-Si surface is oxidized in HNO_3 at 80°C and etched in buffered HF. Mainly organic rests are oxidized during the cleaning step at 80°C in a solution of H_2O_2 and NH_4OH. Residuals of metal contaminations form volatile chlorides during the second cleaning step at 80°C in a solution of HCl and H_2O_2. Finally, cleaned silicon wafers are rinsed in deionized water.

The purity of deionized water, of all the chemicals and of the surrounding ambience is very important in order to avoid recontamination of cleaned c-Si surfaces. For this reason, c-Si solar cells with very high η have to be processed in specially designed clean rooms, none of the handling tools are allowed to contain metal and only ultra-clean chemical solutions may be used.

Silicon surfaces have to be structured or textured for light trapping (see Chapter 2). Processes of anisotropic etching in alkaline solutions such as potassium hydroxide (KOH) are applied for texturing c-Si surfaces of c-Si solar cells (Bean, 1978). As an alternative, the starting porous silicon formation in HF-containing solutions can also be used in order to roughen a c-Si surface (see, for example, Smith and Collin, 1992).

The etch rate of c-Si in alkaline solutions has a minimum in the ⟨111⟩ direction since there is only one free bond per silicon surface atom available (Figure 7.19(b)). In contrast, the etch rate of c-Si is much higher in the ⟨100⟩ direction since there are two free bonds per silicon surface atom available (Figure 7.19(a)) (Valle Rios *et al.*, 2016). As a consequence, facets of the most stable c-Si(111) surface remain after etching of c-Si in alkaline solutions (Figure 7.20(a)).

On c-Si wafers oriented in the ⟨100⟩ direction, facets of c-Si(111) surfaces form inverted pyramids, which are excellent for light trapping in c-Si solar cells. SiO₂ is stable in alkaline solutions. Therefore, SiO₂ masks can be applied for etching ordered arrays of inverted pyramids on silicon wafers oriented in the ⟨100⟩ direction or of pyramids oriented in the ⟨100⟩ direction (Bean, 1978) (Figure 7.20(b)). Incidentally, light trapping

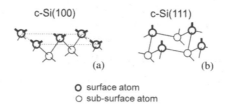

Figure 7.19. Bond configurations at the c-Si(100) (a) and c-Si(111) (b) surfaces.

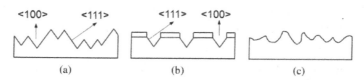

Figure 7.20. Schematic cross-sections of c-Si(100) surfaces structured with random pyramids (a), with inverted pyramids etched with a SiO₂ mask (b) and of a roughened c-Si surface (c).

by scattering at inverted pyramids was and is important for reaching world records in the η of c-Si solar cells.

Anisotropic etching in alkaline solutions is not very useful for mc-Si solar cells. Alternatives are based on isotropic roughening of silicon surfaces (Figure 7.20(c)) by stain etching in solutions containing HNO_3, sodium nitrite and HF (Archer, 1960) or by the formation of macropores during electrochemical treatment in fluoride-containing organic electrolytes (Ponomarev and Levy-Clement, 1998).

7.4.2 The c-Si/SiO₂ interface

SiO_2 is an insulator with a wide band gap of about 9.3 eV (Weinberg *et al.*, 1979), and SiO_2 is chemically stable in most acidic, alkaline and other solutions except HF. Therefore, SiO_2 is an excellent electrically inactive protection layer for c-Si solar cells.

The c-Si/SiO₂ interface played a decisive role for the invention of the MOSFETs (metal oxide semiconductor field-effect transistors), the heart of modern electronics (Kahng, 1960). The reason for this is that electronic devices can be excellently passivated (see Chapter 3) with c-Si/SiO₂ interfaces. Over many years, the surfaces of world record c-Si solar cells were passivated with a c-Si/SiO₂ interface as well (Zhao *et al.*, 1998). Recently, c-Si/SiO₂ interfaces became important for solar cells based on liquid-phase crystallization (Haschke *et al.*, 2016).

SiO_2 layers are grown on silicon wafers by thermal oxidation of c-Si at temperatures of the order of 1000°C. The growth rates of SiO_2 are, for example, about 1.7 or 20 nm/min at 1100°C in dry or wet oxygen, respectively (Deal and Grove, 1965). The reactions can be described by the following equations for dry and wet oxidation:

$$Si + O_2 \rightarrow SiO_2 \tag{7.33}$$

$$Si + 2H_2O \rightarrow SiO_2 + 2H_2 \tag{7.34}$$

The oxidation process is limited by diffusion of oxidizing species such as O_2^- through the growing SiO_2 layer for layer thicknesses above 20–30 nm (Deal and Grove, 1965).

In the following, the passivation mechanisms at c-Si/SiO₂ interfaces will be described. Figure 7.21 shows a schematic bond configuration at a c-Si/SiO₂ interface. Si–Si bonds are replaced by Si-O bonds at the c-Si/SiO₂

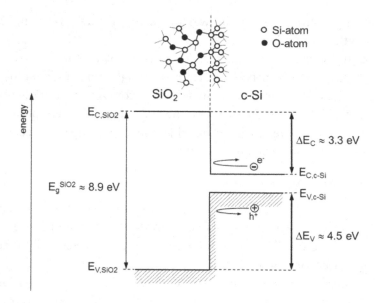

Figure 7.21.   Schematic bond configuration and band diagram at a c-Si/SiO$_2$ interface.

interface. There is a very thin SiO$_x$ transition layer at the c-Si/SiO$_2$ interface where the oxidation state of Si atoms varies gradually from Si$^{(0)}$ (oxidation state of Si atoms in c-Si) and Si$^{(1+)}$ via Si$^{(2+)}$ and Si$^{(3+)}$ to Si$^{(4+)}$ (oxidation state of Si atoms in SiO$_2$) (Himpsel *et al.*, 1988). The energies of bonding and anti-bonding states of oxidized Si atoms are deep in the valence and conduction bands of c-Si, respectively. This means that oxidized Si surface atoms do not have surface states in the band gap of c-Si.

The energies of the valence and conduction band edges of SiO$_2$ are much lower or higher, respectively, than for c-Si. The related band offsets ΔE_C and ΔE_V at the c-Si/SiO$_2$ interface are defined as the differences between the conduction band edges of c-Si ($E_{C,\text{c-Si}}$) and SiO$_2$ (E_{C,SiO_2}) and between the valence band edges of c-Si ($E_{V,\text{c-Si}}$) and SiO$_2$ (E_{V,SiO_2}), respectively. The values of ΔE_C and ΔE_V are very large for c-Si/SiO$_2$ interfaces and amount to −3.3 and 4.5 eV, respectively. The values of the band offsets can vary over about 0.2 eV depending on oxidation technology and on orientation of the silicon wafer (see, for example, Keister *et al.*, 1999; Ribeiro *et al.*, 2009). Therefore, the c-Si/SiO$_2$ interface provides huge barriers for free electrons and holes in the conduction and valence bands of c-Si.

The values of bond lengths and bond angles are different in c-Si and in SiO_2. This causes stress at c-Si/SiO_2 interfaces, which can only partially relax by local re-arrangement of atoms and of chemical bonds in the interface region. The SiO_x transition layer plays an important role for the local relaxation of stress at c-Si/SiO_2 interfaces. The relaxation of remaining stress at c-Si/SiO_2 interfaces is realized by omitting some of the chemical $\equiv$Si–O bond at the c-Si/SiO_2 interface. This leads to the formation of unsaturated or dangling bonds (Si-dbs), the most prominent defects at c-Si/SiO_2 interfaces (Figure 7.22). Thermally oxidized c-Si/SiO_2 interfaces cannot be free of Si-dbs. Si-dbs at Si atoms with three silicon back bonds can be recombination active. Si-dbs pointing into the $\langle 111 \rangle$ direction form so-called P_b recombination centers (Poindexter *et al.*, 1981) with defect states in the middle of the band gap of c-Si (Caplan *et al.*, 1979). In contrast, Si-dbs at silicon dimers at reconstructed Si(100) surfaces are not recombination active (Dittrich *et al.*, 2002). Si-dbs at Si atoms with oxygen back bonds can form trap states (Caplan *et al.*, 1979).

The density of P_b defect centers at c-Si/SiO_2 interfaces can be minimized by controlling orientation, surface roughness and oxidation regime. In general, the lower the surface roughness and the lower the oxidation rate, the lower the density of P_b centers at the c-Si/SiO_2 interface. The lowest density of P_b defect centers has been reached for thermal oxidation of c-Si(100) wafers in dry oxygen (Poindexter *et al.*, 1981). In contrast, much higher densities of P_b defect centers were observed after

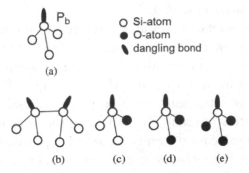

Figure 7.22. Configurations of recombination active (a) and not recombination active silicon dangling bonds (b–e).

oxidation of c-Si(111) wafers (Caplan *et al.*, 1979). This is another reason why c-Si(100) wafers are preferred for c-Si solar cells.

The surface recombination velocity increases with increasing density of surface states (D_{it}) and is proportional to the capture cross-sections of electrons (σ_{se}) and holes (σ_{sh}) (see Chapter 3). Values of D_{it} of the order of 10^{11}–10^{12} eV^{-1}cm^{-2} can be reached by thermal oxidation of c-Si (Poindexter *et al.*, 1981). As a good approximation, the σ_{se} and σ_{sh} of P_b defect centers are about 10^{-16} and 10^{-14} cm^2, respectively (see Aberle *et al.*, 1992 and references therein). The surface recombination velocity remains still as high as 10^3–10^4 cm/s after thermal oxidation of c-Si surfaces. However, surface recombination velocities below 10–100 cm/s are required for c-Si solar cells with very high η.

7.4.3 *Hydrogen passivation of silicon dangling bonds*

Hydrogen passivation is a general concept that is applied whenever the density of Si-dbs should be reduced or kept as low as possible. The density of Si-dbs at c-Si/SiO$_2$ interfaces can be further reduced by hydrogen passivation, during which Si-dbs react with hydrogen atoms and form Si–H bonds. Usually hydrogen passivation of c-Si/SiO$_2$ interfaces is performed during post annealing in so-called forming gas, which contains hydrogen and NO$_2$.

Figure 7.23 shows the principle of hydrogen passivation of a Si-db at a c-Si/SiO$_2$ interface. The density of Si-dbs at c-Si/SiO$_2$ interfaces can be further reduced by hydrogen passivation to values even below 10^{10} cm^{-2}, i.e. surface recombination velocities between 10 and 100 cm/s and even less can be realized (Stephens *et al.*, 1994).

Hydrogen passivation takes also place at c-Si surfaces coated with SiN$_x$:H or a-Si:H layers where hydrogen is introduced together with the reactant gas during layer deposition. Si-dbs are passivated in a-Si:H thin-film solar cells with hydrogen as well (see Chapter 9). Furthermore, the preparation of ultra-clean hydrogen-passivated c-Si surfaces by wet chemical treatments in acidic HF-containing solutions is an important pretreatment for subsequent thermal oxidation or plasma-enhanced deposition steps of passivation layers on c-Si surfaces at low temperatures. The D_{it} at hydrogenated c-Si(111) surfaces can be reduced to very low values up to 10^{10} eV^{-1}cm^{-2}. Extremely low surface recombination velocities of

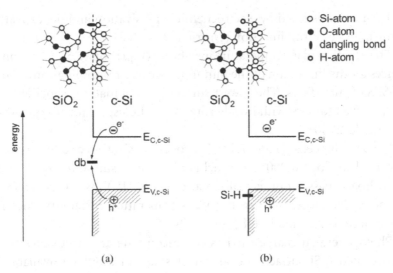

Figure 7.23. Schematic bond configuration and band diagram of the c-Si surface before (a) and after (b) hydrogen passivation of a silicon dangling bond (Si-db).

about 0.7 cm/s have been demonstrated for c-Si surfaces immersed in HF (Yablonovich *et al.*, 1986a).

The oxidation state of silicon atoms at the c-Si/SiO$_2$ interface does not change during hydrogen passivation (Himpsel *et al.*, 1988). Annealing in vacuum at T above 400–500°C leads to an increase of the density of Si-dbs at the c-Si/SiO$_2$ interface, i.e. $\equiv$Si–H interface bonds can dissociate into Si-dbs and hydrogen (Stathis, 1995).

$$\equiv \text{Si–H} \rightarrow \ \equiv \text{Si} \cdot + \text{H}^0 \tag{7.35}$$

Hydrogen passivated c-Si surfaces are also unstable at high negative electric fields (negative-bias-temperature instability, see, for example, Helms and Poindexter, 1994). The dissociation of hydrogen from $\equiv$Si–H bonds is enhanced in the presence of holes at the c-Si/SiO$_2$ interface leading to the following electrochemical reaction:

$$\equiv \text{Si–H} + \text{A} + h^+ \rightarrow \ \equiv \text{Si} \cdot + \text{AH}^+ \tag{7.36}$$

The component A in Equation (7.36) is usually related to water molecules, which can occur at c-Si/SiO$_2$ interfaces (Helms and Poindexter, 1994). The correlation between the fixed positive charge and the D_{it} has also been demonstrated by electrochemical passivation and depassivation cycling under electron injection (Dittrich *et al.*, 2001).

The negative-bias-temperature instability (Equation (7.36)) has consequences for the limitation of the minimum surface recombination velocity at c-Si/SiO$_2$ interfaces. The lowest surface recombination velocities have been achieved for c-Si with low density of free holes, i.e. for *n*-type doped c-Si (Aberle, 2000).

In HIT solar cells (Figure 7.18), hydrogen passivation of c-Si surfaces is performed at *T* below 200°C (Taguchi *et al.*, 2005; Tsunomura *et al.*, 2009) and hydrogen passivation is realized at c-Si/a-Si:H(i) instead of c-Si/SiO$_2$ interfaces. Therefore, the negative-bias-temperature instability does not limit surface passivation in HIT solar cells.

Photogenerated charge carriers can recombine at defect states at the hydrogenated c-Si surface and at defect states in a-Si:H(i) available by tunneling. Consequently, only an interface region with a thickness of a couple on nanometers at the a-Si:H(i) side is relevant for recombination in HIT solar cells. If taking into account a density of Si-dbs in a-Si:H(i) of $3 \cdot 10^{15}$ cm^{-3} and less (Jackson and Amer, 1982), electronic states with an a real density less than 10^9 cm^{-2} are available for recombination at the a-Si:H(i) side of the charge-selective contacts of HIT solar cells. The D_{it} at hydrogenated c-Si surfaces can be reduced to values of the order of and below 10^{12} eV^{-1}cm^{-2} (see, for example, Rauscher *et al.*, 1995). Since the complete area of the c-Si absorber is passivated with a-Si:H(i) and since the density of free charge carriers is very low in a-Si:H(i), the surface recombination rate (see Chapter 3) is strongly reduced in HIT solar cells due to hydrogen passivation. This is key for reaching record values of V_{OC} in c-Si solar cells.

7.4.4 Passivation with c-Si/SiN$_x$:H and c-Si/a-Si:H interfaces

The band gap of silicon nitride (Si$_3$N$_4$) is about 5.4 eV (Miyazaki *et al.*, 2003). Si$_3$N$_4$ is chemically very stable. The ΔE_C and ΔE_V at c-Si/Si$_3$N$_4$ interfaces are about 2.8–2.4 and 1.5–1.9 eV, respectively (Miyazaki *et al.*, 2003; Higuchi *et al.*, 2007), i.e. barriers for electrons and holes can be used for passivation of c-Si surfaces with Si$_3$N$_4$.

Hydrogenated silicon nitride can be deposited on c-Si surfaces by plasma-enhanced deposition processes containing hydrogen (also denoted

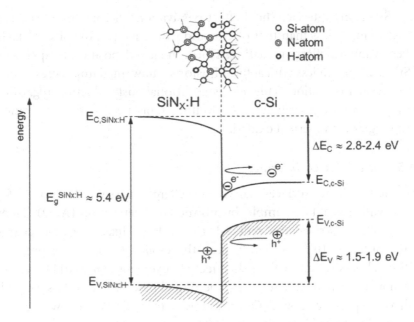

Figure 7.24. Bond configuration and band diagram at the c-Si/SiN$_x$:H interface.

as SiN$_x$:H), which enables hydrogen passivation of remaining Si-dbs. Furthermore, a fixed positive charge is formed in SiN$_x$:H near the c-Si/SiN$_x$:H interface after plasma-enhanced deposition at temperatures below 500°C (Leguijt *et al.*, 1996). Figure 7.24 shows a schematic bond configuration and the band offsets at a c-Si/Si$_3$N$_4$ interface.

The fixed positive charge introduces an electric field in c-Si, thereby attracting free electrons and repelling free holes. Very low surface recombination velocities can be reached at c-Si/SiN$_x$:H interfaces due to the low D_{it} and due to the built-in electric field caused by the fixed positive charge. A surface recombination velocity as low as 4 cm/s has been demonstrated even for standard *p*-type doped c-Si wafers covered with SiN$_x$:H (Lauinger *et al.*, 1996).

The passivation of c-Si solar cells with a multi-functional SiN$_x$:H layer has important technological advantages. First, three measures of surface passivation are realized within one deposition step, namely the (i) creation of barriers for free electrons and holes due to large ΔE_C and ΔE_V, (ii) passivation of remaining Si-dbs by hydrogen and (iii) creation of an electric field by positive fixed charge in the SiN$_x$:H layer near the c-Si/SiN$_x$:H interface (field-effect passivation). Second, SiN$_x$:H has a refractive index less than that of c-Si and, consequently, the SiN$_x$:H layer serves as an antireflection coating. Third, the low deposition temperature of SiN$_x$:H is much less critical for diffusion of unwanted impurities in c-Si than thermal oxidation. This allows additional cost reduction regarding the clean room atmosphere and handling tools needed for passivation technologies with thermal oxidation.

7.4.5 *The c-Si/Al$_2$O$_3$ interface*

Crystalline silicon surfaces can be excellently passivated with Al$_2$O$_3$ layers synthesized, for example, by atomic layer deposition (ALD) (Hoex *et al.*, 2008). Besides the reduced D_{it}, a large fixed negative charge of the order of 10^{12}–10^{13} cm^{-2} near the c-Si/Al$_2$O$_3$ interface provides additional passivation by the field effect (Dingemans *et al.*, 2011). Surface recombination velocities at the *p*-type doped base side for c-Si solar cells with a base passivated by Al$_2$O$_3$ are as low as for c-Si solar cells with a base passivated by SiO$_2$ (Schmidt *et al.*, 2008).

Figure 7.25 shows a schematic of the bond configuration and of the band diagram at the c-Si/Al$_2$O$_3$ interface. The band gap of very thin Al$_2$O$_3$ layers grown by ALD depends sensitively on the deposition parameters and can range between 6.2 and 7 eV (Bersch *et al.*, 2008). The oxidation state of Si atoms changes at the c-Si/Al$_2$O$_3$ interface, i.e. there is an ultra-thin SiO$_x$ layer at the c-Si/Al$_2$O$_3$ interface. The extension and the variation of the oxidation state of Si atoms at and near the c-Si/Al$_2$O$_3$ interface are important for ΔE_C and ΔE_V which can vary between 2.1 and 2.8 eV and between 2.9 and 3.7 eV, respectively (Bersch *et al.*, 2008).

The values of σ_{se} and σ_{sh} are about $7 \cdot 10^{-15}$ and $4 \cdot 10^{-16}$ cm^2, respectively, and the D_{it} is about $1 \cdot 10^{11}$ eV^{-1}cm^{-2} at the c-Si/Al$_2$O$_3$ interface (Werner *et al.*, 2012). Due to the large fixed negative charge near the c-Si/Al$_2$O$_3$ interface, surface recombination of electrons is strongly suppressed. Since σ_{se} is much larger than σ_{sh}, the effective surface recombination velocity at highly *p*-type doped c-Si/Al$_2$O$_3$ interfaces is less by about one to three orders of magnitude in comparison to the passivated highly *p*-type doped c-Si/SiO$_2$ or c-Si/a-SiN$_x$:H interfaces (Hoex *et al.*, 2008). Even surface recombination velocities at highly *p*-type doped emitters can be as low as 100–300 cm/s in case of passivation with Al$_2$O$_3$.

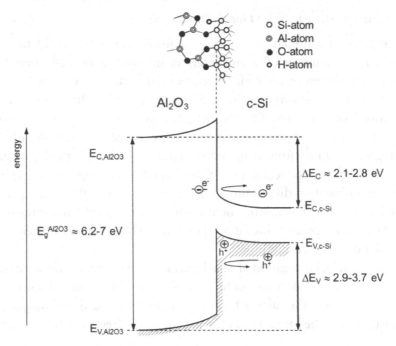

Figure 7.25. Bond configuration and schematic band diagram at the c-Si/Al$_2$O$_3$ interface.

An energy conversion efficiency of 23.2% has been demonstrated for c-Si solar cells with an *n*-type base passivated with SiO$_2$ and a highly *p*-type doped emitter passivated with Al$_2$O$_3$ (Benick *et al.*, 2008).

The surface of *n*-type doped c-Si can also be well passivated with an Al$_2$O$_3$ layer (Hoex *et al.*, 2008). For example, effective surface recombination velocities below 0.8 cm/s have been achieved with plasma ALD on *n*-type doped c-Si surfaces (Dingemans *et al.*, 2010).

The high thermal stability of passivated c-Si/Al$_2$O$_3$ interfaces enables local treatments of c-Si solar cells at high temperatures even after the deposition of the passivation layer. For example, Al$_2$O$_3$ passivation layers can be locally opened by laser ablation (laser ablation means the removal of surface atoms from a substrate by strong short laser pulses). Furthermore, opened areas can be doped and contacted without increasing the effective surface recombination velocity. This opened technological opportunities for mass production of c-Si solar cells with very high η at low costs.

7.4.6 *Back surface field and local ohmic contacts*

The surface recombination velocity at ohmic contacts should be very large since a splitting of Fermi-levels is impossible by definition (see Chapter 5). Therefore, the surface recombination rate can only be reduced by reducing the density of photogenerated minority charge carriers at the ohmic contact. A barrier repelling photogenerated minority electrons and attracting majority holes is necessary to reduce the density of photogenerated electrons at the ohmic contact at the *p*-type doped base of a c-Si solar cell. The region of a related barrier should be wide enough to avoid recombination due to tunneling of electrons into the highly *p*-type doped ohmic contact region. The introduced controlled barrier region is called back surface field since it is related to the back contact of a classical c-Si solar cell (Figure 7.26).

The principal realization of a back surface field is possible for the ohmic contact of the base of a conventional c-Si solar cell because the effective density of states at the valence band edge is about 2–3 order of magnitude larger than p_0. Therefore, the potential energy of free electrons increases

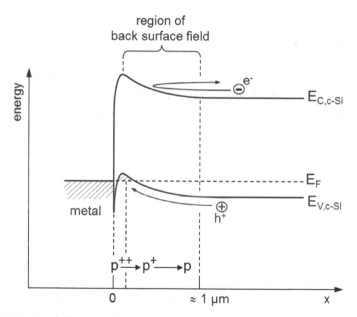

Figure 7.26. Band diagram close to the ohmic contact at the *p*-type doped base of a c-Si solar cell with back surface field.

by about 0.12 . . . 0.18 eV towards the highly p-type doped region near the ohmic contact. The barrier height is further increased due to very high doping (denoted by p^{++}) near the ohmic contact of the order of 10^{20} cm^{-3}. The region of the back surface field is adjusted to an extension of about 1 μm by a concentration gradient of acceptors.

The ohmic contact area has to be strongly reduced (see also Chapter 5) in c-Si solar cells with a pn-junction and with very high η (local back surface field). The area between local ohmic contacts is passivated, for example, with a c-Si/SiO$_2$ interface (Figure 7.27). Over decades, the concept of locally diffused ohmic back contacts (PERL — passivated emitter rear locally diffused) was key for reaching highest η (Wang *et al.*, 1990).

For mass production of c-Si solar cells with local ohmic back contacts, an Al back-contact layer can be deposited directly onto the passivation layer (Figure 7.28(a)). A strong short laser pulse induces local melting of the Al surface layer, subsequent local damage of the passivation layer, local alloying between Si and Al and some diffusion of Al into c-Si (Figure 7.28(b)). As result, a laser fired local ohmic back contact is formed (Schneiderlöchner *et al.*, 2002; Moors *et al.*, 2012).

Laser ablation can be used for opening well-defined windows in a passivation layer deposited on the p-type doped base (Figure 7.28(c)). The local back surface field and the highly doped regions can be formed by conventional and/or laser-assisted diffusion. Finally, the Al back-contact layer is deposited, for example, by printing (Moors *et al.*, 2012) (Figure 7.28(d)). The high thermal stability of c-Si/Al$_2$O$_3$ interfaces is an advantage of this kind of technology.

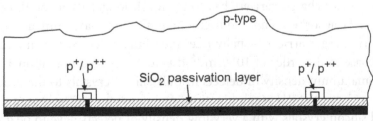

Figure 7.27.   Schematic cross-section of the back contact of a c-Si solar cell with a local back surface field.

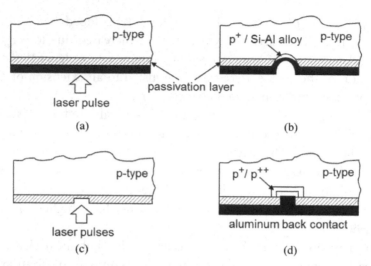

Figure 7.28. Schematic cross-sections of *p*-type doped c-Si coated with a passivation and aluminum layers before (a) and after formation of laser-fired local p^+-type doped c-Si/Si-Al alloy/Al contacts (b), of *p*-type doped c-Si coated with a passivation layer during opening of a window in the passivation layer by laser ablation (c) and after diffusion of a local back surface field and deposition of the Al back-contact layer (d).

7.5 Summary

The absorption length is of the order of 100 μm for a large part of photons in the sun spectrum with energies above the band gap of c-Si. For this reason silicon wafers with a thickness of 150–300 μm are demanded for c-Si solar cells with high η.

Recombination of photogenerated charge carriers in the thick c-Si absorber should be as low as possible. The diffusion length of photogenerated minority charge carriers has to be much longer than the thickness of the c-Si absorber. This requirement limits the maximum density of majority charge carriers (usually free holes introduced by boron doping) in the base to the order of 10^{16} cm^{-3} due to Auger recombination as well as the maximum density of defects in bulk silicon crystals to the order of 10^{11} cm^{-3} due to SRH recombination. As a consequence, homogeneously doped silicon crystals with a very low density of defects have to be grown for c-Si solar cells.

Purification by distillation of SiHCl$_3$ (precursor for silicon crystal growth) and refinement by segregation at the liquid–solid interface are applied for getting homogeneously doped and very pure silicon crystals.

Etching and cleaning of c-Si surfaces are crucial for reaching high η since defects caused by wafering have to be removed, c-Si surfaces have to be structured for light trapping and surface contaminations must be removed or avoided.

Photogenerated charge carriers have to be collected at the highest possible potential difference between charge-collecting electron and hole contacts. For this demand, the emitter is doped with a density of dopants (phosphorus for n-type doping) as high as 10^{19} cm^{-3}. Technologies of thermal diffusion, ion implantation, thermal annealing or laser processing can be applied for the preparation of the emitter. As an alternative, doped a-Si:H/a-Si:H(i)/c-Si hetero-junctions are applied as charge-selective and passivating contacts in c-Si solar cells (HIT concept) as well.

For c-Si solar cells with very high η, sophisticated measures of passivation of c-Si surfaces are required for reducing the density of recombination active surface defects to values below 10^{10} cm^{-2}. Silicon surfaces can be passivated with c-Si/SiO$_2$, c-Si/SiN$_x$:H, c-Si/a-Si:H(i) or c-Si/Al$_2$O$_3$ interfaces. The surface recombination rate is additionally reduced by fixed positive or negative charge near c-Si/SiN$_x$:H and c-Si/Al$_2$O$_3$ interfaces, respectively, as well as by the back surface field at ohmic contacts. In c-Si solar cells with a pn-homo-junction, the passivation layer has to be opened locally for contacting the base and the emitter with ohmic contacts.

There are numerous types of c-Si solar cells which are mainly distinguished by their architecture (for example, buried-contact solar cells, back-contacted solar cells, local-contact solar cells, emitter-wrap-through solar cells, etc.) and/or by their technology or passivation concepts (for example, screen-printed c-Si solar cells, passivated emitter and rear locally diffused c-Si solar cells, passivated emitter and rear contact c-Si solar cells, hetero-junction with intrinsic thin layer, etc.). Each concept has individual physical and technological advantages and disadvantages, important for the production costs per watt peak of c-Si solar cells and modules, and which are especially important for the further reduction of the energy needed for the production of a c-Si solar cells with high η.

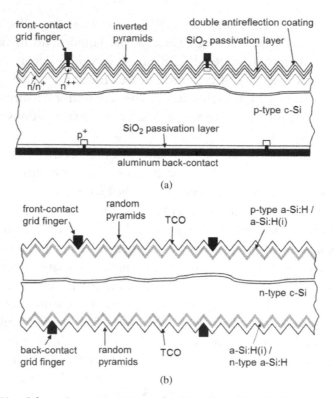

Figure 7.29. Schematic cross-sections of c-Si solar cells with highest η based on a *pn*-homo-junction (PERL: passivated emitter and rear locally diffused) (a) and on a-Si:H(i)/c-Si hetero-junctions (HIT: hetero-junction with intrinsic thin layer) (b). In the PERL concept, most of the c-Si surface area is passivated with c-Si/SiO$_2$ opened locally for ohmic contacts (a). In the HIT concept, the complete c-Si surface area is passivated with c-Si/a-Si:H(i) and the *n*-type doped c-Si absorber is sandwiched between two charge-selective contacts (b). Inverted (a) and random (b) pyramids are etched on the c-Si(100) absorbers for light trapping. A double antireflection coating and an aluminum rear contact as back reflector are applied for minimization of optical losses in the PERL concept (a). For minimization of optical losses in the HIT concept, a transparent conducting oxide (TCO) layer is optimized as an antireflection coating and as a layer for total internal reflection (b).

The basic experience obtained during the optimization of c-Si solar cells with η above 24% is summarized in Figure 7.29 for the PERL (a) and HIT (b) concepts. Different measures for minimization of optical, recombination and resistive losses have been successfully applied for both concepts. The reduction of the thickness of c-Si solar cells combined with

Table 7.1. Values of the thickness of the c-Si absorber, short circuit current density, open circuit voltage, fill factor and energy conversion efficiency for different c-Si solar cell concepts. References are given.

	d_{c-Si} (μm)	I_{SC} (mA/cm^2)	V_{OC} (V)	FF (%)	η (%)	Reference
HIT	165	42.3	0.744	**83.8**	**26.3**	Yoshikawa *et al.* (2017)
PERL	260	**42.2**	0.706	82.8	24.7	Zhao *et al.* (1999)
HIT	98	39.6	**0.750**	83.2	24.7	Taguchi *et al.* (2014)
PERL (thinning)	47	37.9	0.699	81.1	21.5	Wang *et al.* (1996)
Kerfless epitaxy	43	38.14	0.682	77.4	20.1	Kapur *et al.* (2013)
Exfoliation	25	33.6	0.580	76.7	14.9	Rao *et al.* (2012)
LPE on p^+	16.8	27.2	0.660	82.2	14.7	Werner *et al.* (1993)
LPC, back-contacted	15	31.7	0.652	77.0	15.9	Sonntag *et al.* (2017)
SOI, back-contacted	10	29.0	0.623	76.0	13.7	Jeong *et al.* (2013)

Notes: PERL passivated emitter rear locally diffused; HIT hetero-junction intrinsic thin film; LPE liquid phase epitaxy; LPC liquid phase crystallized on glass; SOI silicon on insulator.

a reduction of surface recombination gives the opportunity of further increasing of η of c-Si solar cells (see Chapter 4).

Record c-Si solar cells with different thicknesses are compared in Table 7.1. Until 2014, the thickest c-Si solar cell (260 μm) made by the PERL concept had the highest η (Zhao *et al.*, 1999). However, the highest open circuit voltage (V_{OC}) equal to 0.75 V and the highest fill factor have been reached with a 98 μm thin c-Si solar cell with an area of 101.8 cm^2 made by the HIT concept (Taguchi *et al.*, 2014). This demonstrates the importance of the reduction of the thickness of c-Si solar cells for increasing V_{OC}. However, additional optical losses in HIT solar cells are caused by free carrier absorption and by fundamental absorption in the TCO layer (see Chapter 9). Optical losses related to the TCO and to shading by front contacts have been avoided in large interdigitated back-contacted HIT solar cells so that η above 25% has been reached (25.6%: Masuko *et al.*, 2014, 26.3%: Yoshikawa *et al.*, 2017). With regard to manufacturing, the production of interdigitated back-contacted solar cells still demands a relatively large number of complex processing steps. The number of processing steps can be reduced, for example, by implementing tunnel-interdigitated back contacts (η above 22.5% was demonstrated, Tomasi *et al.* (2017)).

The high potential of very thin c-Si solar cells has been shown by model experiments for a 16.8 μm thin c-Si epitaxial layer without any measures of antireflection coatings or light trapping ($\eta = 14.7\%$; Werner *et al.*, 1993), a 10 μm thin c-Si layer grown on an insulator that was removed in the area of the solar cell ($\eta = 13.7\%$; Jeong *et al.*, 2013) and a 15 μm thin c-Si layer grown directly on glass by liquid phase crystallization ($\eta = 15.9\%$; Sonntag *et al.*, 2017). Incidentally, liquid phase crystallization on glass (Haschke *et al.*, 2016) is a technology which allows for a strong reduction of the energy for the production of PV modules and which can be up-scaled at low costs.

Kerfless exfoliation (Rao *et al.*, 2012) and kerfless transfer of epitaxial c-Si layers (Kapur *et al.*, 2013) can be realized on large areas and η of 14.9% and 20.1% were obtained on 25 and 43 μm thin c-Si solar cells, respectively. However, high η can be hardly expected for the exfoliation technology with an intimate c-Si/metal contact due to generation of stress at high temperature.

The solar energy conversion efficiency of conventional c-Si PV modules decreases with increasing temperature by a factor of about -0.45%/K. Operation temperatures of 40–60°C are easily reached under full illumination of PV modules in summer so that η is reduced by about 10% in comparison to operation at 25°C. A further advantage of HIT solar cells is that the temperature dependence of η is reduced to -0.25%/K (Taguchi *et al.*, 2005) which is caused by the lower thermal generation rate at the a-Si:H side of the charge-selective contact.

The energy payback time and the energy payback factor are important parameters of c-Si solar cells in the context of sustainable renewable energy production. The energy payback time of conventional c-Si solar cells is about 1.5 years in the south of Italy (Wild-Scholten *et al.*, 2010). The degradation rates of c-Si PV modules are about 0.5% per year on average (Jordan *et al.*, 2016) and c-Si PV systems working well longer than 30 years are known. Therefore, the energy payback factor of c-Si solar cells and modules is larger than 10–15 with respect to warranty for PV modules given by producers. The energy payback time is further reduced by reducing the thickness of the wafers and by the consequent application of low-temperature processes and/or local laser heating in mass production of c-Si solar cells. A very strong reduction of the energy payback time can be expected if implementing liquid phase crystallization in mass production (Haschke *et al.*, 2016). The energy payback factor of PV modules can be even higher than 50 depending on the operation regime and on the total operation time.

7.6 Tasks

T7.1: n-type doped base and p-type doped emitter

Calculate the density of free electrons in an *n*-type doped base and ascertain the thickness of a *p*-type doped emitter of a c-Si solar cell.

T7.2: Silver required for c-Si based photovoltaics

Bus bars are screen-printed from a silver (Ag) containing paste in conventional industrial processes of c-Si solar cells. The specific conductivity and the density of Ag are about $6 \cdot 10^7$ S/m and 10.5 g/cm^3. Estimate the mass of Ag which would be required for a complete support of electricity for mankind produced only with c-Si solar cells and discuss the importance of replacement of Ag in c-Si solar cells. About $2 \cdot 10^4$ tons of Ag are produced globally in one year (Brooks, 2011).

T7.3: Boron doped c-Si(100) wafers

Why are boron-doped c-Si(100) wafers used for the production of conventional c-Si solar cells?

T7.4: Diffusion of phosphorus in c-Si

Calculate the thickness of a phosphorus emitter after diffusion from an inexhaustible source for 1000 s at 900°C.

T7.5: Diffusion of copper in c-Si

Calculate the spread of copper after diffusion from a localized copper contamination of one monolayer for 10, 100 and 1000 s at 900°C.

T7.6: V_{OC} of PERL and HIT c-Si solar cells

Explain why a higher V_{OC} can be reached for c-Si solar cells made by the HIT concept than those made by the PERL concept.

T7.7: Importance of kerfless wafer transfer technologies

Discuss the potential of kerfless wafer transfer technologies for realizing high energy payback factors with c-Si solar cells under the conditions of low-cost mass production.

8

Solar Cells Based on III–V Semiconductors

The absolute highest solar energy conversion efficiencies (η) are reached for single- and multi-junction solar cells based on III–V semiconductors illuminated at non-concentrated and highly concentrated sunlight. This Chapter illuminates properties of III–V semiconductors and their interfaces which are decisive for very high values of η. Binary, ternary and quaternary III–V semiconductors are introduced and variations of band gaps with stoichiometry are described. High diffusion length of minority charge carriers in comparison with the optical absorption length as well as n-type and p-type doping over a wide variable range can be achieved with III–V semiconductors. Barrier layers, passivation layers and tunneling junctions can be realized at hetero-junctions between III–V semiconductors. The principle of epitaxy of III–V semiconductor layers is explained for the growth of sophisticated layer systems realizing functions of optical absorption, charge separation, charge transport, surface passivation and ohmic contacts for mono-lithically stacked multi-junction solar cells based on III–V semiconductors. Formation principles of ohmic metal contacts with III–V semiconductors are given. Architectures are derived for tandem, triple-junction and quadruple-junction solar cells based on III–V semiconductors.

8.1 III–V Semiconductor Family

8.1.1 *Binary III–V semiconductors*

Semiconductors of the III–V family are built from elements of the third (aluminum — Al, gallium — Ga and indium — In) and fifth (nitrogen — N,

III	IV	V
B 5	C 6	N 7
boron	*carbon*	*nitrogen*
Al 13	Si 14	P 15
aluminum	*silicon*	*phosphorus*
Ga 31	Ge 32	As 33
gallium	*germanium*	*arsenic*
In 49	Sn 50	Sb 51
indium	*tin*	*antimony*

Figure 8.1. Part of the periodic table with Al, Ga and In of the third group and N, P, As and Sb of the fifth group which form the family of III–V semiconductors.

phosphorus — P, arsenic — As and antimony — Sb) groups of the periodic table (Figure 8.1). Atoms are covalently bonded in III–V semiconductors.

The 12 binary III–V semiconductors contain only one kind of atoms of the third and fifth groups of the periodic table ($A^{III}B^{V}$, A or B corresponds to Al, Ga, In or to N, P, As, Sb, respectively).

The band gap (E_g) of III–V semiconductors decreases with increasing atom numbers of the components. For example, the E_g of the III–V semiconductors with the lowest (AlN) and highest (InSb) atom numbers are 6.2 and 0.18 eV, respectively. The band gaps of the binary III–V semiconductors (GaN, InN, AlP, GaP, InP, AlAs, GaAs, InAs, AlSb, GaSb and InSb) are visualized in Figure 8.2 in relation to the metal atom. The most prominent III–V semiconductors are GaAs (1.42 eV) and InP (1.34 eV) with band gaps in the optimum range for photovoltaic solar energy conversion (see Chapter 6). The band gaps of GaP, AlAs and InAs are 2.26, 2.16 and 0.36 eV, respectively (Vurgaftman *et al.*, 2001 and references therein).

The temperature dependence of E_g has been described by the following empirical equation (Varshni, 1967b):

$$E_g = E_{g0} - \frac{\alpha \cdot T^2}{T + \beta} \tag{8.1}$$

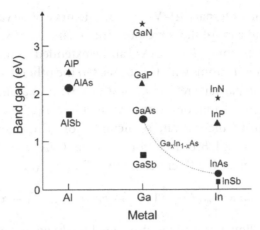

Figure 8.2. Visualization of the band gaps of binary III–V semiconductors at room temperature. The dashed line schematically shows the behavior of the band gap for alloying between GaAs and InAs.

The band gaps at 0 K (E_{g0}) are equal to 1.519, 2.24, 0.417, 2.35 and 1.424 eV for GaAs, AlAs, InAs, GaP and InP, respectively. The constants α and β are equal to 0.5405 meV/K and 204 K for GaAs, to 0.7 meV/K and 530 K for AlAs, to 0.276 meV/K and 93 K for InAs, to 0.577 meV/K and 372 K for GaP and to 0.363 meV/K and 162 K for InP (Vurgaftman *et al.*, 2001).

8.1.2 *Ternary and quaternary III–V semiconductors*

Atoms of the third or fifth groups of the periodic table can be partially replaced by other atoms of the third or fifth groups of the periodic table. The replacement of given atoms in a compound by atoms with similar chemical properties is also called alloying.

In ternary III–V semiconductors, atoms of the third (A^{III}) or fifth (B^{V}) groups of the periodic table are partially replaced by other atoms of the third (C^{III}) or fifth (D^{V}) groups of the periodic table ($A_x^{III} C_{1-x}^{III} B^{V}$ or $A^{III} B_y^{V} D_{1-y}^{V}$, where the composition parameters x and y range between 0 and 1).

In quaternary III–V semiconductors, both atoms of the third and fifth groups of the periodic table are partially replaced by other atoms of the third and fifth groups of the periodic table ($A_x^{III} C_{1-x}^{III} B_y^{V} D_{1-y}^{V}$, where the composition parameters x and y range between 0 and 1).

The band gap of ternary III–V semiconductors can be varied smoothly between the band gaps of the corresponding binary III–V semiconductors by alloying. For example, E_g of InAs can be extended to higher values by partially replacing In atoms with Ga atoms. On the other hand, E_g of GaAs can be reduced by partially replacing Ga atoms with In atoms.

The band gap of a compound semiconductor $A_x B_{1-x}$ is a superposition of the band gaps of the separate semiconductors $E_g(A)$ and $E_g(B)$. The semiconductors A and B can be, for example, GaAs and InAs with the corresponding ternary III–V semiconductor $Ga_x In_{1-x}As$.

$$E_g(A_x B_{1-x}) = x \cdot E_g(A) + (1 - x) \cdot E_g(B) - x \cdot (1 - x) \cdot C \qquad (8.2)$$

Coefficient C in Equation (8.2) is also called the bowing parameter which is given in the unit of the band gap. The value of C, for example, is equal to 0.477 eV for $Ga_x In_{1-x}As$ (Vurgaftman *et al.*, 2001).

It should be taken into account that Equation (8.2) is valid for the band gaps at the same symmetry points or valleys of the band structure of the given semiconductors (Vurgaftman *et al.*, 2001). Incidentally, the band structure describes the dependence of the energy of electrons on the wave vector. For interested readers, the band structures of some III–V semiconductors are given in, amongst other sources, Chelikowsky and Cohen (1976).

For GaAs and InAs, both band gaps are related to the so-called Γ-valley. Therefore, the dependence of E_g on x depends only on one bowing parameter of $Ga_x In_{1-x}As$ (Figure 8.3).

For AlAs and GaP, the band gaps are related to the so-called X-valley. The transition energies are 3.0 and 2.9 eV for AlAs and GaP in the Γ-valleys and 1.9 and 1.37 eV for GaAs and InAs in the X-valleys, respectively. The bowing parameters are different for the transition energies in the different valleys and so the transition energies are equal for the Γ- and X-valleys for a certain composition (also called intercrossing point). The intercrossing points are reached for $Al_x Ga_{1-x}As$ and $Ga_x In_{1-x}P$ at compositions of 0.45 and 0.64, respectively. The bowing parameter of $Al_x Ga_{1-x}As$ is equal to 0.055 for x larger than 0.45 (X-valleys) and equal to $-0.127 + 1.31x$ for x less than 0.45 (Γ-valleys) (Vurgaftman *et al.*, 2001). At the intercrossing point, the band gap of $Al_{0.45}Ga_{0.55}As$ is about

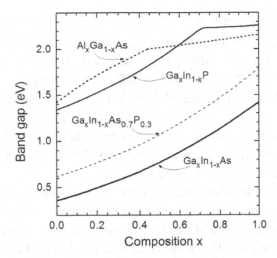

Figure 8.3. Dependence of the band gap on the composition parameter x for the ternary compounds $Ga_xIn_{1-x}P$ (thin solid line) and $Ga_xIn_{1-x}As$ (thick solid line), for $Al_xGa_{1-x}As$ (dotted line) and for the quaternary compound $Ga_xIn_{1-x}As_{0.7}P_{0.3}$ (thin dashed line).

1.95 eV. For $Ga_xIn_{1-x}P$ the bowing parameters are 0.65 for x less than 0.64 (Γ-valleys) and 0.20 for x larger than 0.64 (X-valleys) (Vurgaftman *et al.*, 2001). Figure 8.3 shows the dependence of the band gap on the composition for $Al_xGa_{1-x}As$ and $Ga_xIn_{1-x}P$.

In quaternary compounds, the band gap depends on two composition parameters x and y. The band gap of $Ga_xIn_{1-x}As_yP_{1-y}$ can be calculated, for example, by using the following equation (Nahory *et al.*, 1978)

$$E_g(x, y) = (1.35 + 0.668 \cdot x - 1.17 \cdot y + 0.758 \cdot x^2 + 0.18 \cdot y^2$$
$$- 0.069 \cdot x \cdot y - 0.322 \cdot x^2 \cdot y + 0.03 \cdot x \cdot y^2) \cdot eV \quad (8.3)$$

The dependence of E_g on the composition parameter x is plotted for $Ga_xIn_{1-x}As_{0.7}P_{0.3}$ in Figure 8.3.

A great advantage of alloying of III–V semiconductors is that practically any band gap relevant for photovoltaic solar energy conversion can be obtained. This property is key for the realization of multi-junction solar cells (Chapter 6) with highest solar energy efficiency (η).

8.1.3 *Optical absorption of III–V semiconductors*

Figure 8.4 shows the optical absorption spectra of some binary III–V semiconductors (InAs, InP, GaAs, GaP and AlAs), $Al_{0.3}Ga_{0.7}As$ and germanium (Ge)). The absorption coefficient increases to values larger than 10^4 cm^{-1} within the intervals between E_g and $E_g + 0.1$ eV for InP and GaAs, E_g and $E_g + 0.3$ eV for InAs and Ge and E_g and $E_g + 0.6$ eV for GaP and AlAs. Absorption of photons with energies below E_g can be neglected for epitaxially grown layers of III–V semiconductors due to the very low density of defects with states in the band gap.

The optical absorption spectra of ternary and quaternary III–V semiconductors behave very similar to those of the binary III–V semiconductors. The absorption coefficient of ternary and quaternary III–V semiconductors also increases steeply within an interval between E_g and $E_g + 0.1$ eV when E_g is dominated by transitions in the Γ-valley, for example, for $Ga_xAl_{1-x}As$ with $x < 0.45$ (Adachi, 1989; Aspnes *et al.*, 1986). Absorber layers with large absorption coefficients (α) for photons with energies relatively close to the band gap can be realized practically for the whole region of band

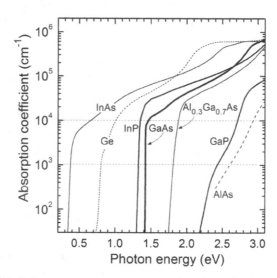

Figure 8.4. Optical absorption spectra of InAs, GaAs, GaP, AlAs, $Al_{0.3}Ga_{0.7}As$ and Ge. Data taken from Dixon and Ellis (1961), Turner *et al.* (1964), Aspnes and Studna (1983), Aspnes *et al.* (1986), Adachi (1989) and Braunstein *et al.* (1958). The dashed lines mark the absorption lengths of 10 and 1 μm.

gaps important for single-junction and multi-junction solar cells (see Chapter 6).

From the optical point of view, III–V semiconductors are well suited for stacked multi-junction solar cells with integrated series connection due to the variability of band gaps, high absorption coefficients for photons with energies above E_g and excellent transparency for photons with energies below E_g.

The optical absorption spectra and the thicknesses of the absorber layers are key for reaching high quantum efficiencies (QE) in multi-junction solar cells. The limitation of QE by optical absorption can be obtained for each of the stacked solar cells in a multi-junction solar cell by analyzing the light absorbed in the layers with the larger band gap and the light transmitted to the following layer with the lower band gap. In the following, this approach will be illustrated for a hypothetic triple-junction solar cell. The QE of the top, middle and bottom cells (QE_{top}, QE_{middle}, QE_{bottom}) are estimated considering only the definition of QE (Equation (1.41)) and transmission losses (Equation (2.14)). This means that losses caused by reflection, interference and recombination are not taken into account. In this very simplified case, QE_{top}, QE_{middle}, QE_{bottom} depend only on the related spectra of α (α_{top}, α_{middle}, α_{bottom}, respectively) and on the thicknesses of the absorbers (d_{top}, d_{middle}, d_{bottom}, respectively).

$$QE_{top} = 1 - \exp\left(-\alpha_{top} \cdot d_{top}\right) \tag{8.4'}$$

$$QE_{middle} = 1 - QE_{top} - \exp(-\alpha_{middle} \cdot d_{middle}) \tag{8.4''}$$

$$QE_{bottom} = 1 - QE_{top} - QE_{middle} - \exp(-\alpha_{bottom} \cdot d_{bottom}) \tag{8.4'''}$$

Equations (8.4')–(8.4''') correspond to the maximum QE (QE_{max}). The set of these equations has to be extended by one or two equations in for quadruple-junction of five-junction solar cells, respectively, following the same principle.

For a triple-junction solar cell with high η, the band gaps of the bottom, middle and top cells can be, for example, 0.7, 1.35 and 1.86 eV, respectively (see Figure 6.12(b)). These band gaps correspond, for example, to the ternary III–V semiconductors $Ga_{0.43}In_{0.57}As$, $Ga_{0.93}In_{0.07}As$ and $Ga_{0.5}In_{0.5}P$, respectively.

As an idealized approximation, the spectra of high absorption coefficients are proportional to the square root of the difference between the photon energy (E_{ph}) and the band gap (Equation (8.5)) For the given example, a proportionality factor (α_0) close to that of GaAs was used ($5 \cdot 10^4$ cm^{-1}).

$$\alpha = \alpha_0 \cdot \sqrt{E_{ph} - E_g} \tag{8.5}$$

Figure 8.5 shows the corresponding absorption spectra for the top, middle and bottom cells of the hypothetic triple-junction solar cell (a) and the spectra of QE$_{max}$ for thicknesses of the absorbers of 1 μm (b) and 3 μm (c).

For the example shown in Figure 8.5, the short-circuit current (I_{SC}) density was calculated for each of the cells by using equation (1.42″). The top, middle and bottom cells are connected by an integrated series

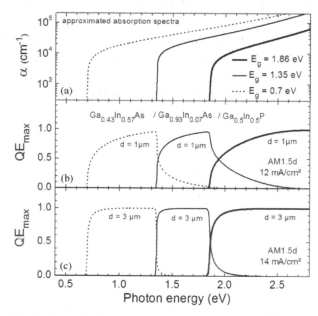

Figure 8.5. Idealized optical absorption spectra (a) and spectra of the maximum QE ((b) and (c)) for a hypothetic triple-junction solar cell. The band gaps of the absorbers of the top, middle and bottom cells (thick solid, thin solid and dashed lines, respectively) were 0.7, 1.35 and 1.86 eV, respectively. Calculations were performed using Equations (8.4′)–(8.4‴) and (8.5) for thicknesses of the absorbers of 1 μm (b) and 3 μm (c).

connection for the triple-junction solar cell. The I_{SC} density was limited by the top cell for the given example and amounted to 12 and 14 mA/cm^2 for thicknesses of the absorbers of 1 and 3 μm, respectively. Current matching in stacked multi-junction solar cells with integrated series connection (see Chapter 6) is optimized by varying the thicknesses of the absorber layers.

8.2 Hetero-Junctions of III–V Semiconductors

8.2.1 *Type I and type II semiconductor hetero-junctions*

Semiconductors with different band gaps are brought into contact with each other in III–V semiconductor solar cells. Contacts between semiconductors with different band gaps are known as semiconductor hetero-junctions.

Barriers for electrons and holes are formed at semiconductor hetero-junctions depending on the energies of the conduction and valence band edges of the two semiconductors at the contact. In the following, the band gaps of the semiconductors with the lower and higher band gaps are denoted by E_{g1} and E_{g2}, respectively. Similarly, the energies of the corresponding conduction and valence band edges are denoted by E_{C1}, E_{C2}, E_{V1} and E_{V2}.

The conduction and valence band offsets (ΔE_C and ΔE_V, respectively) are defined as the differences between E_{C2} and E_{C1} and between E_{V2} and E_{V1}.

$$\Delta E_C \equiv E_{C2} - E_{C1} \qquad (8.6')$$

$$\Delta E_V \equiv E_{V2} - E_{V1} \qquad (8.6'')$$

Type I and type II semiconductor hetero-junctions are distinguished with regard to the signs of ΔE_C and ΔE_V. For type I semiconductor hetero-junctions, the signs of ΔE_C and ΔE_V are different. This means that barriers exist at the hetero-junction for both free electrons and holes in the semiconductor with the lower band gap (Figure 8.6(a)). Type I semiconductor hetero-junctions are applied for surface passivation of photovoltaic absorbers with wide-gap semiconductors (see Chapter 3).

The signs of ΔE_C and ΔE_V are identical for type II semiconductor hetero-junctions. At the side of the semiconductor with the lower band gap, a barrier is formed for free electrons whereas the potential energy of free holes in the semiconductor with the lower band gap is reduced in the case of positive signs of ΔE_C and ΔE_V (Figure 8.6(b)). In contrast,

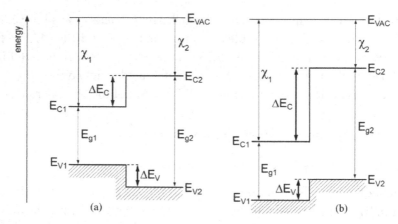

Figure 8.6. Schematic band diagrams of type I (a) and type II (b) semiconductor hetero-junctions. The signs of the conduction and valence band offsets are positive in the case of the type II semiconductor hetero-junction.

a barrier is formed for free holes whereas the potential energy of free electrons in the semiconductor with the lower band gap is reduced at a type II semiconductor hetero-junction in the case of negative signs of ΔE_C and ΔE_V. Therefore type II semiconductor hetero-junctions can serve as charge-selective contacts in solar cells.

In the case that local electric fields are absent at the interface of a semiconductor hetero-junction, i.e. the vacuum level (E_{VAC}) is constant across the interface (Figure 8.6), ΔE_C can be obtained from the difference between the electron affinities (see Chapter 5) of both semiconductors (χ_1 and χ_2).

$$\Delta E_C = -(\chi_2 - \chi_1) \tag{8.7}$$

The valence band offset can be calculated from the difference between the band gaps and the conduction band offset at the semiconductor hetero-junction.

$$\Delta E_V = (E_{g2} - E_{g1}) - \Delta E_C \tag{8.8}$$

The electron affinity can be replaced by the difference between the ionization energy (see Chapter 5) and the band gap. Similarly to Equation (8.7), ΔE_V corresponds to the negative difference between the ionization energies of both semiconductors.

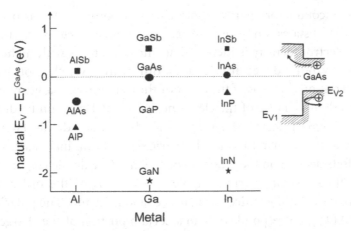

Figure 8.7. Visualization of the natural valence band alignment of binary III–V semiconductors in comparison to GaAs (values taken from Wei and Zunger (1998).

The valence band offset defined by Equation (8.8) is also called the natural ΔE_V, which has been analyzed theoretically (Wei and Zunger, 1998). The trends of the natural ΔE_V can be visualized by comparing it for the 12 binary III–V semiconductors in relation to GaAs (Figure 8.7). The lowest natural ΔE_V is obtained for the GaSb/InSb hetero-junction ($\Delta E_V = -0.01$ eV) followed by GaAs/InAs hetero-junction ($\Delta E_V = 0.05$ eV). Large values of ΔE_V, i.e. large natural barriers for free holes, were found, for example, for the GaAs/AlAs hetero-junction ($\Delta E_V = 0.51$ eV).

The values of the natural ΔE_C can be calculated from the values of the natural ΔE_V and the differences of the corresponding band gaps (Equation (8.9)). Therefore the nature of a natural binary III–V semiconductor hetero-junction can be found by combining the values of ΔE_V (Figure 8.7) with the values of E_g (Figure 8.2). For example, the natural GaAs/AlAs hetero-junction is a type I hetero-junction, whereas the natural GaAs/InP hetero-junction is a type II hetero-junction (see also task T8.1 at the end of this chapter).

8.2.2 Role of interface dipoles and interface states for band offsets

The density of interface atoms at a hetero-junction is of the order of $5 \cdot 10^{14}$ cm^{-2}. Atoms at a hetero-junction can carry a polarization charge different to that of bulk atoms due to different electronegativity of atoms in both semiconductors. Furthermore, chemical bonds can be distorted and charged defect states can be present at semiconductor hetero-junctions.

The electron density is increased at interface atoms with higher electronegativity, but reduced at interface atoms with lower electronegativity. The polarization charge is of the order of the elementary charge multiplied with the relative change of the electronegativity. Polarization by distorted chemical bonds leads to the formation of interface dipole layers and therefore to the formation of local electric fields. Resulting changes in the band offsets depend on the direction of the interface dipole (Figure 8.8).

The thickness of interface dipole layers (δ) is of the order of two bond lengths or of the lattice constant, i.e. about 0.5 nm. The polarization charge (ΔQ_{pol}) corresponds only to a certain portion of the charge of an electron in a covalent chemical bond. The drop of the potential energy across a dipole layer (ΔE_{dipole}) can be estimated by using the equation of the parallel plate capacitor.

$$\Delta E_{dipole} = q \cdot \frac{\delta}{\varepsilon \cdot \varepsilon_0} \cdot \Delta Q_{pol} \tag{8.9}$$

Band offsets at intrinsic semiconductor hetero-junctions depend on the orientation of the crystal (Grant *et al.*, 1983). This is not surprising

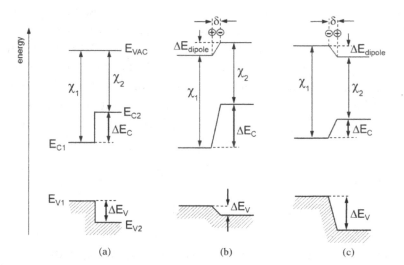

Figure 8.8. Band diagrams of an ideal type I hetero-junction (a) and of type I hetero-junctions with interface dipoles of opposite orientations (b) and (c).

if taking into account, for example, the different densities of atoms of the third and fifth groups of the periodic table at differently oriented surfaces.

Interface dipoles can be engineered by introducing, for example, additional interfacial doping layers of sub-monolayer thickness as demonstrated, for example, for a AlGaAs/GaAs hetero-junction (Capasso *et al.*, 1985). The distance between the sheets of donor and acceptor layers at the interface is only several nm for this type of interface dipole engineering.

Electronic defect states can arise at semiconductor hetero-junctions due to imperfections. Interface states are partially charged due to trapping. Charge trapped at interface states causes an additional modification of interface dipole layers at hetero-junctions (Figure 8.9). Depleted barrier layers arise at both sides of doped semiconductor hetero-junctions with large densities of interface states.

Fermi-level pinning (see Chapter 5) becomes possible at semiconductor hetero-junctions for high densities of interface states. Fermi-level pinning arises, for example, due to the removal of atoms of the third or fifth groups of elements in the periodic table from their regular lattice positions at surfaces of III–V semiconductors (Spicer *et al.*, 1979). Incidentally,

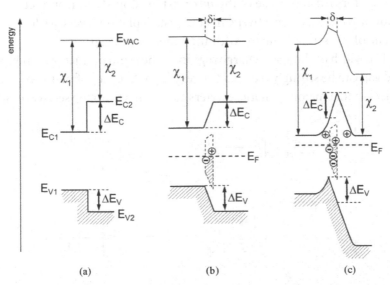

Figure 8.9. Band diagrams of an ideal type I hetero-junction (a) and of type I hetero-junctions with Fermi-level pinning caused by interface states for undoped and *n*-type doped type I hetero-junctions (b) and (c).

unpinning of the Fermi-level at surfaces of III–V semiconductors is possible as well as demonstrated by alternating adsorption of donor or acceptor molecules at GaAs surfaces (Alperovich *et al.*, 1994).

One of the most important advantages of III–V semiconductors is that the formation of interface states can be almost completely avoided at hetero-junctions formed by lattice-matched epitaxy, for example, at $GaAs/Al_xGa_{1-x}As$ hetero-junctions. The discovery of the $GaAs/Al_xGa_{1-x}As$ hetero-junction (Alferov *et al.*, 1969; 1968) triggered a burst of development of electronic devices including, for example, semi-conductor lasers (Alferov *et al.*, 1970), high-efficiency solar cells (Alferov *et al.*, 1971) and high electron mobility transistors (Mimura *et al.*, 1980).

8.2.3 *Doped type I and type II semiconductor hetero-junctions*

The band diagrams of type I semiconductor hetero-junctions at interfaces of two *n*-type or two *p*-typed doped semiconductors are drawn in Figures 8.10(a) and 8.10(b), respectively. Majority charge carriers move from the semiconductor with the higher band gap into the semiconductor with the lower band gap. Therefore, accumulation layers of electrons or holes are formed at the side of the *n*- or *p*-type doped semiconductor with the lower band gap, respectively. In contrast, depletion layers are formed at the side of the semiconductor with the higher band gap.

The width of the space charge region of the semiconductor with higher band gap can be strongly reduced for very high densities of majority charge carriers so that majority charge carriers can tunnel from the semiconductor

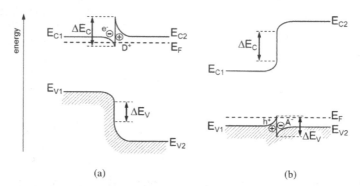

(a)   (b)

Figure 8.10. Schematic band diagrams of type I hetero-junctions *n*-type (a) or *p*-type (b) doped at both sides of the type I hetero-junction.

with the lower band gap into the semiconductor with the higher band gap (see Chapter 5). At the same time, minority charge carriers photogenerated in the semiconductor with the lower band gap can hardly overcome the large barrier at the doped type I semiconductor hetero-junction. The barrier is equal to the sum of the corresponding band offset (ΔE_C for p-type and ΔE_V for n-type doping) and the band bending at the sites of the semiconductors with the higher and lower band gaps.

The highly doped semiconductor with the higher band gap can form an ohmic contact with metals (see Chapter 5). At the same time, the highly doped semiconductor with the higher band gap serves as a barrier layer for minority charge carriers photogenerated in the semiconductor with the lower band gap. Therefore, type I semiconductor hetero-junctions with highly doped semiconductors of a higher band gap than the photovoltaic absorber are nearly ideal contact systems for majority charge carriers if, of course, an extremely low density of interface states is assumed at the type I semiconductor hetero-junction. This behavior is one of the keys for reaching very high solar energy conversion efficiencies with III–V semiconductors.

Ohmic or recombination contacts can be formed at highly doped pn-junctions (see Chapter 5). Solar cells based on III–V semiconductors with different band gaps can be stacked on each other and connected in series by introducing highly pn-doped semiconductor hetero-junctions between the separate solar cells (stacked integrated series connection). Whether the highly pn-doped semiconductor hetero-junction is of type I or of type II is not really important for the formation of recombination contacts (Figure 8.11).

The formation of highly pn-doped semiconductor hetero-junctions for recombination contacts is decisive for the realization of multi-junction concentrator solar cells based on III–V semiconductors with extremely high solar energy conversion efficiency.

8.3 Epitaxial Growth of III–V Semiconductors

8.3.1 *Principle of epitaxy with III–V semiconductors*

Epitaxy is a deposition principle enabling the growth of doped III–V semiconductor layers and layer systems with very low densities of bulk

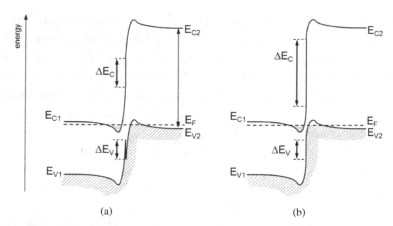

Figure 8.11. Band diagrams of highly doped recombination *pn*-contacts at type I (a) and type II (b) hetero-junctions.

and interface defects. The complex combination of III–V semiconductor absorber layers with different band gaps, with charge-selective pn-junctions, with highly doped barrier layers between photovoltaic (PV) absorbers and ohmic contacts, and with highly doped recombination contact layer systems is possible only with epitaxy.

Epitaxy means the growth of a crystal layer on a substrate crystal that provides orientation and growth direction for the growing crystal layer. For epitaxial crystal growth, the difference between the lattice constants (a_{lc}) of the substrate crystal and of the crystal of the growing layer should not be too large. The normalized difference between a_{lc} of the substrate and the growing crystal layer is called the lattice mismatch ($\Delta a_{lc}/a_{lc}$). In general, the $\Delta a_{lc}/a_{lc}$ between two compound crystals with $a_{lc}(A_xB_{1-x})$ and $a_{lc}(C_yD_{1-y})$ can be calculated by the following equation:

$$\frac{\Delta a_{lc}}{a_{lc}} = 2 \cdot \left| \frac{|a_{lc}(A_xB_{1-x}) - a_{lc}(C_yD_{1-y})|}{a_{lc}(A_xB_{1-x}) + a_{lc}(C_yD_{1-y})} \right| \tag{8.10}$$

Figure 8.12 shows a schematic interface between two binary crystals with the same a_{lc}, i.e. $\Delta a_{lc}/a_{lc}$ is equal to zero. The crystal lattice of the substrate crystal is continued ideally in the growing crystal layer. Homoepitaxy is the growth of an epitaxial layer on a crystal of the same material, for example, the growth of a GaAs layer on a GaAs crystal.

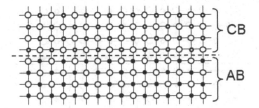

Figure 8.12. Schematic of an ideal interface between two binary crystals AB and CB with the same lattice constant.

Growth conditions can be optimized for epitaxial layers so that the density of bulk defects can be much lower in epitaxial layers than in large crystals of III–V semiconductors. This is crucial for realizing a long lifetime and therefore a long diffusion length of photogenerated charge carriers in the absorber (see Chapter 3). The homoepitaxial growth of a buffering III–V semiconductor layer with a very low density of defects between a substrate crystal and a layer system of a solar cell is very important for reaching very high η.

The interface between two crystals is free from interface defect states if both crystal layers have the same a_{lc}. Therefore, surface recombination can be neglected at interfaces between epitaxial layers with the same a_{lc} or with a very low $\Delta a_{lc}/a_{lc}$. This is the case for interfaces, for example, between epitaxial $Al_x Ga_{1-x} As$ and GaAs layers.

Stress arises in growing epitaxial layers if the a_{lc} of the substrate and growing crystals are different (Figure 8.13(a)). If the $\Delta a_{lc}/a_{lc}$ is not very large, interfacial stress can be compensated by slight distortion of chemical bonds and atoms in the interface region. Distortion of chemical bonds from their equilibrium causes fluctuations in the periodic lattice potential of the crystals and therefore the formation of exponential tail states at the conduction and valence band edges in the interface region. This type of defect states becomes relevant for surface recombination at concentrated sunlight. Further, distortion of chemical bonds causes electric polarization in the interface region and therefore the formation of an interface dipole layer.

The $\Delta a_{lc}/a_{lc}$ between the substrate and growing crystals can be very large and consequently, for compensation of stress, some atoms have to be omitted in the interface region in addition to distortion of chemical bonds (Figure 8.13 (b)). The complete removal of interface atoms from

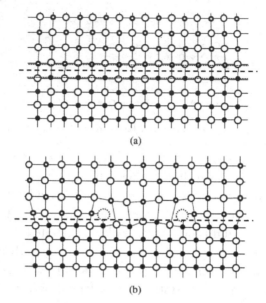

(a)

(b)

Figure 8.13. Schematic interface configuration between two binary crystals AB and CB with slightly different lattice constants (a) and with very different lattice constants (b). The narrow dashed lines in (a) characterize the distortion of atoms in the interface region from their lattice positions. The large dashed circles in (b) correspond to missing atoms at the interface of the semiconductor with the higher lattice constant.

certain lattice positions starts when the $\Delta a_{lc}/a_{lc}$ becomes larger than a certain critical lattice mismatch, which also depends on the thickness of the epitaxial layer.

Systems of III–V semiconductor layers with various E_g are required for high-efficiency solar cells. However, not all III–V semiconductors can be combined with each other for epitaxial growth due to their $\Delta a_{lc}/a_{lc}$. The band gaps of several semiconductors — including binary III–V semiconductors and also Si and Ge — are plotted versus their lattice constants in Figure 8.14. The a_{lc} of GaAs, GaP, InAs and InP are 0.56532, 0.54505, 0.60583 and 0.58697 nm, respectively, at room temperature (see (Vurgaftman *et al.*, 2001) for the calculation of the temperature dependent a_{lc} of III–V semiconductors).

The a_{lc} of Ge, GaAs and AlAs are very similar and, as such, epitaxial GaAs and Al$_x$Ga$_{1-x}$As layers can also be grown on Ge crystals. The a_{lc} of GaP and AlP are also very close to each other, although their band gaps

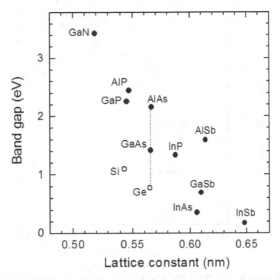

Figure 8.14. Band gaps of binary III–V semiconductors and Si and Ge plotted against their lattice constants. The vertical line marks epitaxy of GaAs and AlAs on Ge.

are rather large for PV solar energy conversion at maximum η. For most combinations of binary III–V semiconductors, the $\Delta a_{lc} / a_{lc}$ is too large for epitaxial growth with low density of defects.

Lattice constants can be smoothly changed for compound crystals by alloying. The a_{lc} of an alloy $A_x B_{1-x}$ is a linear superposition of the composition parameter x and the a_{lc} of the compounds ($a_{lc}(A)$ and $a_{lc}(B)$) as expressed by the Vegard's law (Vegard, 1921)

$$a_{lc}(A_x B_{1-x}) = x \cdot a_{lc}(A) + (1 - x) \cdot a_{lc}(B) \qquad (8.11)$$

For a ternary III–V semiconductors, for example, A and B can correspond to GaAs and InAs. The band gap and the a_{lc} of ternary III–V semiconductors can be calculated as a function of the composition parameter by using Equations (8.2) and (8.10). As a result, the band gap can be plotted as a function of the lattice constant for any alloy system of III–V semiconductors. This is shown for the ternary III–V semiconductors $Ga_x In_{1-x} As$, $Ga_x In_{1-x} P$ and $AlAs_x Sb_{1-x}$ and for the quaternary semiconductor $Ga_x In_{1-x} As_{0.7} P_{0.3}$ in Figure 8.15.

Lattice-matched epitaxy is possible along vertical lines of constant lattice constants. GaAs (line (1) in Figure 8.15) and InP (line (2) in

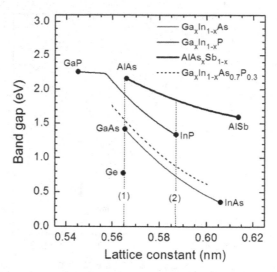

Figure 8.15. Dependence of the band gap on the lattice constant for the ternary III–V semiconductors $Ga_xIn_{1-x}As$, $Ga_xIn_{1-x}P$ and $AlAs_xSb_{1-x}$ (thin, middle and thick solid lines, respectively) and for the quaternary III–V semiconductors $Ga_xIn_{1-x}As_{0.7}P_{0.3}$ (dashed line). The vertical lines mark lattice-matched epitaxy with GaAs (1) and InP (2).

Figure 8.15) are important III–V semiconductors for epitaxy. The band gap between AlAs and GaAs can be smoothly varied by alloying within the $Al_xGa_{1-x}As$ system without changing the a_{lc}. The a_{lc} of Ge is 0.56575 nm (Dismukes *et al.*, 1964). A perfect lattice matching is obtained for $Ga_{0.99}In_{0.01}As$ and $Ga_{0.51}In_{0.49}P$ epitaxial layers on Ge crystals which correspond to the combination of band gaps of 1.4 (middle cell), 1.92 (top cell) and 0.7 (bottom cell made from Ge) eV, respectively, for triple-junction solar cells. However, the combination of these band gaps does not correspond to the optimum for reaching the maximum η (see Chapter 6). A necessary broader variation of the band gaps especially between those of Ge and GaAs is hardly possible at the lattice constant of GaAs.

The ternary III–V semiconductors $Ga_{0.46}In_{0.56}As$ and $AlAs_{0.44}Sb_{0.56}$ have the same a_{lc} as InP and band gaps of 0.73 and 1.9 eV, respectively. Therefore, a combination of the band gaps of 1.36 (middle cell), 0.73 (bottom cell) and 1.7–1.9 (top cell, variation due to uncertainty in the bowing parameter) eV is possible by lattice-matched epitaxy on InP (line (2) in Figure 8.15).

The high variability of band gaps within a lattice-matched epitaxy is also needed for the realization of solar cells with more than three junctions (Szabo *et al.*, 2008). In ideal case, E_g can be tuned over extended ranges at constant a_{lc}. This is possible with quaternary compounds. For example, the a_{lc} of $Ga_xIn_{1-x}As_yP_{1-y}$ compounds can be calculated from the a_{lc} of GaAs, GaP, InAs and InP (Nahory *et al.*, 1978):

$$a_{lc}(x, y) = xy \cdot a_{lc}(GaAs) + x(1 - y) \cdot a_{lc}(GaP) + (1 - x)y$$
$$\times a_{lc}(InAs) + (1 - x)(1 - y) \cdot a_{lc}(InP) \qquad (8.12)$$

The a_{lc} of a quaternary compound can be kept constant for certain combinations of x and y. Figure 8.16 shows the dependence of the composition parameter x on the composition parameter y for keeping the a_{lc} constant at that of InP and the corresponding dependence of E_g that ranges between 0.73 and 1.36 eV.

The lattice-matched growth of $Ga_xIn_{1-x}As$ with a band gap of about 1.1 eV on InP is impossible but desirable for multi-junction solar cells from some technological point of view. For example, a tandem solar cell

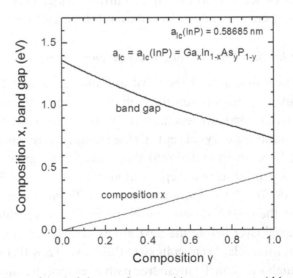

Figure 8.16. Dependence of the metal composition parameter x (thin solid line) and of the band gap (thick solid line) on the composition parameter y for the quaternary III–V semiconductors $Ga_xIn_{1-x}As_yP_{1-y}$ for lattice matched epitaxy on InP. Values were calculated using Equations (8.3) and (8.12).

with a $Ga_{0.35}In_{0.65}P$ top cell and a $Ga_{0.82}In_{0.18}As$ bottom cell (Dimroth *et al.*, 2000) has a combination of band gaps (1.67 and 1.18 eV, respectively) corresponding to the theoretical optimum (see Chapter 6). However, the a_{lc} is quite different for both crystals (0.59853 nm for $Ga_{0.82}In_{0.18}As$ and 0.55972 nm for $Ga_{0.35}In_{0.65}P$). In contrast to lattice-matched epitaxy of III–V semiconductor layers, the combination of lattice-mismatched layers would allow a higher degree of freedom for optimal choice of band gaps in multi-junction solar cells with extremely high η (King *et al.*, 2007).

Additional defects are introduced into epitaxial layers in lattice-mismatched or so-called metamorphic solar cells. It is possible to reduce the density of defects to a certain minimum by introducing, for example, misoriented substrate crystals with composition graded buffers in which stress is reduced (Dimroth *et al.*, 2000; King *et al.*, 2000). Misorientation means that the crystal is cut under a certain angle in relation to the surface at which the epitaxial layer is grown. In this way, additional steps are introduced at the surface of the substrate crystal so that stress or strain can relax in a thin interfacial layer. The relaxation of stress is very important in order to avoid the propagation of extended defects such as threading dislocations into active absorber layers during further crystal growth as well as during operation of the solar cell over its lifecycle (King *et al.*, 2000).

8.3.2 *Molecular beam epitaxy of III–V semiconductors*

Molecular beams of atoms of the third and fifth groups of the periodic table are formed by evaporation through blends from effusion cells into ultra-high vacuum (pressure less than 10^{-8} Pa). The molecular beams are directed onto a substrate crystal kept at a high temperature. Atoms reaching the hot crystal are incorporated into the lattice of the growing epitaxial layer. A typical temperature for epitaxial growth of GaAs is 590°C (Cho and Arthur, 1975). The evaporation rate is sufficiently low that atoms condensing at the crystal surface can be perfectly incorporated into the growing epitaxial layer before the next atoms arrive. For this reason, the growth rate of molecular beam epitaxy (MBE) is very low (0.1–1 nm/s).

In ultra-high vacuum, the mean free path of evaporated atoms is much longer than the distance between the effusion cells and the substrate crystal. This means that evaporated atoms do not interact with each other or with rest gas molecules before reaching the substrate surface. In an ultra-high vacuum chamber, the density of rest gas molecules is much lower than the density of evaporated atoms and extremely pure elemental effusion cells can be produced. Therefore, very pure epitaxial layers can be grown by MBE. For example, the background doping of undoped GaAs layers grown by MBE is less than 10^{14} cm^{-3}, which is two orders of magnitude lower than the maximum density of defects that can be tolerated for the minimum lifetime for SRH recombination in GaAs (see Figure 3.7).

Figure 8.17 shows a general setup for MBE growth of III–V semiconductor layers. Cryogenic traps around the heated sample holder additionally improve the ultra-high vacuum. A mass spectrometer is connected to control the components of the rest gas in the ultra-high vacuum chamber. Effusion cells for evaporation of, for example, Al, Ga, In, P, As, Sb, and Be and S (for p- and n-type doping, respectively) can be closed separately with shutters. The effusion cells, an electron gun and a RHEED (reflection high-energy electron diffraction) detector are all pointed towards the sample holder.

For MBE growth, semiconductor crystals kept in ultra-high vacuum at high temperature. The energy of pure surfaces is minimized by pushing out certain surface atoms from the lattice position of the bulk crystal. In this way, two-dimensional ordered surface structures are formed at very clean

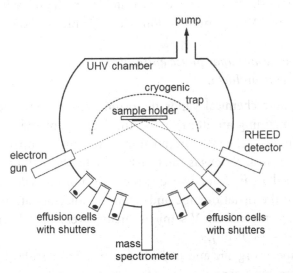

Figure 8.17. Principle setup for epitaxial layer growth by MBE (UHV–ultra-high vacuum).

semiconductor surfaces in ultra-high vacuum. The geometry of the two-dimensional surface structures depends on the orientation of the crystal and on temperature. The formation of two-dimensional ordered surface structures is called surface reconstruction. Surface reconstruction and phase diagrams of surface reconstruction can be investigated by RHEED (see, for example, Däweritz and Hey, 1990).

The epitaxial layer-by-layer growth can be well monitored by analyzing surface reconstruction since the surface becomes rough during the growth of a new mono-layer (reduction of the RHEED signal) and smooth during completion of a growing mono-layer (maximum of the RHEED signal). A very precise control of the growing layer thickness has been reached by connecting computer-controlled shutters of effusion cells with RHEED monitoring of epitaxial layer growth.

Surface chemical bonds and reconstruction influence the optical properties of surfaces of III–V semiconductors. The change of optical properties caused by a variation of surface chemical bonds can be monitored very sensitively (see, for example, Samuelson *et al.*, 1992) by measuring the variation of reflected polarized light (the reflection difference — RD) (Aspness *et al.*, 1988).

Extremely pure epitaxial layers of III–V semiconductors with a defined number of mono-layers and with defined stoichiometry of each mono-layer can be grown by MBE in combination with RHEED and RD.

8.3.3 *Metal organic vapor phase epitaxy of III–V semiconductors*

In metal organic chemical vapor deposition (MOCVD), metal organic molecules containing an atom of the third group of the periodic table and hydride molecules containing an atom of the fifth group of the periodic table are diluted in a carrier gas and transported to a heated substrate crystal where the metal organic and hydride precursor molecules decompose. The core atoms of the metal organic and hydride molecules are incorporated into a growing epitaxial III–V semiconductor layers (metal organic vapor phase epitaxy — MOVPE).

Typical metal organic and hydride precursor molecules for MOCVD of III–V semiconductors are, for example, trimethyl gallium (TMG — $Ga(CH_3)_3$) and arsine (AsH_3). Metal organic and hydride precursor

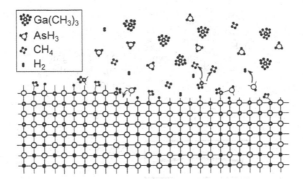

Figure 8.18. Scheme of trimethyl gallium (Ga(CH$_3$)$_3$) and arsine (AsH$_3$) arriving at the surface of a growing GaAs epitaxial layer. Precursor molecules decompose at the hot crystal surface and elements of the third and fifth groups are incorporated into the lattice of the growing epitaxial layer.

molecules decompose near the hot crystal surface by different reactions. As an example, Figure 8.18 shows a scheme of TMG and AsH$_3$ arriving at the surface of a growing epitaxial GaAs layer.

TMG and AsH$_3$ start to decompose near the hot surface via reactions such as

$$AsH_3 \rightarrow = AsH + H_2 \qquad (8.13)$$

$$Ga(CH_3)_3 + H_2 \rightarrow = GaCH_3 + 2CH_4 \qquad (8.14)$$

Activated radicals such as $=$AsH and $=$GaCH$_3$ form bonds with reactive sites at the surface of the growing epitaxial layer and continue to decompose. The overall reaction equation of deposition of GaAs by MOCVD is

$$Ga(CH_3)_3 + AsH_3 \rightarrow GaAs + 3CH_4 \qquad (8.15)$$

Metal organic precursor molecules such as dimethylzinc are used for doping of III–V semiconductor layers during MOVPE.

A certain amount of carbon, hydrogen and oxygen is incorporated into the growing crystal besides the desired elements of the third and fifth groups of the periodic table. The amount of incorporated carbon, hydrogen and oxygen can be reduced to a minimum by choosing appropriate deposition conditions. The incorporation of carbon can be used for *p*-type doping of III–V semiconductors. Carbon has a low diffusion coefficient in III–V

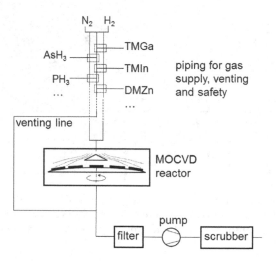

Figure 8.19.   Schematic MOCVD setup.

semiconductors so that highly doped p-type doping is possible with carbon (see, for example, Kuech *et al.*, 1988).

The growth rate of epitaxial III–V semiconductor layers is proportional to the gas flow in the MOCVD reactor (Figure 8.19). Usual growth rates are of the order of 1–2 nm/s for MOVPE. Gas flows can be well controlled in large-scale production. MOCVD reactors are operated at reduced pressure of the order of 100 mbar.

Hydrides such as AsH_3 and PH_3 are highly toxic gases and metal organics are highly pyrophoric, i.e. they react violently with oxygen and moisture (see, for example, Shenai-Khatkhate *et al.*, 2004 and references therein). Therefore, safety measures including gas detection, gas protection and destruction removal of unreacted molecules are very important for the production of III–V semiconductors by MOCVD. Processes of catalytic decomposition, combustion and adsorption are used for the destruction removal of unreacted molecules. This process is called scrubbing.

Figure 8.19 shows a schematic of an MOCVD setup, including the supply of carrier and hydride gases, the supply of metal organic molecules, the run and venting pipes, the MOCVD reactor with the sample holder, the filters for particles and phosphorus, the processing pump and the scrubber for decomposing unreacted metal organic molecules. The power for substrate heating is coupled into a massive carbon susceptor from a high-frequency power supply (not shown in the figure). Numerous substrates can be processed at the same time. Homogeneous deposition is achieved by optimizing the design of the gas flow in the MOCVD reactor and by planetary motion of the substrates during deposition.

8.3.4 *Epitaxial lift-off of III–V semiconductors*

The increase of the diffusion length of photogenerated charge carriers, the reduction of the absorber layer thickness and the realization of a very low surface recombination rate are decisive for reaching very high η (see Chapter 4). It is useful to combine GaAs solar cells with back reflectors and/or photonic structures for light trapping in order to reduce the absorber layer thickness (see, for example, Miller *et al.*, 2012). For this purpose, a GaAs solar cell has to be transferred from the substrate crystal to a substrate enabling advanced optical design.

The transfer of GaAs solar cells is possible with the so-called epitaxial lift-off, making use of a sacrificial AlAs layer which is etched away during the lift-off process (Figure 8.20). The etch rate of AlAs in hydrofluoric acid (HF) is several orders of magnitude larger than the etch rate of GaAs in HF (Yablonovich *et al.*, 1987). This property is used to lift-off epitaxial GaAs layers and to recycle the GaAs substrate crystal many times for subsequent epitaxy of GaAs layers (Voncken *et al.*, 2004).

The highest η of a solar cell with a single band gap has been reached by using epitaxial lift-off of GaAs (Kayes *et al.*, 2011). In the epitaxial lift-off technology of GaAs solar cells, the back contact is deposited onto the

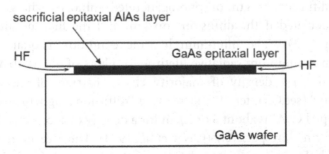

Figure 8.20. Schematic of etching a sacrificial epitaxial AlAs layer for lifting off a GaAs epitaxial layer. Prior to the epitaxial lift-off, the GaAs epitaxial layer has to be fixed at a handle or at another semiconductor wafer such as c-Si or InP (wafer bonding, not shown).

complete epitaxial layer system of the GaAs solar cell together with a flexible handle prior to the epitaxial lift-off (Kayes *et al.*, 2011). Instead of a flexible handle, another wafer can be bonded so that epitaxial layers with rather different a_{lc} can be combined, for example, in a quadruple-junction solar cell with two top cells grown on a GaAs wafer and two bottom cells grown on an InP wafer (Dimroth *et al.*, 2014).

Stacks of many epitaxial sacrificial AlAs layers and epitaxial GaAs solar cells can be grown and etched simultaneously (Yoon *et al.*, 2010). This facilitates ways to assemble flexible lightweight photovoltaic modules with very high η at moderate costs (Yoon *et al.*, 2010).

8.4 Single- and Multi-Junction Solar Cells of III–V Semiconductors

8.4.1 Design of the pn-junction of solar cells of III–V semiconductors

III–V semiconductors can be doped *n*-type and *p*-type over a wide range. In III–V semiconductors, the mobility of free electrons is higher than the mobility of free holes by a factor of 5–10 due to the low effective mass of electrons and the high effective mass of holes. Typical values of electron and hole mobilities are 3000 cm^2/Vs and 200 cm^2/Vs, respectively, for GaAs doped with donors or acceptors with densities of about $5 \cdot 10^{17}$ cm^{-3} (Sze, 1981). The diffusion coefficient of free electrons and holes can be obtained from the corresponding mobility by using the Einstein equation (Equation (9.13)).

The diffusion length of photogenerated minority charge carriers can be calculated if the diffusion constant and the lifetime are known (Equation (3.5)). The minimum lifetime condition (Equation (3.6)) resulted in a minimum radiative lifetime for GaAs of about 10 ns, which corresponds to a density of majority charge carriers of the order of $5 \cdot 10^{17}$ cm^{-3} (see Chapter 3). The measured diffusion length of electrons in *p*-type doped GaAs is about 8 or 5 μm for a density of acceptors of $3 \cdot 10^{17}$ or $8 \cdot 10^{17}$ cm^{-3}, respectively (Casey *et al.*, 1973). The measured diffusion length of holes in *n*-type doped GaAs is about 1.5 μm for a density of donors of $1 \cdot 10^{18}$ cm^{-3} (Casey *et al.*, 1973). The diffusion potential of a *pn*-junction of a GaAs solar cell ranges between 1.38 and 1.40 V.

The sum of the thicknesses of the n-type and p-type doped regions of a solar cell should be larger by a factor of three than the optical absorption length. More than 95% of incoming photons with energy above the band gap of GaAs is absorbed within a GaAs layer with a thickness of 3 μm (see Chapter 2). Therefore, a GaAs solar cell with high η usually contains an n-type doped layer with a thickness of about 0.5 μm and a p-type doped layer with a thickness of about 2.5 μm.

The absorption lengths and the diffusion coefficients of electrons and holes are of a similar order in III–V semiconductors. Therefore doping and extension of n-type and p-type doped regions are rather similar for solar cells based on III–V semiconductors.

8.4.2 *Architecture and band diagram of single-junction solar cells*

Figure 8.21 shows an example for the layer structure of a GaAs solar cell with very high η grown on a highly n-type doped wafer of GaAs(100).

First, a so-called buffer layer with a thickness of the order of 1 μm was grown on top of the GaAs(100) wafer in order to reduce the density of extended defects. The buffer layer is highly n-type doped as well. Second, a highly n-type doped $Al_xGa_{1-x}As$ layer was grown on top of the defect-free buffer layer to form the ohmic contact with the n-type doped GaAs layer and with a barrier for holes photogenerated in the n-type doped GaAs layer. Third, the n-type and p-type doped absorber layers were grown. Fourth, a

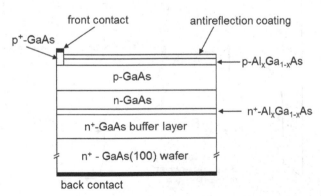

Figure 8.21. Schematic cross section of a GaAs solar cell grown on a highly n-type doped GaAs wafer. Incidentally, the GaAs wafer can be reused if replacing it by a flexible handle within an epitaxial lift-off.

p-type doped $Al_xGa_{1-x}As$ passivation layer was grown on the p-type doped GaAs layer. Fifth, the p-type doped $Al_xGa_{1-x}As$ buffer was partially opened by a mesa etching and the p-type doped GaAs layer was contacted with a highly doped GaAs layer. The $Al_xGa_{1-x}As$ buffer layer was coated with a silicon nitride antireflection coating. Finally, metal contacts were deposited and ohmic contacts were formed by annealing steps between 100°C and 300°C.

A similar architecture shown in Figure 8.21 can be realized in combination with an epitaxial lift-off. In this case, the (reusable) GaAs wafer in the figure has to be replaced by the back contact and the flexible handle. In a process with an epitaxial lift-off, the epitaxial layers are grown in the opposite order since the back contact is deposited on the front side of the wafer after epitaxy and the solar cell is flipped and coated with the front contact and antireflection coating after the lift-off (Keyes *et al.*, 2011).

The idealized band diagram of the solar cell shown in Figure 8.21 is depicted in Figure 8.22 for the cross section across the antireflection coating

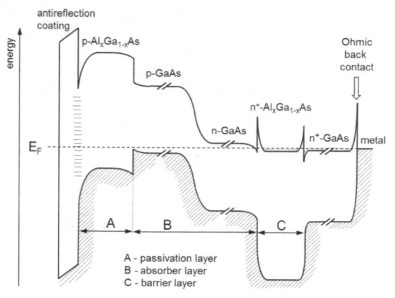

Figure 8.22. Idealized band diagram of a GaAs solar cell in the dark across the antireflection coating, the passivation, absorber and barrier layers and the highly n-type doped GaAs wafer with the ohmic back contact.

and the following passivation, absorber and barrier layers and the highly *n*-type doped GaAs wafer. The band gap of the antireflection coating is much larger than the band gap of the absorber layer. Numerous defects at the interface between the antireflection coating and the passivation layer are not important for the performance of the solar cell since the p-$Al_x Ga_{1-x}As$ passivation layer keeps photogenerated electrons and holes away from defects at the interface between the antireflection coating and the passivation layer (see Chapter 3).

Electrons can be transferred from the *n*-type doped region of the GaAs absorber through the highly *n*-type doped $Al_x Ga_{1-x}As$ barrier layer to the ohmic back contact whereas holes are reflected at the interface between the *n*-type doped region of the GaAs absorber and the barrier layer. Holes can be transferred from the *p*-type doped region of the GaAs absorber through the highly *p*-type doped GaAs layer to the ohmic front contact.

The combination of absorber layers with *pn*-homo-junctions as charge-selective contacts and hetero-junctions for formation of barrier and passivation layers is applied in all kinds of solar cells based on III–V semiconductors.

Solar cells based on GaAs can also be prepared on *p*-type doped GaAs wafers. In this case, the thickness of the *p*-type doped GaAs absorber layer is increased to about 3 μm concomitant with a reduction of the density of acceptors to $5 \cdot 10^{16}$ cm^{-3} whereas the thickness of the *n*-type doped GaAs layer is reduced to about 0.1 μm concomitant with an increase of the density of donors to about $1–2 \cdot 10^{18}$ cm^{-3} (Lu *et al.*, 2011).

8.4.3 *Ohmic metal contacts with III–V semiconductors*

The formation of ohmic metal contacts with low contact resistance demands a highly doped semiconductor and a low barrier height at the metal–semiconductor interface (see Chapter 5). A strong reduction of the barrier height at the interface between a metal and differently doped III–V semiconductors is practically impossible due to Fermi-level pinning during covering a clean surface of a III–V semiconductors with one given metal such as gold (see, for example, Spicer *et al.*, 1979; Walukiewicz, 1988).

The principle of the formation of ohmic contacts at III–V semiconductor/metal interfaces (see Figure 8.23) is based on the formation of vacancies in the III–V semiconductors and of a stable intermetallic phase of a metal

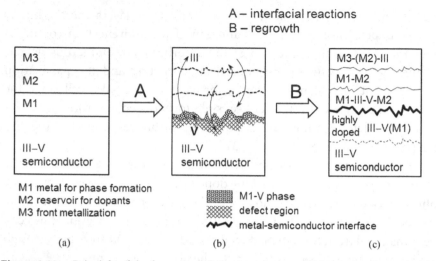

Figure 8.23. Principle of the formation of ohmic contacts at III–V semiconductor/metal interfaces. (a) stack of deposited metal layers (M1–M3) needed for (i) the formation of a stable M1-V phase, for example, NiAs or PdAs, (ii) doping of the III–V semiconductors, for example, Ge as acceptor and Be as donor, (iii) front metallization, for example, Au. (b) contact region with M1-V phase, defect region and diffusion processes during interfacial reactions. (c) contact region after regrowth of the crystal of the III–V semiconductors.

(denoted by M1 in Figure 8.23) with the element of the fifth group of the periodic table. During interfacial reactions, the intermetallic phase and a defect layer are formed near the surface of the III–V semiconductors and atoms of dopants (denoted by M2 in Figure 8.23) diffuse into the defect region of the III–V semiconductors. During the regrowth, the defect region is well recrystallized and the atoms of dopants are incorporated into the crystal lattice. On top of the metallization layer (denoted by M3 in Figure 8.23), a contact wire can be bonded. In addition, diffusion processes lead to the formation of different alloys in the layer system.

The interface at ohmic metal contacts with III–V semiconductors is rough due to interfacial chemical reactions and several phase transitions and due to inter-diffusion of atoms between layers. The thickness of the complete ohmic contact layer system is of the order of 100–200 nm. After the formation of the ohmic contact, electroplating is applied for increasing the thickness of the contact pad so that external metal wires can be bonded to the ohmic metal contact of the solar cell.

Layer systems such as Ni/Ge/Au (Kim and Holloway, 1997) or Pd/Ge/Ti/Pt/Au (Zide *et al.*, 2006) are deposited onto *n*-type doped III–V semiconductors and annealed in order to form ohmic contacts with very low contact resistance. At the beginning of annealing, arsenic (As) diffuses, for example, from GaAs into the nickel (Ni) or palladium (Pd) layers and stable intermetallic compounds such as NiAs are formed at the interface between the metal and GaAs. This process leads to a deficiency of As and to damage of the crystal lattice in the contact region of GaAs. During the following regrowth of GaAs in the contact region, Ge atoms are incorporated into the crystal lattice of GaAs. The incorporation of Ge atoms into the lattice of GaAs leads to a highly *n*-type doped interface region between the metal and the III–V semiconductors.

Ohmic metal contacts can be formed on *p*-type doped GaAs, for example, with an Au/BeAu/Cr contact system (Algora *et al.*, 2001). Beryllium atoms are acceptors in GaAs crystals. Beryllium atoms are incorporated into the surface region of the GaAs layer during the formation of the ohmic contact. The ohmic contact is formed between the metal layer and a highly *p*-type doped GaAs(Be) layer at the contact.

The Fermi-energy at III–V semiconductor/metal interfaces is often pinned near the middle of the band gap of the semiconductor. Therefore, the barrier heights at III–V semiconductor/metal contacts usually increase with increasing band gap of the III–V semiconductors. The barrier height at a wide gap III–V semiconductors/metals contact can be reduced by introducing a very thin layer of a highly doped III–V semiconductors with a low band gap on top of the III–V semiconductors with a high band gap (Figure 8.24). By this way, ohmic metal contacts with very low contact resistances have been formed on III–V semiconductors (see, for example, Kim and Holloway, 1997). For example, a highly doped germanium layer can be grown on top of a GaAs to form an ohmic GaAs/metal contact (Devlin *et al.*, 1980). The contact resistance of ohmic GaAs/metal contact has been reduced to 10^{-7} Ω cm^2 by implementing a highly doped Ge (Stall *et al.*, 1981), InAs (Kumar *et al.*, 1989) or graded InGaAs (Mehdi *et al.*, 1989) contact layers between the GaAs absorber and the metal contact.

The contact resistance of ohmic metal contacts formed on GaAs with Au/Ge can be as low as 10^{-5} Ω cm^2 (Kim and Holloway, 1997). This value has to be related to the tolerable series resistance of solar cells (see Chapter 1)

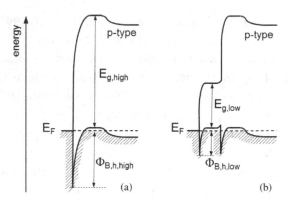

Figure 8.24. Schematic band diagrams of contacts between a metal and a highly p-type doped semiconductor with a high band gap ($E_{g,\text{high}}$) (a) and a low band gap ($E_{g,\text{low}}$) (b) for contacting a moderately p-type doped semiconductor with a high band gap. The barrier height for holes has been reduced from $\Phi_{B,h,\text{high}}$ to $\Phi_{B,h,\text{low}}$ in addition to a reduction of the width of the space charge region. Fermi-level pinning around the middle of the band gap at the metal–semiconductor contact and a type I hetero-junction between the semiconductors with high and low band gaps are assumed.

based on III–V semiconductors and operated under concentrated sunlight by taking into account a reduced area of the front contact for minimized shading. For comparison, a GaAs solar cell with an open-circuit voltage (V_{OC}) of 1.107 V and a short-circuit current (I_{SC}) density of 29.6 mA/cm^2 under operation at air mass (AM) 1.5 (Kayes *et al.*, 2011) would require a contact resistance of about $3 \cdot 10^{-6}$ Ω cm^2 for operation at a concentration factor of 1000 if assuming a reduction of the area of the front contact by 100 times. This makes the contact resistance very important for solar cells based on III–V semiconductors operated under highly concentrated sunlight.

8.4.4 *Tandem solar cells with III–V semiconductors*

In comparison to single-junction solar cells, the η can be increased with tandem solar cells due to the reduction of thermalization losses while the theoretical limit of η is about 44% for a combination of band gaps of 1.1 and 1.7 eV (see Chapter 6).

GaAs and Al$_x$Ga$_{1-x}$As have nearly identical lattice constants (Figure 8.15) and, as such, lattice-matched epitaxial layer growth can be used for fabrication of tandem solar cells. The band gap of GaAs is 1.42 eV and the corresponding optimum band gap of the top cell is 1.91 eV with regard to Figure 6.12. A band gap of 1.91 eV is obtained for $Al_{0.36}Ga_{0.64}As$ with regard to Figure 8.3. The maximum η of a tandem solar cell with band gaps of 1.91 and 1.42 eV for the top and bottom cells, respectively, is about 40% (see Figure 6.15).

Figure 8.25 shows a schematic cross section of a $GaAs/Al_{0.36}Ga_{0.64}As$ tandem solar cell grown by MOVPE on an n-type doped GaAs wafer with a layer system used by Takahashi *et al.* (2005). For reaching a very high η of 28.85% under illumination at AM1.5, the authors (i) increased the lifetime of holes in n-type doped $Al_{0.36}Ga_{0.64}As$ layers from about 1.7 ns to 4.1 ns by exchanging Si for Se donors and (ii) reduced the thermal diffusion of acceptors into the highly n-type doped region of the tunneling junction by exchanging Zn for carbon acceptors, which have a low diffusion coefficient at increased temperatures (Takahashi *et al.*, 2005). This example demonstrates the importance of a dedicated increase of the lifetime of

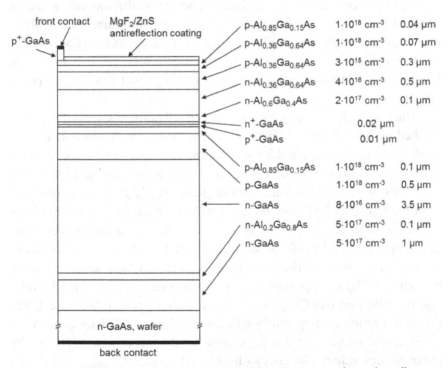

Figure 8.25. Schematic cross-section of a $GaAs/Al_{0.36}Ga_{0.64}As$ tandem solar cell grown on an n-type doped GaAs wafer. After Takahashi *et al.* (2005).

minority charge carriers in the absorber and the reduction of the tunneling resistance of the recombination contact.

A GaAs/Al$_{0.36}$Ga$_{0.64}$As tandem solar cell consists of at least 12 epitaxial layers including the n-type and p-type doped absorber layers, barrier layers, passivation layers and the highly doped layers of the tunneling contact (see also Bedair *et al.*, 1979). Layers with thicknesses between 10 nm and 3.5 μm and a doping range between $3 \cdot 10^{15}$ cm^{-3} and $5 \cdot 10^{19}$ cm^{-3} have been grown. In addition, highly doped contact layers were formed for ohmic contacts. Finally, a MgF$_2$/ZnS double antireflection coating and ohmic contact layer systems were deposited.

The ternary compound Ga$_{0.49}$In$_{0.51}$P has a lattice constant equal to that of GaAs and a band gap of 1.93 eV which is very close to the optimum band gap for a tandem solar cell with a GaAs bottom cell. A flexible GaAs/Ga$_{0.5}$In$_{0.5}$P tandem solar cell was built by so-called inverted epitaxy on GaAs and by applying epitaxial lift-off (Kayes *et al.*, 2014). Inverted epitaxy means that the layer structure of the top cell is grown first and GaAs bottom cell is grown after. Under illumination at AM1.5, this flexible tandem solar cell had an η as high as 30.8% (Kayes *et al.*, 2014). Incidentally, the GaAs/Ga$_{0.5}$In$_{0.5}$P tandem solar cell had a very low temperature coefficient (see Chapter 2) of only -0.1%/K (Kayes *et al.*, 2014).

The very high η of flexible GaAs/Ga$_{0.5}$In$_{0.5}$P tandem solar cell has been reached by controlling the thicknesses of each layer very precisely and by implementing an optimized back reflector. The optimized back reflector allowed for reabsorption of a large part of photons emitted inside the solar cells by radiative recombination (Kayes *et al.*, 2014). The optically optimized so-called photon recycling (Figure 8.26) is key for reaching highest η for solar cells limited by radiative recombination (Miller *et al.*, 2012). In tandem solar cells, photons generated by radiative recombination (secondary photons) in the top and bottom cells can be reabsorbed in the bottom cell. Light trapping of secondary photons can be realized by total internal reflection (see Chapter 2) at the interface between the top cell and the antireflection coating and by reflection at the metallic back contact.

A combination of band gaps close to the maximum of η can be obtained for tandem solar cells with a Ga$_{0.83}$In$_{0.17}$As ($E_{g,\text{bottom}} = 1.18$ eV) bottom cell and a Ga$_{0.35}$In$_{0.65}$P top cell ($E_{g,\text{top}} = 1.67$ eV). The III–V

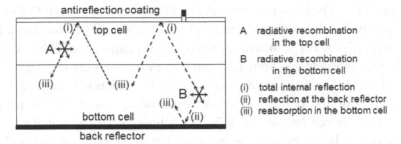

Figure 8.26. Principle of photon recycling in a tandem solar cell by light trapping of photons generated by radiative recombination (secondary photons) in the top cell (A) and in the bottom cell (B). Secondary photons can undergo total internal reflection at the interface between the top cell and the antireflection coating (i), reflection at the back reflector (ii) and reabsorption in the bottom cell (iii). Total internal reflection is limited by the scape within the critical angle (see Chapter 2).

semiconductor layers have the same lattice constant for both solar cells. However, there are no crystals with an appropriate lattice constant on which the layer system for the tandem solar cell can be grown by lattice-matched epitaxy. Dimroth *et al.* showed that very efficient $Ga_{0.83}In_{0.17}As$ solar cells can be produced on a misoriented GaAs(100) wafer covered with an In-graded buffer layer (Dimroth *et al.*, 2000). Incidentally, the number of epitaxial layers increases for such metamorphic growth of tandem solar cells due to the additional In-graded buffer layer.

A strong advantage of tandem solar cells is the increased photovoltage in comparison to a single-junction solar cell. For example, V_{OC} increased from 1.047 V for a GaAs solar cell to 2.42 V for an $Al_{0.36}Ga_{0.64}As$/GaAs tandem solar cell (Takahashi *et al.*, 2005). Incidentally, the increased photovoltage of tandem solar cells is sufficient to provide the energy required for water splitting by electrolysis (at least 1.23 eV minimum, see, for further discussions, the review by Walter *et al.*, 2010). An efficiency of solar energy to hydrogen conversion of 18% has been reached with an integrated system including a $Ga_{0.83}In_{0.17}As$/$Ga_{0.35}In_{0.65}P$ tandem solar cell and a cell for electrolysis of water with a proton exchange membrane (Dimroth, 2006).

8.4.5 *Triple-junction concentrator solar cells*

Conventional triple-junction solar cells make use of Ge, which has an appropriate E_g (about 0.7 eV, Figure 8.4) for the bottom cell and an appropriate a_{lc} for lattice-matched epitaxy of GaAs and $Ga_{0.5}In_{0.5}P$

(Figure 8.15). The band gaps of GaAs (1.42 eV) and $Ga_{0.5}In_{0.5}P$ (1.95 eV) are well suitable for the middle and bottom cells in a triple-junction solar cell, respectively. For comparison, the optimum band gaps of the middle and top cells would be about 1.15 and 1.75 eV, respectively (Figure 6.12). Therefore, in the case of perfectly lattice-matched epitaxy (Figure 8.27(a)), the I_{SC} density of the top cell would limit the I_{SC} density of the triple-junction solar cell (see also task T8.6). A little lattice mismatch can be tolerated for lattice-matched epitaxy so that band gaps for the middle

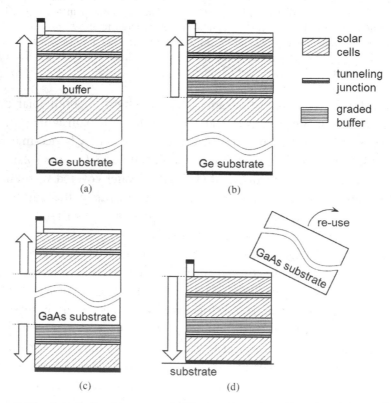

Figure 8.27. Principal architectures of triple-junction solar cells with very high η grown on Ge substrates by lattice-matched (a) and metamorphic (b) epitaxy, grown on a GaAs substrate by lattice-matched epitaxy of the middle and top cells at one side and metamorphic epitaxy of the bottom cell at the other side of the GaAs substrate (c) and grown on a GaAs substrate by inverted metamorphic epitaxy followed by transfer onto a substrate and subsequent epitaxial lift-off of the GaAs substrate (d). The middle and top cells are made from $Ga_xIn_{1-x}As$ and $Ga_xIn_{1-x}P$, respectively. The bottom cells are made from Ge ((a) and (b)) and from $Ga_xIn_{1-x}As$ ((c) and (d)). The arrows mark the growth direction.

and top cells could be reduced to about 1.4 ($Ga_{0.99}In_{0.01}As$) and 1.9 ($Ga_{0.49}In_{0.51}P$) eV, respectively, that resulted in an η of 40.1% for the lattice-matched triple-junction solar cell under illumination with concentrated sunlight (King *et al.*, 2007).

Metamorphic or lattice-mismatched epitaxy on Ge crystals (Figure 8.27(b)) provides the opportunity to reduce the band gaps of the middle and top cells to come closer to the optimum values of band gaps in triple-junction solar cells (King *et al.*, 2000). For band gaps of the middle and top cells of 1.3 ($Ga_{0.92}In_{0.08}As$) and 1.8 ($Ga_{0.44}In_{0.56}P$) eV, respectively, an η of 40.7% was reached for the lattice-mismatched triple-junction solar cell under illumination with concentrated sunlight (King *et al.*, 2007).

More than 20 epitaxial layers are required for the production of a triple-junction solar cell. The bottom cell consists of a p-type doped Ge base and an n-type doped Ge emitter. On top of the Ge emitter, a nucleation layer, an n-type doped $Ga_xIn_{1-x}As$ buffer layer and the first tunneling junction are grown in case of lattice-matched epitaxy (Cotal *et al.*, 2009). In case of metamorphic epitaxy, an additional layer system is included for reducing strain and extended defects. The layer structures of the following middle and top solar cells are similar for lattice-matched and metamorphic epitaxy and include layers for absorption, barrier layers, the second tunneling junction and passivation layers (Cotal *et al.*, 2009).

The values of η under illumination at AM1.5 are quite similar for lattice-matched (32%, King *et al.*, 2007) and metamorphic (31.3%, King *et al.*, 2007) epitaxial growth of the layer systems of triple-junction solar cells on Ge crystals. A very high η of more than 40% was reached for the first time with triple-junction concentrator solar cells (King *et al.*, 2007) and has been further increased for metamorphic epitaxy to 44% (Solar Junction, 2012).

A bottom cell with a band gap of 1.0 eV has an optimum combination with band gaps of a top cell and middle cell of 1.9 ($Ga_{0.5}In_{0.5}P$) and 1.4 eV ($Ga_{0.99}In_{0.01}As$), respectively (Figure 6.12). The maximum η in the limit of radiative recombination is about 50% for a triple-junction solar cell with $E_{b,\text{bottom}} = 1.0$, $E_{g,\text{middle}} = 1.4$ and $E_{g,\text{top}} = 1.9$ eV (Figure 6.15). For this combination of band gaps, the top cell and the middle cell can be grown on a GaAs crystal or on a germanium crystal by lattice-matched epitaxy. A band gap of 1.0 eV can be realized with a layer of $Ga_{0.68}In_{0.32}As$ which has a high lattice mismatch with GaAs.

A layer of $Ga_{0.68}In_{0.32}As$ can be grown on a GaAs substrate by metamorphic epitaxy with a system of graded buffer layers for reducing strain and for reducing the density of extended defects (see, for example, Cotal *et al.*, 2009). The transparency of a thin GaAs substrate is very high for photons with energy below the band gap of GaAs and so the GaAs substrate can be used for lattice-matched epitaxial growth of the $Ga_{0.99}In_{0.01}As$ middle and of the $Ga_{0.5}In_{0.5}P$ top cell at one side and of the $Ga_{0.68}In_{0.32}As$ bottom cell at the other side of the GaAs substrate (Figure 8.27(c)).

It is possible to start the epitaxial growth with the $Ga_{0.5}In_{0.5}P$ top cell and the $Ga_{0.99}In_{0.01}As$ middle cell by lattice-matched epitaxy on a GaAs substrate and to finish the layer growth by metamorphic epitaxy of the $Ga_{0.68}In_{0.32}As$ bottom cell (inverted metamorphic growth). In this case, the solar cell can be transferred to a mechanically stable substrate such as a silicon wafer by using an epitaxial lift-off so that the GaAs substrate can be reused (Figure 8.27(d)). Under illumination at AM1.5, values of V_{OC} larger than 3 V and an η as high as 37.9% have been be reached for a triple junction solar cell with band gaps of 1.0, 1.4 and 1.9 eV for the bottom, middle and top cells, respectively (Sasaki *et al.*, 2013). Under illumination with concentrated sunlight, η reached 44.4% (Sharp, 2013).

8.4.6 *Quadruple- and five-junction solar cells*

The further increase of the number of solar cells with different band gaps allows a further increase of the η of multi-junction solar cells. With regard to Figures 6.15 and 6.12, the maximum η in the limit of radiative recombination is about 54% for a quadruple-junction solar cell with band gaps of about 0.7, 1.0, 1.4 and 1.9 eV, respectively, illuminated at AM1.5. Inverted metamorphic epitaxy (Figure 8.27(d)) has been extended to the growth of a second $Ga_xIn_{1-x}As$ bottom cell with a band gap lower than 0.9–0.8 eV (NREL, 2014; unfortunately, details are missing). With inverted metamorphic epitaxy, an η of 45.7% under concentrated sunlight was demonstrated (NREL, 2014).

Taking into account Figure 8.15, the two top cells with the higher band gaps and the two bottom cells with the lower band gaps of a quadruple-junction solar cell can be grown by lattice-matched epitaxy on a GaAs and an InP substrates, respectively. For example, a solar cell based on

Ga$_{0.43}$In$_{0.57}$As with a band gap of 0.73 eV followed by a solar cell based on Ga$_{0.2}$In$_{0.8}$As$_{0.43}$P$_{0.57}$ with a band gap of 1.03 eV were grown on an InP substrate (Szabo *et al.*, 2008). Such two bottom cells were combined with two top cells based on GaAs and Ga$_{0.5}$In$_{0.5}$P solar cells grown on a GaAs substrate (Dimroth *et al.*, 2014). The two bottom cells grown on an InP substrate and the two top cells grown on a GaAs substrate were connected with each other by wafer bonding (Figure 8.28) and the GaAs substrate was removed by an epitaxial lift-off (Dimroth *et al.*, 2014). With the combination of the optimal band gaps of a quadruple-junction solar cell, an η of 46.3% was achieved under concentrated sunlight with a concentration factor of 340 (Steiner *et al.*, 2016).

The combination of the growth of solar cells by lattice-matched epitaxy on GaAs and InP substrates and their connection by wafer bonding followed by an epitaxial lift-off of the GaAs substrate allows for a further increase of the number of solar cells in multi-junction solar cells. To date, three top cells grown on a GaAs substrate were combined with two bottom cells

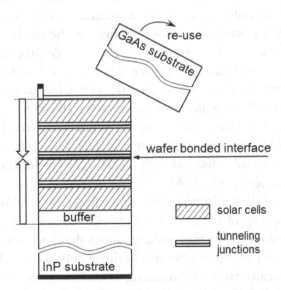

Figure 8.28. Principal architecture of a quadruple-junction solar cell prepared by lattice-matched epitaxy of two top cells based on GaAs and Ga$_x$In$_{1-x}$P grown on a GaAs substrate and of two bottom cells based on Ga$_x$In$_{1-x}$As$_y$P$_{1-y}$ and In$_x$Ga$_{1-x}$As grown on an InP substrate. The arrows denote the growth directions. The two bottom and the two top cells were connected with each other by wafer bonding and the GaAs substrate was removed by epitaxial lift-off.

grown on an InP substrate to a five-junction solar cell and resulted in an η of 38.8% under illumination at AM1.5 (Chiu *et al.*, 2014).

8.5 Summary

Binary, ternary and quaternary III–V semiconductors are practically ideal photovoltaic absorbers. Their band gaps can be varied over the whole range of the sun spectrum, and their diffusion lengths of minority charge carriers can be much longer than the optical absorption length. Additionally, III–V semiconductors can be doped over a wide range so that *pn*-junctions for practically ideal charge-selective contacts with very low densities of defects at the contact can be realized and ohmic contacts can be formed with metal contacts. Hetero-junctions that are free from interface defects can be formed between layers of different III–V semiconductors by epitaxial growth due to engineering of lattice constants of ternary and quaternary III–V semiconductors. This allows for the implementation of (i) passivation layers without introducing remarkable surface recombination at hetero-junctions and (ii) barrier layers for minimizing the density of minority charge carriers at ohmic contacts. Therefore, single-junction solar cells based on III–V semiconductors include *n*-type and *p*-type doped absorber layers of the same band gap, barrier and passivation layers made from III–V semiconductors with a higher band gap than the absorber layer and metal layer systems for forming ohmic contacts to the two external metal leads. The absolutely highest η of a single-junction solar cell illuminated at AM1.5 was reached for a GaAs absorber with $Al_xGa_{1-x}As$ barrier and passivation layers (28.8%, Green *et al.*, 2016).

Layers of III–V semiconductors are transparent for photon energies below the band gap of the given III–V semiconductors. This means that solar cells based on III–V semiconductors with different band gaps can be stacked on top of each other without additional optical losses. Band gaps of absorber layers can be tuned for reaching the optimum combination for the maximum η of multi-junction solar cells. At the same time, lattice constants of III–V semiconductors can be engineered in such a way that sophisticated layer systems including different absorber layers, barrier layers, passivation layers and contact layers can be grown with a minimum of defects in the different absorbers and at the numerous interfaces. The properties of the tuning of band gaps, of ideal transparency for photon energies below the band gap and of the realization of tunneling contacts between highly doped layers of III–V semiconductors are additional properties, which are decisive for the realization of very efficient stacked multi-junction solar cells. An η as large as 38.8% has been reached for a five-junction solar cell illuminated at AM1.5 (Chiu *et al.*, 2014).

Table 8.1 compares basic characteristics of different single-junction and multi-junction solar cells based on III-semiconductors.

The η of solar cells can be increased under illumination with highly concentrated sunlight. However, illumination of solar cells with highly concentrated sunlight demands very low series resistances for minimizing resistive losses. The resistance of layers of III–V semiconductors is very low and can be neglected under illumination with highly concentrated

Table 8.1. Short circuit current density (I_{SC}), open circuit voltage (V_{OC}), fill factor (FF) and solar energy conversion efficiency (η) for different single-junction and multi-junction solar cells based on III–V semiconductors for given concentration factors of sunlight (X).

	X (suns)	I_{SC} (mA/cm^2)	V_{OC} (V)	FF (%)	η (%)	Reference
GaAs	1	29.68	1.122	86.5	28.8	Green *et al.* (2016)
GaAs/AlGaAs	1	13.44	2.42	88.7	28.85	Takahashi *et al.* (2005)
GaAs/GaInP	1	14.3	2.547	84.7	30.8	Kayes *et al.* (2014)
GaInP/GaAs/ GaInAs	1	14.27	3.065	86.7	37.9	Sasaki *et al.* (2013)
Five-junction	1	9.564	4.213	85.2	38.8	Chiu *et al.* (2014)
GaInP/GaInAs/Ge	240	3830	2.911	87.5	40.7	King *et al.* (2007)
GaInP/GaAs/ GaInAs	302	—	—	—	44.4	Sharp (2013)
GaInP/ GaInAs/Ge	454	7475	2.867	87.2	41.1	Bett *et al.* (2009)
GaInP/GaAs; GaInAsP/GaInAs	508	6500	4.227	85.1	46.0	FhG-ISE (2014)
GaInP/GaAs/GaInNAs	947	—	—	—	44	Solar Junction (2012)
GaAs/wafer	2509	67340	1.184	75	23.1	Algora *et al.* (2001)

sunlight. Very low resistances at III–V semiconductor/metal contacts can be realized with appropriate systems of metal layers taking into account interfacial reactions and epitaxial regrowth after contact formation. Tunneling resistances between highly n-type and p-type doped layers of III–V semiconductors usually limit the series resistance of multi-junction solar cells under illumination with highly concentrated sunlight. Inter-diffusion of dopants at highly doped pn-hetero-junctions leads to a partial compensation of doping and therefore to an increase of the width of the corresponding space charge regions. As consequence, the tunneling resistance increases. Inter-diffusion of dopants at highly doped pn-hetero-junctions has to be minimized by choosing dopants with low diffusion coefficients and by reducing the thermal impact during the subsequent epitaxy and other processing steps.

An η as high as 44.4% has been reached with a $Ga_xIn_{1-x}P/$ $Ga_xIn_{1-x}As/Ge$ triple-junction solar cell illuminated with highly concentrated sunlight with a concentration factor of 302 (Sharp, 2013). To date, the absolutely highest η was achieved with a quadruple-junction solar cell (46% at a concentration factor of 508, FhG-ISE (2014)). Incidentally, the concentration of sunlight requires a focusing lens, which introduces optical losses in concentrator photovoltaic power plants. Under optimized optical conditions, a 43% sunlight to electricity conversion efficiency has been demonstrated for a concentrator solar cell with the highest η (Steiner *et al.*, 2016). Incidentally, direct solar-to-hydrogen conversion at activated surfaces of tandem solar cells based on III–V semiconductors is a very promising alternative for the efficient storage of solar energy (May *et al.*, 2015).

8.6 Tasks

T8.1: Type I and type II semiconductor hetero-junctions

Establish whether the natural GaP/InP, InN/AlAs and GaAs/InAs interfaces form type I or type II semiconductor hetero-junctions.

T8.2: Influence of an interface dipole on the band offset

Estimate the change of the potential energy at a semiconductor hetero-junction with a polarization charge of 10% of the elementary charge per unit cell of the lattice. Discuss the consequences for the natural valence and conduction band offsets at the GaAs/AlAs and GaAs/InAs interfaces.

T8.3: Tuning of band gaps in lattice-matched epitaxy

Calculate the minimum and maximum band gaps that can be realized with lattice-matched epitaxy of $Ga_x In_{1-x} As_y P_{1-y}$ at a lattice constant of 0.5800 nm.

T8.4: Absorber resistance of solar cells based on III–V semiconductors

Estimate the resistance of a GaAs absorber in a GaAs solar cell and discuss the role of the resistance of the absorber in solar cells based on III–V semiconductors operated under highly concentrated sunlight.

T8.5: Epitaxial layers in tandem solar cells with III–V semiconductors

Explain the function of each epitaxial layer in a $GaAs/Al_{0.36}Ga_{0.64}As$ tandem solar cell as shown in Figure 8.24.

T8.6: Maximum efficiency of a lattice-matched triple-junction solar cell

Ascertain the maximum efficiency within the limit of radiative recombination of a lattice-matched triple-junction solar cell with $E_{g,\text{bottom}} = 0.7$ eV, $E_{g,\text{middle}} = 1.38$ eV, $E_{g,\text{top}} = 1.86$ eV illuminated at AM1.5. Use the nomograms given in Chapter 6. Compare the result with the maximum η of a triple-junction solar cell with optimum band gaps.

T8.7: Lattice mismatch for lattice-matched epitaxy

Ascertain the lattice mismatch for lattice-matched epitaxy realized for a triple-junction solar cell on a Ge substrate with a $Ga_x In_{1-x} As$ middle cell and a $Ga_x In_{1-x} P$ top cell with band gaps of 1.38 and 1.86 eV, respectively.

T8.8: Lattice mismatch for lattice-mismatched epitaxy

Ascertain the lattice mismatch for the growth of a triple-junction solar cell a $Ga_x In_{1-x} As$ ($E_g = 1.28$ eV) middle and a $Ga_x In_{1-x} P$ ($E_g = 1.80$ eV) top cells on c-Ge ($a_{lc} = 0.56575$ nm) substrate. Compare with the lattice mismatch for an inverted metamorphic triple-junction solar cell including a $Ga_xIn_{1-x}As$ bottom cell with $E_g = 1.0$ eV grown on GaAs.

T8.9: Type II hetero-junctions for charge separation

Is it useful to apply type II hetero-junctions instead of *pn*-homo-junctions as charge-selective contacts in multi-junction solar cells?

T8.10: Finger grid of concentrator solar cells

Estimate the distance between grid fingers at te front contact of a triple-junction solar cell operated at concentrated sunlight with a concentration factor of 2000.

9

Thin-Film Solar Cells

Thin-film solar cells and photovoltaic (PV) thin-film modules are based on homogeneous layers of PV absorbers, which have a high absorption coefficient and which can be formed on foreign substrates at reduced temperatures. A thin-film absorber has a high quality if the lifetime of photogenerated charge carriers is longer than the transport time to the charge-selective contact. The concept of thin-film photovoltaics and two examples of typical thin-film deposition techniques are given at the beginning of this chapter. Transparent conducting oxides (TCO) are key for thin-film PVs. Resistive and optical losses caused by TCO layers in thin-film solar cells are illuminated. The main part of this chapter is devoted to principle properties of thin-film absorbers and their charge-selective contacts. The classes of amorphous hydrogenated silicon (a-Si:H) and micro-crystalline silicon (μc-Si:H), of chalcogenide (cadmium telluride: CdTe; chalcopyrites: $CuIn_{1-x}Ga_x(Se_{1-y}S_y)_2$; kesterites: $Cu_2ZnSn(Se_{1-x}S_x)_4$) and of hybrid organic–inorganic metal halide perovskites (for example, methylamine lead iodide: $CH_3NH_3PbI_3$) thin-film solar cells are distinguished. A-Si:H and CdTe are directly deposited due to passivation of defects by hydrogen and due to congruent sublimation of CdTe, respectively. Metal layers are transformed to chalcopyrites and kesterites during selenization and/or sulfurization. Hybrid organic–inorganic metal halides are prepared from solvents. Very high solar energy conversion efficiencies were achieved for thin-film solar cells based on CdTe, chalcopyrites and hybrid organic–inorganic metal halide. A-Si:H, hybrid organic–inorganic metal halide perovskite and kesterite absorbers are produced from earth abundant elements.

9.1 Concept of Thin-Film Photovoltaics

9.1.1 *About the architecture of thin-film solar cells and modules*

Thin-film absorbers usually have a thickness of 0.5–5 μm. Therefore, the absorption length of thin-film absorbers should be of the order of 0.1–2 μm. Thin-film absorber layers are deposited directly onto or formed on a foreign substrate, such as glass sheets, metal foils of polyimide foils. Properties of foreign substrates can be adjusted according to application demands including costs, weight and flexibility of solar cells and photovoltaic (PV) modules.

Figure 9.1 shows a principle cross-section of a thin-film solar cell. The back-contact layer is deposited onto the foreign substrate and the thin-film absorber and the charge-selective contact layers are deposited onto the back-contact layer. Charge-selective contact layers at thin-film absorbers have a thickness of only several tens of nm. Therefore, the front-contact layer should completely cover the absorber and charge-selective contact layers. Depending on whether the thin-film solar cell is illuminated through the front-contact (strate configuration) or through the back-contact (super-strate configuration) the front- or back-contact layers, respectively, have to be transparent in the absorption range of the absorber. Transparent contact layers are realized with transparent conducting oxides (TCOs).

Thin-film absorbers can be based on hydrogenated amorphous and microcrystalline silicon (a-Si:H, μc-Si:H), chalcopyrites (CuIn$_{1-x}$Ga$_x$(Se$_{1-y}$S$_y$)$_2$), kesterites (Cu$_2$ZnSn(Se$_{1-x}$S$_x$)$_4$), hybrid organic–inorganic metal halide perovskites such as methyl amine lead iodide (CH$_3$NH$_3$PbI$_3$) or cadmium telluride (CdTe). These absorbers have high absorption

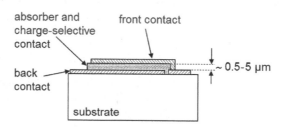

Figure 9.1. Principle cross-section of a thin-film solar cell.

coefficients and can be deposited with sufficient diffusion or drift lengths on appropriate substrates coated with back-contact layers. Charge-selective contact layers are called buffer layers in thin-film solar cells with chalcopyrite, kesterite or cadmium telluride absorber.

In thin-film technology, the maximum processing temperature is often limited by the foreign substrate, for example, by the softening temperature of glass. In this way, a foreign substrate can limit the formation temperature of absorber layers, contacts and contact layers. Furthermore, the foreign substrate and the back-contact layer are sources for uncontrolled impurities, which can penetrate into the absorber layer and contact regions. Both limited processing temperatures and penetration of impurities have a tremendous influence on the density of defects in thin-film absorbers and therefore on the diffusion or drift lengths. Impurities can cause a reduction or an increase of the solar energy conversion efficiency (η) of thin-film solar cells depending on specific interactions between impurity atoms and host atoms in absorber layers and/or contact regions.

Thin-film deposition technologies are optimized for a homogeneous covering of substrates with thin metal, absorber and TCO layers on areas as large as several m². This enabled a lateral integrated series connection of thin-film solar cells in a monolithic process of PV thin-film modules. For this purpose, the large area of the deposited back-contact is separated into numerous back-contact stripes determining the areas of the separate solar cells. The separation of the back-contact stripes is called the P1 cut (Figure 9.2(a)). The P1 cut is usually realized with a focused laser beam (laser scribing). The absorber and charge-selective contact layers are deposited in the subsequent step. The absorber and charge-selective contact layers are separated with a narrow overlap between two neighboring solar cells during the P2 cut (Figure 9.2(b)). After the P2 cut, the front-contact layer is deposited. The front-contact layer is separated into front-contact stripes during the P3 cut in such a way that the front contact of each solar cell is connected with the back contact of the following solar cell (Figure 9.2(c)). Depending on the absorber layer, the P2 and P3 cuts can be performed by laser or mechanical scribing. Incidentally, the area of the P1, P2 and P3 cuts and the area between the P1 and P3 cuts are lost for photocurrent generation. Conventional P1, P2 and P3 cuts have a width of the order of 100 μm.

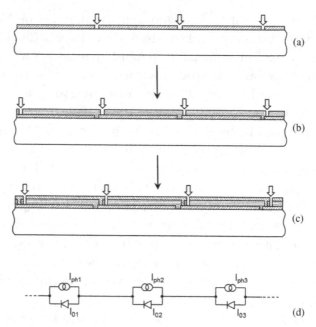

Figure 9.2. Separation of back contacts (a), absorbers layers (b) and front contacts (c) for lateral integrated series connection of thin-film solar cells. The arrows mark the corresponding P1 (a), P2 (b) and P3 (c) cuts, and the dashed lines mark the alignment between them. The equivalent circuit of thin-film solar cells connected in series is given in (d).

The values of the photovoltage of thin-film solar cells are added for integrated series connection in a PV thin-film module (Figure 9.2(d)). The solar cell with the lowest photocurrent limits the photocurrent of the whole PV thin-film module. Therefore, the current-matching condition demands identical photocurrents for all solar cells in a PV thin-film module in the ideal case. As a consequence, all solar cells in a PV thin-film module must be very homogeneous.

The current-matching condition requires thin-film depositions techniques capable of producing very homogeneous absorber and TCO layers with respect to photogeneration, optical losses and resistive losses. The homogeneity of optical losses over a thin-thin PV module, for example, is limited by the thickness of the TCO layer, which should be controlled with an accuracy of only a few nanometers. Resistive losses are mainly caused by the resistance of the TCO layers and local shunts.

Thin-film technologies are scalable so that the area of thin-film PV modules can easily be varied from several cm^2 and less (low-power mobile applications) to several m^2 (high-power energy production). Furthermore, PV thin-film modules can be combined with additional functionalities given by the choice of the substrate. This broadens degrees of freedom for applications of thin-film solar cells, for example, in consumer goods and architecture.

9.1.2 *Sputtering and plasma-enhanced chemical vapor deposition*

The concept of thin-film photovoltaics is closely related to the development of technological tools of thin-film deposition. In thin-film photovoltaics, most of thin-film deposition techniques are related to sputtering (Thornton, 1974), physical vapor deposition (PVD) and plasma-enhanced chemical vapor deposition (PECVD) (van Sark, 2002). These technologies allow for the deposition of homogeneous metal contact layers (PVD, sputtering), TCO layers (sputtering (Minami *et al.*, 1985)) and/or absorber layers (sputtering, PECVD) on large areas, such as PECVD of a-Si:H (for example, Otero *et al.*, 2012).

Figure 9.3 shows a principle setup for sputtering based on a vacuum chamber, an electric power supply, a sputter target and a substrate. Energetic ions are used for the transfer of a material from a sputter

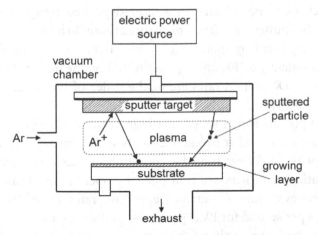

Figure 9.3. Principle setup for sputtering.

target onto a substrate at well-controlled low substrate temperature of the order of 100–400°C. Usually, argon (Ar) atoms are ionized for sputtering. Trajectories of ions and electrons can be enhanced in a magnetron so that ionization of argon is very efficient. Argon ions are accelerated in a high electric field towards the sputter target where argon ions kick out or sputter target atoms. Sputtered target atoms have a high kinetic energy and move towards the substrate where they form a growing layer. Sputtered target atoms passing the plasma can be ionized and undergo numerous chemical reactions before reaching the substrate.

Constant or variable electric fields are applied in direct current (DC) or radio frequency (RF, usually 13.56 MHz) sputtering from conducting or insulating sputter targets, respectively. Sputter targets used for the production of thin-film solar cells consist of (i) metals such as molybdenum, copper or indium, (ii) metal oxides such as ZnO or (iii) sulfides such as ZnS. The rates of sputtering are specific for each element in mixed or multi-component sputter targets. Oxygen as a reactive species is added to the Ar gas in reactive ion sputtering for the deposition of TCO layers such as ZnO:Al. Reactive selenium or sulfur species can also be added to the argon gas for depositing sulfide or selenide layers. For example, $CuInSe_2$ has been sputtered by Thornton *et al.* in an Ar/H_2Se working gas (Thornton *et al.*, 1988).

The stoichiometry of sputter targets can be varied by the alloying of metals or by combining different sputter targets. Practically any stoichiometry can be adjusted in a layer deposited by sputtering since deposition by sputtering is far from chemical equilibrium.

Layers deposited by sputtering are multi-crystalline. The deposition rate depends mainly on the argon pressure and on the power of the electric power source. Deposition rates are of the order of hundreds of nm per minute for sputtering.

A PVD process is performed in a high vacuum chamber with a base pressure below 10^{-4} mbar or less. Elements such as metals, sulfur or selenium are evaporated from heated evaporation sources. The stoichiometry and deposition sequences are controlled by opening or closing shutters covering the evaporation sources. Deposition rates are of the order of nanometers per second for PVD processes. Sulfide- and/or selenide-based chalcopyrite or kesterite absorber layers can be deposited by PVD, such as $CuInS_2$ by a coevaporation process (Scheer *et al.*, 1993). Metal layers

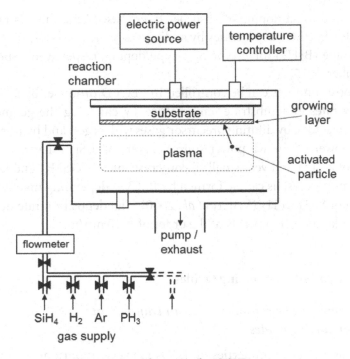

Figure 9.4. Schematic of a PECVD setup.

deposited by PVD or sputtering can be transformed by post-treatment steps into chalcopyrite absorber layers, such as by sulfurization of Cu/In layers (Klenk *et al.*, 2005).

PECVD is used for the deposition of a-Si:H and μc-Si:H thin-film absorbers. A PECVD setup contains a gas supply system besides the electric power source and the high vacuum chamber with the substrate (Figure 9.4). The gas supply system is connected with several gas sources such as Ar as a carrier gas and silane (SiH_4) as the main precursor gas. The gas flow is measured with a flowmeter. A precise temperature control is important for the quality of growing a-Si:H or μc-Si:H layers.

Precursor molecules are activated in the plasma by electron impact excitation, dissociation and ionization and undergo different chemical reactions in the plasma and at the surface of a substrate (Perrin, 1995). The chemical surface reactions are related to bonding of $-SiH_3$ at reactive surface sites and to the transformation of $\equiv$Si-SiH_3 surface species to $=$Si$=SiH_2$, $-Si\equiv SiH$ and $\equiv$Si-Si$\equiv$ bond configurations under the release of hydrogen.

Substitutional doping of absorber layers based on a-Si:H can be well controlled in PECVD processes by adding gases such as phosphine (PH_3) or diborane (B_2H_2) for *n*-type or *p*-type doping, respectively (Spear and Le Comber, 1975).

Deposition rates can be controlled in PECVD processes by modifying gas flow rates and substrate temperature, by changing the geometry of PECVD reactors, by diluting precursor gases with argon and hydrogen and by the power of the RF power source. Layers of a-Si:H, layers of mixed phases of a-Si:H and very small silicon crystallites (μc-Si:H) and layers of silicon micro-crystals can be formed by PECVD depending mainly on the dilution of SiH_4 in H_2 (Vetterl *et al.*, 2000). The deposition rate of a-Si:H absorber layers by PECVD is of the order of 1–10 nm/min.

9.2 Transparent Conducting Oxides

9.2.1 *Doping and electron mobility in transparent conducting oxides*

Optical and electric properties of TCOs are very important for the solar energy conversion efficiency (η) of thin-film solar cells. Some metal oxides have large forbidden band gaps so that most of sunlight can be transmitted. For example, the band gaps of ZnO, SnO_2 and In_2O_3 are 3.2 (Thomas, 1960), 3.6 (Summit *et al.*, 1964) and 2.98 (Haines and Bube, 1978) eV, respectively. Undoped transparent metal oxides are insulators. The electric conductivity of transparent metal oxides can be increased by many orders of magnitude by intrinsic and extrinsic doping. Intrinsic doping is caused by deviations in the stoichiometry. An excess of oxygen causes *p*-type doping and a deficiency of oxygen leads to *n*-type doping (see, for example, Kim *et al.*, 1986). Defined intrinsic doping demands a well-controlled partial pressure of oxygen over several orders of magnitude during processing at high temperature. Therefore, extrinsic doping is much more suitable for practical reasons.

Knowledge about the density and mobility of free electrons (n_0, μ_n) in TCOs is important for minimizing resistive losses in thin film solar cells. Free carrier absorption causes additional optical losses in TCO layers depending on the doping.

Metal atoms are oxidized and oxygen atoms are reduced (oxidation state O^{2-}) in metal oxides. Host metal atoms can be replaced by metal atoms in a higher oxidation state. For example, zinc is in the oxidation state Zn^{2+} in ZnO. Zinc atoms can be replaced by aluminum atoms, which are in the oxidation state Al^{3+} (Minami *et al.*, 1985). The additional positive charge at the metal side is compensated by a free electron in the conduction band of ZnO. Aluminum doped ZnO:Al is the most important TCO for thin film photovoltaics.

In SnO_2, oxygen atoms can be replaced by fluorine atoms, which are in the oxidation state F^- (Bhardwaj *et al.*, 1981). The missing negative charge is compensated by a free electron in the conduction band of SnO_2:F (also called FTO — fluorine doped tin oxide). FTO coatings are used for standard glass in architecture for reflecting infrared light but transmitting visible light. It can be also applied in solar cells. Incidentally, FTO can be simply produced by spray pyrolysis on glass sheets (see, for example, Zhang *et al.*, 2011).

In In_2O_3:Sn (also called — indium tin oxide, ITO) a part of indium atoms, which are in the oxidation state In^{3+} are replaced by tin atoms, which are in the oxidation state Sn^{4+} (Haines and Bube, 1978). The additional positive charge at the metal side is compensated by a free electron in the conduction band of ITO, which is the standard TCO for thin-film displays.

The n_0 can be as high as 10^{21} cm^{-3} in TCO layers. The μ_e can be as high as 130 cm^2/(Vs) in moderately doped ZnO crystals; see for example, Wagner and Helbig (1974). The μ_e decreases with increasing n_0 in highly doped TCOs or semiconductors due to scattering at ionized impurities. For example, μ_e is up to 22 cm^2/(Vs) in highly doped sputtered ZnO:Al layers (Minami *et al.*, 1985).

Sputtered TCO layers are polycrystalline. Therefore, the electron mobility is limited not only by transport processes within one grain (intra-grain transport) but also by charge transfer across grain boundaries (inter-grain electron transfer), as depicted in Figure 9.5.

A part of the free electrons from the bulk of a TCO crystallite can be trapped at interface states in grain boundaries. The negative charge of electrons trapped at grain boundaries leads to an increase of the potential energy of free electrons at the conduction band edge. As consequence, a barrier between two neighboring grains is formed. The barrier height

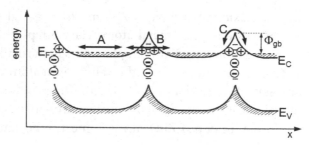

Figure 9.5. Band diagram across charged grain boundaries in a TCO layer. A, B and C denote intra-grain transport, inter-grain electron transfer by tunneling and inter-grain electron transfer by thermionic emission, respectively.

depends on the density of interface states at grain boundaries (of the order of 10^{12}–10^{13} cm^{-3}). The negative charge of electrons trapped at grain boundaries is compensated by positive charge in the space charge regions near the grain boundaries.

Electrons transferred from one to the other grain have to overcome the barrier at the grain boundary by thermionic emission or by tunneling (see Chapter 5). Therefore, μ_e increases with increasing grain size in TCO layers. Tunneling becomes important for n_0 larger than 10^{19} cm^{-3}. The μ_e usually increases with increasing tunneling rate for n_0 up to about 10^{20} cm^{-2} and decreases for higher n_0 due to the increase of scattering at ionized impurities in the bulk of a grain. Typical values of μ_e in a TCO layer are of the order of 10–40 cm^2/(Vs) for n_0 between 10^{21} and 10^{20} cm^{-3} (Minami et al., 1985).

9.2.2 Resistance of transparent conducting oxide layers

The specific resistance (ρ) of an electron conductor is given by

$$\rho = \frac{1}{q \cdot n_0 \cdot \mu_n} \tag{9.1}$$

For conventional TCO layers ($\mu_e = $ 5–40 cm^2/(Vs) and $n_0 = 10^{21}$–10^{20} cm^{-3}), ρ is of the order of 10^{-3} Ωcm, about three orders of magnitude larger than the specific resistance of copper.

The resistance of the TCO layer gives the major contribution to the series resistance of thin-film solar cells. Figure 9.6 shows a TCO stripe. The height (H, of the order of 0.5–1 μm) and the width of the TCO layer (L) have to be adjusted in accordance with the tolerable series resistance

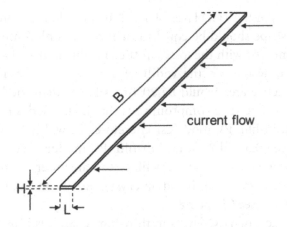

Figure 9.6. TCO stripe of a thin film solar cell.

(see Chapter 1) of a thin-film solar cell. Typically, thin-film solar cells in thin film PV modules have the shape of a long stripe (length B of the order of 1 m) with a width of about 0.5–2 cm. The resistance of a TCO stripe (R_{TCO}) can be calculated by the following equation:

$$R_{TCO} = \frac{1}{q \cdot n_0 \cdot \mu_n} \cdot \frac{L}{B \cdot H} \tag{9.2}$$

The short-circuit current of a solar cell is equal to the density of the short circuit current (I_{SC}) multiplied by the area of the solar cell which is practically equal to the area of the TCO stripe in a thin-film solar cell. With regard to Equation (1.29′) and to the meaning of I_{SC} as a current density, the tolerable series resistance of a TCO stripe ($R_{TCO,tol}$) is

$$R_{TCO,tol} \leq \frac{1}{100} \cdot \frac{V_{OC}}{I_{SC} \cdot B \cdot L} \tag{9.3}$$

With regard to Equations (9.2) and (9.3), an expression for the ratio between L^2 and H can be obtained as a condition for reaching a very low $R_{TCO,tol}$.

$$\frac{L^2}{H} \leq \frac{q \cdot n_0 \cdot \mu_n}{100} \cdot \frac{V_{OC}}{I_{SC}} \tag{9.4}$$

If assuming values of 10^{21} cm^{-3}, 10 cm^2/(Vs), 0.6 V and 0.03 A/cm^2 for n_0, μ_m, V_{OC} and I_{SC}, respectively, the ratio between the squared width and the thickness of the TCO stripe should be less than about 320 cm.

For a thickness of the TCO layer of 1 μm this would mean that the width of the TCO stripe should be equal to or less than only 2 mm. This value has to be compared with the losses in the area due to the P1–3 cuts taking about 0.4 mm away from the width of a TCO stripe, i.e. losses due to the reduced active area would be 20 times larger than the losses due to the series resistance. The compromise is that higher series resistances are tolerated in thin-film PV modules. Therefore, the width of TCO stripes is usually between 0.5 and 2 cm in thin-film PV modules. The resistance of the TCO layer is a major reason why fill factors are lower for thin-film solar cells compared to solar cells based on crystalline silicon (see Chapter 7) or gallium arsenide (see Chapter 8).

The thickness of TCO layers in thin-film solar cells is usually fixed to minimize optical losses. The sheet resistance ($R_\square$) of a TCO layer is defined as ρ divided by H. The unit of $R_\square$ is denoted by $\Omega\square$.

$$R_\square = \frac{\rho}{H} \tag{9.5}$$

TCO layers with $R_\square$ between 2 and 8 $\Omega\square$ are used for thin-film solar cells. As an example, a TCO layer with $R_\square$ of about 2 $\Omega\square$ would be required for a single a-Si:H thin-film solar cell with I_{SC} and V_{OC} of about 0.8 V and 16 mA/cm^2, respectively ($L = 0.5$ cm). However, a value 2 $\Omega\square$ for $R_\square$ is a harsh condition, which demands a thickness of the TCO layer above 1 μm. Incidentally, specially designed contact grids of Ni:Al are deposited onto the TCO layer for reducing the series resistance in small area record thin-film solar cells.

The resistance of ZnO:Al layers increases under wet atmosphere due to the formation of defects compensating the charge of mobile electrons. The degradation of TCO layers of thin-film PV modules is minimized by careful sealing. Accelerated degradation tests (so-called damp heat test, relative humidity of 85%, temperature of 85°C and time of 1000 h) have been developed for testing not only thin-film PV modules.

9.2.3 *Optical losses in transparent conducting oxides*

In a thin-film solar cell the sunlight has to penetrate the TCO layer before the it reaches the PV absorber. The optical transmission of TCO layers has to be as large as possible in the range of the sun spectrum which is absorbed by the thin-film absorber. Therefore, optical losses caused by reflection, free carrier absorption and fundamental absorption in TCO layers have to be minimized for the absorption range of the thin-film absorber. The minimization of optical losses in thin-film solar cells demands simulations with optical models including the spectra of refractive indices and absorption coefficients of all materials involved, as well as light scattering in the near and far fields. Here, the principle role of reflectivity, free charge carriers and absorption of photons with energies above the band gap of a TCO is demonstrated.

The reflection coefficient of a thin-film solar cell depends on the differences between the refraction indices of the ambience (n_{air}), of the TCO layer (n_{TCO}) and of the PV absorber (n_S). A system with reflection between the ambience and the TCO layer and between the TCO layer and the substrate will be considered for simplicity (see also Figure 2.6). The squared refractive index of a transparent material depends on the wavelength (λ) and can be well approximated by the Sellmeier equation

$$n_{TCO}^2 = A + \frac{B \cdot \lambda^2}{\lambda^2 - C^2} + \frac{D \cdot \lambda^2}{\lambda^2 - E^2} \tag{9.6}$$

For ZnO films, the values of the coefficients A, B, C, D and E are 2.0065, $1.5748 \cdot 10^6$, $1 \cdot 10^8$, 1.5868 and 2.606, respectively (for λ given in Å) (Sun *et al.*, 1999). As an example, Figure 9.7 demonstrates the spectra for the refractive index of ZnO and glass. The refractive indices of ZnO and of glass are about 1.9 and 1.5 at longer λ and range between 2.3 and 1.9 and between 1.7 and 1.5 at longer λ, respectively.

The consideration of reflection leads to the following equation for the optical transmission coefficient (T_{otr}) under normal incidence of light.

$$T_{otr} = \frac{1 + r_{air,TCO}^2 \cdot r_{TCO,S}^2 - r_{air,TCO}^2 - r_{TCO,S}^2}{1 + r_{air,TCO}^2 \cdot r_{TCO,S}^2 + r_{air,TCO} \cdot r_{TCO,S} \cdot \cos(\theta)} \tag{9.7$'$}$$

where the reflectivity at the interfaces between air and the TCO layer ($r_{air,TCO}^2$) and the reflectivity between the TCO layer and the substrate ($r_{TCO,S}^2$) are given by the following equations:

$$r_{air,TCO}^2 = \left(\frac{n_{TCO} - n_{air}}{n_{TCO} + n_{air}} \right)^2 \tag{9.7$''$}$$

$$r_{TCO,S}^2 = \left(\frac{n_S - n_{TCO}}{n_S + n_{TCO}} \right)^2 \tag{9.7$'''$}$$

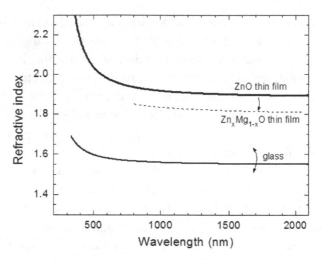

Figure 9.7. Spectra of the refractive index for ZnO thin films and glass. The trend for the reduction of the refractive index by alloying ZnO with MgO and the variability of the refractive index of glass are also shown.

The phase shift (θ) due to the optical path depends on the reciprocal λ of incident light, on the thickness of the TCO layer and on n_{TCO}

$$\theta = \frac{4 \cdot \pi}{\lambda} \cdot n_{\text{TCO}} \cdot H \tag{9.8}$$

According to Equations (9.7′)–(9.8), characteristic interference pattern appear in the transmission spectrum of a TCO layer as shown in Figure 9.8. The transmission of the TCO layer in the given configuration is reduced over the whole spectral range. The reflectivity losses in the TCO layer are between 10% and 25%.

In order to minimize reflection losses in thin-film solar cells, the differences between n_{TCO} and n_{air} (Equation (9.7″)) and/or between n_S and n_{TCO} (Equation (9.7‴)) have to be reduced. This is possible by adding an additional antireflection coating on top of the TCO layer (see Chapter 2) such as MgF$_2$ or by reducing H or n_{TCO} (Equation (9.8)). A reduction of H leads to an increase of $R_\square$ (Equation (9.5)) that shall be compensated by reducing ρ or L of the TCO stripes. However, the reduction of ρ is rather limited for a given TCO. A reduction of n_{TCO} is possible in a certain range by alloying, for example, of ZnO with MgO, the product of which has a

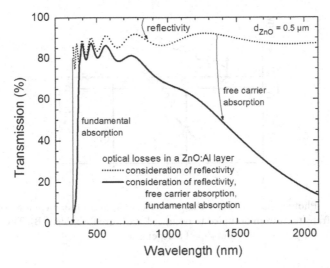

Figure 9.8. Transmission spectra of a ZnO layer (thickness 0.5 μm) deposited on glass under consideration of reflectivity only (dotted line) and of reflectivity, fundamental absorption and free carrier absorption (solid line).

lower refractive index than ZnO (Teng *et al.*, 2000). However, the alloying of different metal oxides often leads to an increase of ρ of the TCO.

Free electrons can be excited to higher energetic states within the conduction band with photons of smaller energies than the band gap of the TCO (free carrier absorption, Figure 9.9).

For free carrier absorption, the absorption coefficient (α) is proportional to n_0 and to λ^2 (Drude model) or λ^3 (Equation (9.9), highly doped ZnO (Weiher, 1966)):

$$\alpha(\lambda, n_0) = \alpha(n_0) \cdot \left(\frac{\lambda}{\lambda_0}\right)^3 \tag{9.9}$$

The absorption coefficient of ZnO amounts to $\alpha_0 = 200\ \text{cm}^{-1}$ at a wavelength of $\lambda_0 = 1\ \mu$m and for $n_0 = 5 \cdot 10^{19}\ \text{cm}^{-3}$ (Weiher, 1966).

The influence of free carrier absorption on optical transmission is demonstrated in the transmission spectrum of a TCO layer shown in Figure 9.8. Losses caused by free carrier absorption become more important with increasing λ. Therefore, the influence of free carrier absorption of TCO layers is less important for thin-film solar cells based on PV absorbers with larger band gaps such as a-Si:H (see also task T9.2).

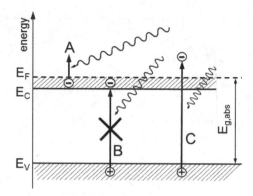

Figure 9.9.   Photoexcitation of free charge carriers (A) and fundamental absorption (C) in a TCO. Transitions from occupied to occupied states are forbidden (B). The absorption gap ($E_{g,\mathrm{abs}}$) is larger than the band gap of a degenerated TCO.

The density of free electrons in TCOs is larger than the effective density of states at the conduction band edge (E_C). Therefore, the Fermi-energy (E_F) is shifted into the conduction band (Figure 9.9). The electronic states between E_C and E_F are occupied. Light is absorbed by excitation of electrons from occupied into unoccupied states. Consequently, the absorption gap of TCO layers is increased by the difference $E_F - E_C$. This effect is known as the Burstein shift (Burstein, 1953). For example, fundamental absorption of ZnO:Al with $n_0 = 3 \cdot 10^{20}$ cm^{-3} sets in at photon energies of about 3.7 eV (see also task T9.3). Therefore, fundamental absorption in the TCO layer does not play an important role for optical losses in thin-film solar cells.

9.3 Amorphous and Microcrystalline Silicon Solar Cells

9.3.1 *Disorder and mobility of charge carriers in amorphous silicon*

Amorphous silicon is a disordered semiconductor, i.e. a unit cell and translational symmetry are missing. The amorphous structure is caused by large variations in bond angles and bond lengths (see Figure 9.10). The probability for the formation of unsaturated or dangling bonds (dbs) is very high in amorphous silicon. Dangling bonds have electronic states in around middle of the band gap, i.e. the Shockley–Read–Hall (SRH)

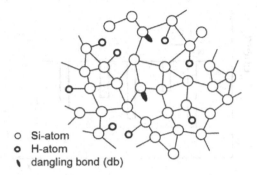

o Si-atom
o H-atom
\ dangling bond (db)

Figure 9.10. Schematic of disordered bond configuration and components in a-Si:H.

recombination rate (see Chapter 3) is at a maximum. The density of dbs in amorphous silicon is strongly reduced by hydrogen passivation. Therefore, silicon thin-film solar cells are based on hydrogenated amorphous silicon (a-Si:H and μc-Si:H).

Each atom in a-Si:H has a specific number of bonds to its nearest neighbors (coordination number). In a relaxed amorphous structure, the preferred coordination number corresponds to the valence of the atoms, i.e. the coordination numbers are 4, 1 and 3 for Si, H and P, for example. Incorporation of atoms with different coordination into amorphous structures is possible. Coordination defects are the elementary defects in a-Si:H. The variability of bond configurations causes a natural variability of the properties of a-Si:H, depending on the content of hydrogen and on preparation parameters. For example the band gap of a-Si:H increases with the increasing content of hydrogen.

The disorder of bond angles and bond lengths causes a random distribution of atom potentials and interatomic distances (Figure 9.11) (Mott, 1985). The disorder potential (V_0) describes the variation of atom potentials. The average interatomic distance is described by a_0. In amorphous semiconductors, the uncertainty of localization (Δx) is of the order of a_0. Therefore, with respect to the Heisenberg uncertainty principle, the uncertainty of the momentum (Δk) in a-Si:H is of the same order as the momentum (k).

$$\Delta k = \frac{\hbar}{\Delta x} \approx \frac{\hbar}{a_0} = k \qquad (9.10)$$

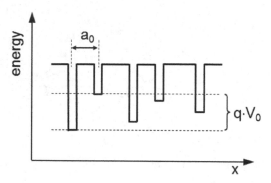

Figure 9.11. Schematic of the random distribution of atom potentials and interatomic distances in a-Si:H.

As a consequence, the momentum is not conserved in a-Si:H and localization in space is most important. Optical transitions are possible between occupied and unoccupied electronic states, which overlap in space. Further, disorder has strong influence on the transport of free charge carriers (Anderson, 1958).

Due to disorder, the motion of free charge carriers in a-Si:H is changed by scattering at each atom passed by a free charge carrier. The mobility of free charge carriers is given by the ratio between the time interval between two scattering events of a moving particle (δt) and the mass of th moving particle ($m^*_{e(h)}$) (Grimsehl, 1990).

$$\mu = q \cdot \frac{\delta t}{m^*_{e(h)}} \tag{9.11}$$

The mobility of free charge carriers in a-Si:H can be estimated by taking into account a scattering length of a_0, which is about 0.25 nm and the maximum velocity, which is the thermal velocity ($v_{th} = 10^7$ cm/s).

$$\mu \approx \frac{q \cdot a_0}{m^*_{e(h)} \cdot v_{th}} \tag{9.12}$$

The field-effect mobility of free electrons, for example, is about 2 cm^2/(Vs) in undoped a-Si:H (Han *et al.*, 2009). This agrees very well with the value estimated from Equation (9.12). For comparison, the electron mobility is about 1000 cm^2/(Vs) in moderately doped silicon crystals (Jacoboni *et al.*, 1977).

The drift mobility of photogenerated charge carriers in electric fields is important for thin-film solar cells based on a-Si:H absorber layers. The drift mobilities of electrons and holes are 2 and 0.02 cm^2/(Vs) at room temperature, respectively. The electron and hole mobilities are controlled by multiple trapping at exponentially distributed defect states leading to dispersive transport in a-Si:H (Tiedje *et al.*, 1981a). Localization of charge carriers at electronic states causes time dependent drift mobilities of photogenerated charge carriers (see, for example, Nebel *et al.*, 1992). The electron drift mobility in a-Si:H depends on the electric field and increases with increasing temperature up to values of 2–5 cm^2/(Vs) (Marshall *et al.*, 1986).

There are numerous electronic states in the band gap of a-Si:H. Defects are part of the chemical equilibrium between different bond configurations in a-Si:H and cannot be avoided in general. This is a principal difference from ideal crystalline semiconductors. Extended and localized states are distinguished in a-Si:H (Figure 9.12). Delocalized states overlap in space so that charge carriers are mobile. The so-called mobility gap (E_μ) is defined as the difference between the lowest energies of delocalized electron and hole states. The E_μ is not really well defined since it also depends on the measurement conditions.

Disorder and stress in bonds cause the so-called exponential tail states at the valence and conduction band edges. The bonding and antibonding

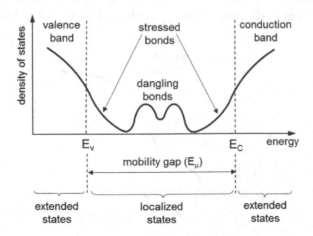

Figure 9.12. Schematic of the distribution of the density of states in a-Si:H.

states of dbs are around the middle of the band gap of a-Si:H (Cody *et al.*, 1981). The high density of electronic states in the band gap of a-Si:H is responsible for low mobilities and very short lifetimes of photogenerated charge carriers in a-Si:H and μc-Si:H.

9.3.2 *Optical absorption and optical band gap in amorphous silicon*

Electrons can be excited from occupied into unoccupied electronic states. Therefore, the distribution of the density of states is reflected in the absorption spectrum. Figure 9.13 shows the schematic absorption spectrum of a-Si:H. It can be separated into three distinct sections. At lower photon energies ($h\nu$), optical absorption is caused by dbs (Cody *et al.*, 1981; Stutzmann *et al.*, 1987). At a certain $h\nu$, α is proportional to the square root of $h\nu$. This part of the absorption spectrum defines the optical band gap (E_{og}) of a-Si:H (Brodsky *et al.*, 1970). For absorption coefficients larger than 10^4 cm^{-1}, the absorption spectrum has been fitted by the following equation (Cody *et al.*, 1981):

$$\sqrt{\alpha \cdot h\nu} = C_{a\text{-Si:H}} \cdot (h\nu - E_{og}) \qquad (9.13)$$

The parameter $C_{a\text{-Si:H}}$ is independent of temperature and amounts to 6.9 (eV μm)$^{-0.5}$ (Cody *et al.*, 1981).

The E_{og} increases with an increasing amount of hydrogen in a-Si:H, for example, from about 1.5 eV (2% of hydrogen) to about 1.8 eV (10% of hydrogen) (van Sark, 2002). The E_{og} is about 1.7 eV (7% of hydrogen) in

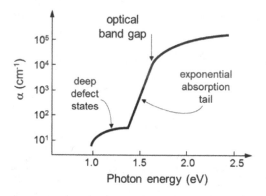

Figure 9.13. Schematic absorption spectrum of a-Si:H.

thin-film solar cells based on a-Si:H. The E_{og} of a-Si:H is controlled by the amount of disorder (Cody *et al.*, 1981). The E_{og} is larger than the E_μ. The absorption length is several hundred nm in a-Si:H absorber layers (Cody *et al.*, 1981).

At $h\nu$ below E_{og}, the absorption coefficient can be described by the following expression:

$$\alpha = \alpha_0 \cdot \exp\left(\frac{h\nu - E_1}{E_t}\right) \qquad (9.14)$$

The parameters α_0 and E_1 are equal to about $1.3 \cdot 10^6$ cm^{-1} and 2.17 eV, respectively (Cody *et al.*, 1981). The characteristic energy describing the exponential increase of α with increasing $h\nu$ is called the tail energy (E_t) and depends on temperature and structural disorder. The value of E_t is of the order of 50–100 meV (Cody *et al.*, 1981). The value of E_{og} increases with decreasing E_t (Cody *et al.*, 1981).

The absorption coefficient can vary between $3 \cdot 10^{-1}$ and 10^2 cm^{-1} in the spectral range below the influence of tail states depending on the density of dbs in a-Si:H (Jackson and Amer, 1982). The density of dbs ranges between $3 \cdot 10^{15}$ and 10^{18} cm^{-3} in a-Si:H (Jackson and Amer, 1982).

9.3.3 *Doping of amorphous silicon*

In relaxed amorphous structures, atoms are usually in their preferred coordination, i.e. four-fold coordinated silicon or threefold coordinated phosphorus, for example. Therefore, it was expected that doping of a-Si:H by substitution of silicon atoms would not work. However, in 1975, Spear and Le Comber found a correlation between the increase of the conductivity of a-Si:H and the increase of the content of PH$_3$ or B$_2$H$_2$ added to the SiH$_4$ gas during deposition of a-Si:H layers (Spear and Le Comber, 1975).

Atoms can appear also in other than preferred coordination due to local chemical reactions. Figure 9.14 shows examples for a P-atom in inactive (three-fold coordination) and active (four-fold coordination) configurations. It has to be considered that an electron leaving a P-atom in four-fold coordination can be trapped at defects. The densities of P-atoms in threefold and four-fold coordination are balanced by chemical reactions involving Si- and P-atoms in different configurations, H-atoms and dbs.

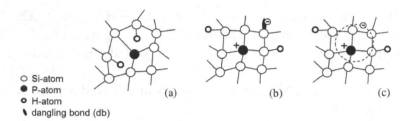

Figure 9.14. Some bond configurations of phosphorus in a-Si:H in inactive configuration (a), defect compensated donor configuration (b) and active configuration (c).

Figure 9.14 suggests that only a part of P- or B-atoms is in active configuration in a-Si:H. The density of P- or B-atoms in active configuration increases with increasing amount of PH_3 or B_2H_2 in the SiH_4 gas. However, the relative amount of P- or B-atoms in active configuration decreases with increasing amount of PH_3 and B_2H_2 in the SiH_4 gas. The density of dbs increases with increasing density of P- or B-atoms in a-Si:H. Therefore, the lowest density of dbs is achieved in undoped a-Si:H (Stutzmann *et al.*, 1987).

The so-called doping efficiency (Stutzmann *et al.*, 1987) of a-Si:H is about 10^{-2}–10^{-3} at high concentrations of P- or B-atoms, i.e. only one from 100 to 1000 P- or B-atoms is in an active configuration. This means that densities of free charge carriers up to about 10^{18} cm^{-3} can be reached in a-Si:H. These high densities of activated dopants are sufficient to form ohmic contacts with a-Si:H due to the high density of defect states in the space charge region (trap assisted tunneling, see Chapter 5).

9.3.4 *The pin concept of amorphous silicon solar cells*

In order to realize a high η, the diffusion length (L) of charge carriers photogenerated in a photovoltaic absorber should be three times larger than the absorption length (see Chapter 3). This implies that an L of more than 1 μm is required for a-Si:H absorbers.

The L of photogenerated charge carriers can be calculated if the diffusion coefficient (D) and the lifetime (τ) of photogenerated charge carriers are known (Equation (3.5)). The diffusion coefficient and the mobility are connected by the Einstein equation:

$$D = \frac{k_B \cdot T}{q} \cdot \mu \qquad (9.15)$$

The longest τ of photogenerated charge carriers in a-Si:H can be estimated by taking into account the minimum possible density of dbs of about 10^{15}–10^{16} cm^{-3} (undoped a-Si:H) and by using Equation (3.21). The resulting τ is about 1–10 ns.

The L can be calculated from μ and τ by combining Equations (3.5) and (9.15):

$$L = \sqrt{\frac{k_B \cdot T}{q} \cdot \mu \cdot \tau} \tag{9.16}$$

The L of photogenerated charge carriers is of the order of 100 nm in undoped a-Si:H and decreases even further for doped a-Si:H. Therefore, the L of photogenerated charge carriers in a-Si:H is not long enough to realize a-Si:H thin-film solar cells with high η.

Photogenerated charge carriers can reach charge-selective contacts by drift in electric fields. The drift length (L_{drift}) of charge carriers photogenerated in a-Si:H can be calculated by considering a certain electric field and by assuming that free charge carriers can drift until they recombine. The value of L_{drift} can be estimated from the product of the lifetime and the average drift velocity of photogenerated charge carriers. The average drift velocity is equal to the product of the drift mobility and the average electric field (ε).

$$L_{drift} = \mu \cdot \tau \cdot \varepsilon \tag{9.17}$$

A homogeneous built-in electric field can be realized in undoped a-Si:H layers if sandwiching the undoped a-Si:H layer between highly p-type and n-type doped a-Si:H layers (Figure 9.15(a)). This is the so-called pin configuration of an a-Si:H thin film solar cell.

The electric field in the undoped a-Si:H absorber can be estimated if taking into account a built-in potential between the p-type and n-type doped regions of the order of 1 V and a thickness of the a-Si:H absorber layer of about 1 μm. Then L_{drift} is of the order of 1 μm for an undoped a-Si:H absorber layer, i.e. a-Si:H absorbers are suitable for thin-film solar cells in the pin configuration. The pin concept has been proposed for thin-film solar based on a-Si:H absorber layers by Carlson and Wronski (1976).

Figure 9.16 shows the schematic band diagram of an a-Si:H solar cell in pin configuration in the dark (a) and the distribution of dbs across the

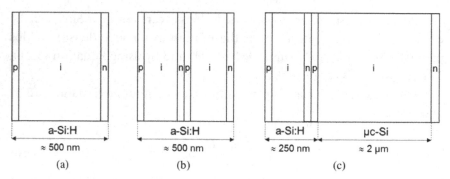

Figure 9.15. Layer structures of silicon thin film absorbers with pin a-Si:H (a), pin–pin tandem a-Si:H (b) and tandem pin–pin a-Si:H/μc-Si:H (c) configuration.

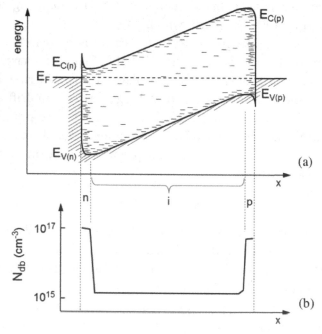

Figure 9.16. Schematic band diagram of an a-Si:H thin film solar cell in pin configuration (a) and densities of dangling bonds (b).

pin structure (b). The density of dbs (N_{db}) is of the order of 10^{17} cm^{-3} in the very thin n-type and p-type doped layers, where N_{db} is of the order of 10^{15} cm^{-3} in the undoped or intrinsic layer (Stutzmann *et al.*, 1987). The built-in potential in the undoped region corresponds, in equivalence to the diffusion potential across a conventional *pn*-junction (see Chapter 4), to the difference of the conduction band edges in the *p*-type and *n*-type doped regions (Equation (4.10)). The potential energy of electrons or holes in the conduction or valence bands, respectively, decreases across the undoped region nearly linearly towards the *n*-type or *p*-type doped layers, respectively. Electrons or holes photogenerated in the undoped layer are accelerated towards the very thin *n*-type or *p*-type dopes layers, respectively, where they become majority charge carriers and are collected at the ohmic contacts.

The maximum built-in potential in a-Si:H solar cells can be estimated if considering an effective density of states of the order of 10^{20}–10^{21} cm^{-3} at the valence and conduction band edges and a saturation limit of P- and B-atoms of the same order. Then, regarding to a doping efficiency of 0.1–1%, the differences between the conduction or valence band edges and the Fermi-energy in the *n*-type or *p*-type doped regions, respectively, correspond to about 0.12–0.18 eV with respect to Equation (2.25). Therefore, the built-in potential is about 1.3–1.4 V in an a-Si:H solar cell with a pin structure. However, the upper limit of V_{OC} has to be lower than the maximum built-in potential in a-Si:H solar cells since a certain electric field has to remain in order to realize carrier collection by drift (see also task 9.4). As a consequence, it is not useful to operate solar cells based on a-Si:H absorber layers under concentrated sunlight, and it is practically impossible to reach very high η with a-Si:H solar cells.

The *p*-type doped contact layer often consists of amorphous silicon carbide (a-SiC:H), which has a larger band gap and enables additional blocking of electrons at the *p*-type doped contact (Konagai, 2011).

The estimation of the minimum diode saturation or thermal generation current density of a-Si:H solar cells is not trivial for a-Si:H solar cells since a-Si:H cannot be treated as an ideal absorber. Furthermore, the E_μ of a-Si:H is lower than the E_{og} and photogeneration due to absorption of blackbody radiation by defect states takes place. These two factors lead, in comparison with an ideal solar cell, to a strong increase of the thermal generation current density, thereby limiting V_{OC}. Therefore, η of an ideal solar cell with a band gap of 1.7 eV in the Shockley–Queisser limit (28%) is much higher than the highest η which can be reached with a-Si:H solar cells in reality (see also Figure 6.6).

9.3.5 Staebler–Wronski effect in amorphous silicon solar cells

The solar energy conversion efficiency of a-Si:H solar cells decreases during the first two to three days of operation of the a-Si:H solar cell and stabilizes after this degradation. The η can be recovered by heating the a-Si:H solar cell to 160–170°C. The effect of reversible degradation of a-Si:H solar cells under illumination is called the Staebler–Wronski effect and was described by Staebler and Wronski in terms of reversible conductivity changes in a-Si:H (Staebler and Wronski, 1977).

The Staebler–Wronski effect is caused by light-induced metastable defects (Stutzmann *et al.*, 1985). The η of a-Si:H solar cells is reduced from about 10% to about 7% due to the Staebler–Wronski effect. Degradation by the Staebler–Wronski effect cannot be avoided in a-Si:H solar cells. The role of the Staebler–Wronski effect for degradation of a-Si:H solar cells can be reduced by increasing the built-in electric field in the undoped a-Si:H absorber. However, the a-Si:H absorber thickness has to be reduced for this purpose, which in turn leads to a decrease of photogeneration in the a-Si:H layer. The reduced photogeneration in a thinner a-Si:H layer can be compensated by stacking two a-Si:H solar cells in series (pin–pin structure of an a-Si:H solar cell, Figure 9.15(b)). It should be mentioned that the thicknesses of the stacked a-Si:H solar cells have to be adjusted in accordance with the current-matching condition.

The origin of the Staebler–Wronski effect is still under discussion and there exist different models for explaining the effect. Figure 9.17 gives an idea of possible mechanisms of photoinduced generation of dbs in a-Si:H (see also the bond breaking model of Stutzmann (Stutzmann *et al.*, 1985)). Hydrogen plays a decisive role for the formation and annealing of dbs in a-Si:H. It is assumed that a photon is absorbed at a Si–Si bond near a H-atom. The Si–Si bond can be weakened by localization of the electron-hole pair at neighboring Si-atoms and the weakened Si–Si bond can break by forming two dbs. Both dbs will recombine within a very short time. However, one of the dbs can be stabilized by a H-atom saturating the other db and leading therefore to an increase of N_{db}.

The Staebler–Wronski effect can be avoided in μc-Si:H consisting of nanoparticles of c-Si with diameters of tens of nm surrounded by a-Si:H (Meier *et al.*, 1994). The a-Si:H provides hydrogen for passivation of surface states at the c-Si nanoparticles. The potential energy of

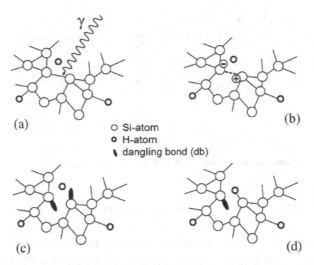

○ Si-atom
● H-atom
❭ dangling bond (db)

Figure 9.17. Possible mechanism of photoinduced generation of a db in a-Si:H consisting of the steps of photon absorption (a), weakening of a Si–Si bond due to localization of a photogenerated electron hole pair (b), breaking of a weakened Si–Si bond (c) and stabilization of one of the dbs with a H-atom (d).

photogenerated electrons and holes is lower in the c-Si nanoparticles so that photogenerated electrons and holes move into the c-Si nanoparticles, where they get delocalized, within a very short time after photogeneration. Recombination of delocalized electrons and holes does not cause bond breaking.

9.3.6 *a-Si:H/μc-Si:H tandem solar cells*

The absorption coefficient of μc-Si:H is higher than the absorption coefficient in c-Si even though the band gaps are the same (Vetterl *et al.*, 2000). This revealed the opportunity to apply a PV absorber with the band gap of c-Si in thin-film solar cells. Solar cells based on μc-Si:H absorbers have also a pin architecture like a-Si:H but do not show degradation due to the Staebler–Wronski effect (Meier *et al.*, 1994). Values of V_{OC} of the order of 0.526 V can be reached in solar cells based on μc-Si:H absorbers (Konagai, 2011).

A tandem solar cell with the combination of band gaps of 1.7 eV (top cell) and 1.1 eV (bottom cell) allows can increase η of silicon thin-film solar cells. The pin–pin concept is applied in a-Si:H/μc-Si:H tandem solar

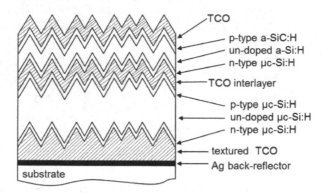

Figure 9.18. Schematic cross-section of a a-Si:H/μc-Si tandem solar cell.

cells (Figure 9.15(c)). It is important to note that the values of I_{SC} reduced from about 17.5 or 25.3 mA/cm^2 for solar cells based on a-Si:H or μc-Si:H absorbers to 14.4 mA/cm^2 for a solar cell based on a-Si:H/μc-Si:H (Konagai, 2011), which is even larger than half of the I_{SC} of the solar cells based on a μc-Si:H absorber, i.e. optical losses can be further reduced in a-Si:H/μc-Si:H multi-layer structures.

Figure 9.18 shows a schematic cross-section of a a-Si:H/μc-Si:H tandem solar cell. Light trapping is optimized by introducing a silver back reflector, a textured TCO layer on top of the silver back reflector and a TCO interlayer. The TCO interlayer has been introduced to increase the reflection of photons with energies above the band gap of a-Si:H, i.e. for optimizing the current matching condition at a thickness of the undoped a-Si:H layer of only 0.3 μm. Tandem solar cells based on a-Si:H/μc-Si:H are also called micromorph (micro-crystalline-amorphous) solar cells. Solar energy conversion efficiencies of the order of 14% have been reached with the a-Si:H/μc-Si:H concept (Konagai, 2011).

9.4 Chalcogenide Solar Cells

9.4.1 *The chalcogenide absorber families*

Chalcogenides are compounds containing elements of the periodic table with the valence of 6. Here, tellurides, selenides and sulfides are considered as chalcogenides. The corresponding elements with the valence of 6, i.e. tellurium (Te), selenium (Se) and sulfur (S), are denoted by X^{VI} in the

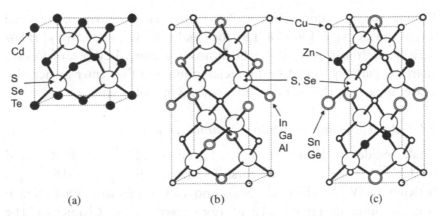

Figure 9.19. Chalcogenides used in thin-film PVs with the structure of cubic zincblende (CdX^{VI}); chalcopyrite (Cu^{I}B^{III}X$_2^{VI}$) and kesterite (Cu$_2^I$C^{II}D^{IV}X$_4^{VI}$), (a), (b) and (c), respectively. X^{VI}, B^{III}, C^{II} and D^{IV} denote elements with the valence of 6 (S, Se), 3 (Fe^{III}, In, Ga, Al), 2 (Zn), and 4 (Sn, Ge), respectively.

following. CdTe is the most prominent chalcogenide in thin-film PV and belongs to the family of A^{II}X^{VI} (A: Zn, Cd, Hg) semiconductors, which have a cubic zincblende structure (Figure 9.19(a)).

Chalcopyrite is a copper-based mineral with the chemical formula CuFeS$_2$. In the chalcopyrite, copper (Cu) and iron (Fe) have the valence of 1 and 3, respectively. Incidentally, Cu and Fe can have the valence of Cu^I and Cu^{II} and of Fe^{II} and Fe^{III}, respectively. Compounds containing Cu^I and elements of the valence 3 and 6 in the stoichiometric ratio 1:1:2, i.e. Cu^{I}B^{III}X$_2^{VI}$ (B^{III} denotes In, Ga or Al), have the structure of chalcopyrite, which is shown in Figure 9.19(b) (see, for example, Jaffe and Zunger, 1984). The most prominent chalcopyrite for PV is CuInSe$_2$.

With regard to the atoms with the valence of 3 in a chalcopyrite, half of these atoms can be replaced by atoms with a valence of 2 and the other half by atoms with a valence of 4. This procedure results in the chemical formula of kesterite, Cu$_2^I$C^{II}D^{IV}X$_4^{VI}$, where C^{II} and D^{IV} denote atoms with the valence of 2, for example, zinc (Zn) atoms, and 4, for example, tin (Sn^{IV}) or germanium (Ge) atoms, respectively. The structure of kesterite is shown in Figure 9.19(c) (see, for example, Persson, 2010). The most prominent kesterite for PV is Cu$_2$ZnSn(Se$_{1-x}$S$_x$)$_4$.

Atoms are bonded by a polar covalent bonding in chalcogenides. The electronegativities of S- or Se-atoms are 2.58 and 2.55, respectively, after

Pauling (see, for example, Allred, 1961). The electronegativity of the metal atoms (for example, 1.9, 1.78, 1.81 for Cu, In and Ga, respectively) is less than the electronegativity of S- or Se-atoms. Therefore, the center of positive charge is partially shifted to the metal atoms and the center of negative charge is partially shifted to S- and Se-atoms in chalcogenides.

9.4.2 Band gaps and absorption in chalcogenides

Chalcogenides with the chemical formulas CdX^{VI}, $Cu^I B^{III} X_2^{VI}$ and $Cu_2^I C^{II} D^{IV} X_4^{VI}$ are semiconductors. The band gap (E_g) of CdTe ranges between 1.5 eV (Adachi *et al.*, 1993) and 1.43 eV (Yamada, 1960), which is in the optimum for PV solar energy conversion (see Chapter 6). The absorption coefficient of CdTe is high (Adachi *et al.*, 1993) so that a thickness of a CdTe layer of several μm is sufficient for light absorption.

The values of E_g are 1.04, 1.53 and 1.68 eV for CuInSe$_2$ (Shay *et al.*, 1973), CuInS$_2$ (Tell *et al.*, 1971) and CuGaSe$_2$ (Shay *et al.*, 1972), respectively. The band gaps of $Cu_2ZnSnSe_4$ and Cu_2ZnSnS_4 are about 1.0 and 1.5 eV, respectively (Persson, 2010). These band gaps are in the range of PV absorbers for high η (see Chapter 6). Incidentally, the stoichiometry between Cu and In or Ga can vary within the chalcopyrite phase so that E_g of a certain chalcopyrite can also vary depending of the Cu:In or Cu:Ga ratio.

The band gap of chalcopyrite thin films can be smoothly tuned by changing the stoichiometry, for example, in $CuIn_{1-x}Ga_xSe_2$ (Dimmler *et al.*, 1987) (Figure 9.20(a)):

$$E_g = 1.018\,eV + 0.575\,eV \cdot x + 0.108\,eV \cdot x^2 \qquad (9.18')$$

It has to be mentioned that the parameters in Equation (9.18′) can vary depending on deposition, probably due to little deviations of the content of copper (Paulson *et al.*, 2003).

The tuning of E_g of chalcopyrites is important for the optimization of η of thin-film solar cells based on chalcopyrite absorbers for two reasons. First, the value of V_{OC} can be increased for larger E_g. Second, a gradient of the content of Ga allows for the reduction of the density of minority charge carriers at ohmic contacts (graded band gap, see Chapter 3) and therefore for a reduction of recombination losses. By changing the stoichiometry, E_g can be varied in other quaternary ($CuIn(Se_{1-x}S_x)_2$) or penternary

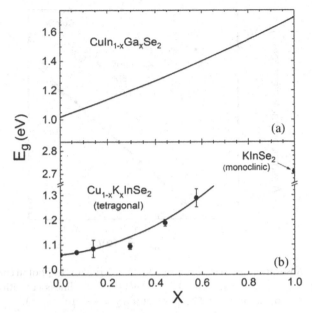

Figure 9.20. Dependence of the band gap of $CuIn_{1-x}Ga_xSe_2$ (a) and tetragonal $Cu_{1-x}K_xInSe_2$ (b) on the stoichiometry following Equations (9.18′) and (9.18″), respectively. The data points given in (b) were obtained after (Muzillo *et al.*, 2016).

chalcopyrite $(CuIn_{1-x}Ga_x(Se_{1-y}S_y)_2)$ or kesterite $(Cu_2ZnSn(Se_{1-x}S_x)_4)$ compounds.

Copper can be partially replaced by alkaline atoms such as potassium (K). For $Cu_{1-x}K_xInSe_2$, the dependence of E_g on the stoichiometry has been investigated (Muzillo *et al.*, 2016) (see the data points in Figure 9.20(b)). For tetragonal $Cu_{1-x}K_xInSe_2$, the data points can be approximated by the following equation:

$$E_g = 1.06\,\mathrm{eV} + 0.1\,\mathrm{eV} \cdot x + 0.5\,\mathrm{eV} \cdot x^2 \qquad (9.18'')$$

The incorporation of K near the ohmic and charge-selective contacts is important in order to reach a very high V_{OC} for solar cells based on $CuIn_{1-x}Ga_xSe_2$ absorbers.

Chalcogenide PV absorbers have large absorption coefficients where α increases strongly within a quite narrow range of E_{ph} above E_g (see, for example, Paulson *et al.*, 2003). Figure 9.21 shows the absorption spectrum of $CuInSe_2$ as a function of E_{ph} reduced by E_g in a double logarithmic

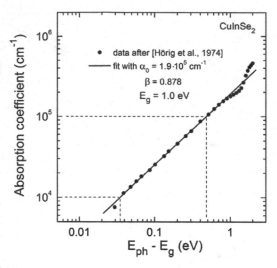

Figure 9.21. Absorption spectrum of CuInSe$_2$ as a function of the photon energy reduced by the band gap. Data points taken after (Hörig *et al.*, 1974), the line is a fit with an empirical power law (power coefficient $\beta = 0.878$, pre-factor $\alpha_0 = 1.9 \cdot 10^5$ cm^{-1}).

scale. The data points were taken from Hörig *et al.* (1974). The absorption coefficient follows an empirical power law for E_{ph} up to 1 eV above E_g. A large α of 10^4 cm^{-1} and a very large α of 10^5 cm^{-1} are reached at E_{ph} of about 35 meV and 0.45 eV above E_g, respectively, i.e. at 1.035 and 1.45 eV. CuInSe$_2$ is the strongest absorber used in thin-film PV.

9.4.3 *Technological aspects of CdTe solar cells*

CdTe sublimates or evaporates in correct stoichiometry as Cd and Te$_2$ (congruent evaporation) and forms only one phase (de Nobel, 1959).

$$2CdTe(s) \leftrightarrow 2Cd(g) + Te_2(g) \qquad (9.19)$$

CdTe is the only semiconductor which can be directly transferred by a very simple sublimation process from a CdTe source onto a substrate within a short time of only 3–5 min. The short time of the deposition of a PV CdTe absorber is a decisive key for the low production costs of CdTe thin-film solar cells. The distance between the source and the substrate is only 2 mm in the CSS (close space sublimation (Britt and Ferekides, 1993; Birkmire and Eser, 1997)) process so that losses of CdTe are very low during the deposition. Furthermore, the CSS process of CdTe deposition is performed in Ar or O_2/He atmosphere at a reduced pressure of about 1000 Pa without need for high vacuum. The deposition temperature of between 550°C and 700°C is not critical and standard glass sheets can be used as substrates (Wu, 2004).

As-deposited CdTe thin films have to be post-treated with $CdCl_2$ in order to reduce recombination losses and to increase the lifetime of photogenerated charge carriers. The $CdCl_2$ is deposited, for example, by spraying a $CdCl_2$/methanol solution onto the CdTe thin film (McCandless and Birkmire, 1991). Annealing of CdTe thin films at about 400°C for 10 min in presence of $CdCl_2$ leads to an increase of grains, to a passivation of grain boundaries and to alloying at the CdS/CdTe interface. The post-treatment of CdTe thin films with $CdCl_2$ is key for high η (Birkmire and Eser, 1997).

Figure 9.22 gives an overview of the basic technology of thin-film solar cells and modules based on CdTe absorber. The amazing points are the simplicity and the short time period of the complete technological process. The production of CdTe solar cells is by far the simplest and fastest technology in PV in general.

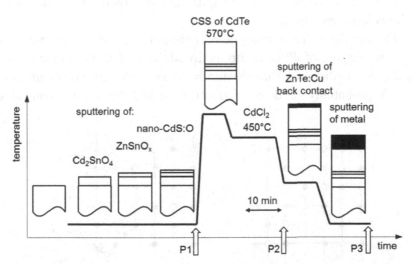

Figure 9.22. Basic technological steps of a CdTe thin-film PV module in time and temperature scales. The arrows mark the P1–3 cuts after deposition of the TCO, after post-treatment with $CdCl_2$ and after final metallization, respectively. For more details see Wu (2004).

Layers for CdTe thin-film PV modules are deposited by sputtering, CSS and spraying, all of which do not demand high energies. Therefore, the application of CdTe thin-film PV modules results in the shortest energy payback time (0.7 years under conditions in Catania, Spain) and the lowest carbon footprint (15 g CO_2-eq/kWh under conditions in Catania, Spain) for PV systems (Wild-Scholten *et al.*, 2010).

Despite the high η and despite the fast and relatively simple technology, CdTe thin-film solar cells are not suitable for a global energy supply due to the limitation of Te resources (George, 2013) (see also task T9.8). Furthermore, broken or degraded CdTe PV modules have to be completely recycled in order to recover Te and to avoid emission of the toxic Cd into the environment.

9.4.4 *Defects, phases and self-compensation*

Intrinsic defects play a decisive role for the electronic properties of chalcogenide absorbers. Positions of atoms in a lattice can be occupied by other atoms (antisite defect), atoms can occupy space between lattice sites (interstitial defect) and atoms can be missed at a lattice site (vacancy defect) as shown schematically in Figure 9.23 for a two-dimensional lattice with two different atoms.

The number of elementary defect configurations in a lattice increases with the number of different atoms. In addition, defects can form complexes, thus further increasing the number of possible defect configurations. A high probability of defect formation makes the precise control of

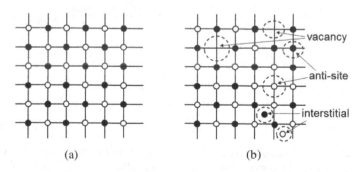

(a) (b)

Figure 9.23. Two-dimensional defect-free lattice with two different atoms (a) and two-dimensional lattice with vacancy, antisite and interstitial defects (b).

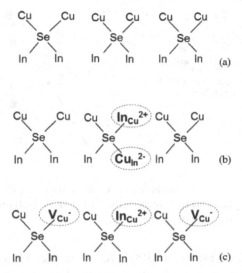

Figure 9.24. Undisturbed configurations of Se atoms in CuInSe$_2$ (a) and configurations with antisite pair (b) and neutral Cu vacancy complex (c) defects in CuInS$_2$.

stoichiometry and phases in chalcopyrite and kesterite thin-film absorbers complicated.

There are two prominent defects in chalcopyrites (Figure 9.24). The antisite defect pair ($Cu_{In}^{2-} + In_{Cu}^{2+}$) consists of a Cu-atom on a lattice site of In (Cu_{In}^{2-}) and of an In-atom on a lattice site of Cu (In_{Cu}^{2+}). The formation energy of the antisite pair defect in chalcopyrite is relatively low and consequently heavy randomization sets in at temperatures much lower than the melting temperature (Zhang *et al.*, 1998). The neutral Cu vacancy complex ($2V_{Cu}^- + In_{Cu}^{2+}$) consists of two Cu vacancies (V_{Cu}^-) and one In-atom on a lattice site of Cu (In_{Cu}^{2+}) compensating the negative charge of the Cu vacancies (Zhang *et al.*, 1998).

The presence of a neutral Cu vacancy complex changes the local stoichiometry of a chalcopyrite. The replacement of Cu atoms by In-atoms can be expressed by the following equation:

$$m \cdot CuInSe_2 + n \cdot In \rightarrow Cu_{m-3n}In_{m+n}Se_{2m} + 3n \cdot Cu \qquad (9.20)$$

The values of the numbers m and n have to be equal to or larger than 3 or 1, respectively. The compounds $Cu_{m-3n}In_{m+n}Se_{2m}$ can form phases (also called ordered vacancy compounds) in the phase diagram

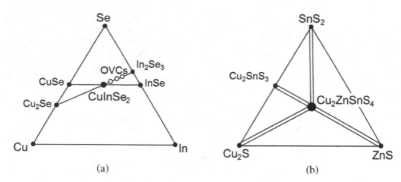

Figure 9.25. Simplified Cu–In–Se phase diagram to form the CuInSe$_2$ and OVC (ordered vacancy compound) phases (a) and pseudo-ternary Cu$_2$S–ZnS–SnS$_2$ phase diagram to form the Cu$_2$ZnSnS$_4$ and Cu$_2$SnS$_3$ phases (b).

of Cu, In and Se (Figure 9.25(a)), for example, CuIn$_5$Se$_8$, CuIn$_3$Se$_5$. Ordered vacancy compounds also semiconductors. The band gaps of ordered vacancy compounds depend sensitively on the preparation and measurement conditions. For example, band gaps of 1.3–1.36 (Philip *et al.*, 2005) and 1.21 (Bodnar, 2008) eV have been reported for CuIn$_5$Se$_8$ and CuIn$_3$Se$_5$.

The formation of the kesterite Cu$_2$ZnSnS$_4$ phase can be illustrated in a pseudo-ternary phase diagram of Cu$_2$S, ZnS and SnS$_2$ (Figure 9.25(b)). Cu$_2$S, ZnS, SnS$_2$ and Cu$_2$SnS$_3$ are secondary phases that can occur during the formation of the kesterite phase. Incidentally, the phase diagrams depend sensitively on stoichiometric deviations (see for more details the review of Kumar *et al.*, 2015) and even off-stoichiometric single phase kesterite can exist (Valle Rios *et al.*, 2016). Furthermore, order-disorder transitions can occur in kesterites and influence on the optical absorption edge (Krämmer *et al.*, 2014). Secondary phases introduce defects at interfaces and additional band gaps into kesterite thin-film absorber.

Changes in the local stoichiometry can induce uncontrolled local phase transitions. The control of phases during the formation of chalcopyrite and kesterite thin-film absorbers is a key technological issue. The formation and transformation of various phases and the distribution of elements have been analyzed for chalcopyrites by Mainz and Klenk (Mainz and Klenk, 2011). Chalcopyrites are usually grown in a Cu rich regime

(Klenk *et al.*, 2005) so that copper sulfide can form on top of a chalcopyrite layer. The highly conductive cupper sulfide layer has to be removed in potassium cyanide solution (KCN) (see, for example, Scheer *et al.*, 1993). It has been shown by neutron scattering that the lowest densities of defects are reached when there is a slight excess of copper in the chalcopyrite layer (Stephan *et al.*, 2011).

The density of each type of defect is determined by a complex of chemical reactions. The equilibrium of chemical reactions involving certain defects and phase transformation can be strongly influenced by additives. For example, layers of $CuIn_{1-x}Ga_xSe_2$ absorbers are usually produced by coevaporation onto substrates of soda lime glass or in presence of additionally deposited sodium increasing the density of acceptors and passivating grain boundaries (see, for example, Ård *et al.*, 2000). The performance of thin-film solar cells based on $CuIn_{1-x}Ga_xSe_2$ improved strongly when sodium was present during the formation of the absorber. The η of a thin-film solar cell above 20% was achieved for the first time with a $CuIn_{1-x}Ga_xSe_2$ absorber deposited on soda lime glass (Jackson *et al.*, 2011). Furthermore, a post deposition treatment of the $CuIn_{1-x}Ga_xSe_2$ absorber with potassium from a very thin deposited KF surface layer allowed for a further increase of η on flexible substrates (Chirila *et al.*, 2013). Recently, an η of 22.6% has been reached by a post deposition treatment of the $CuIn_{1-x}Ga_xSe_2$ absorber with rubidium fluoride (RbF) (Jackson *et al.*, 2016).

Thin-film solar cells based on a $CuIn_{1-x}Ga_xSe_2$ absorber are stable over long period of time despite the large variety of possible reactions of defect formation. Chalcopyrite and kesterite thin-films are excellent PV absorbers in which even the presence of grain boundaries has no influence on η (Abou-Ras *et al.*, 2008) and diffusion lengths of minority charge carriers above $1 \mu m$ are easily reached (Scheer *et al.*, 1995). Therefore, defect reactions and/or local phase transitions can stabilize chalcopyrite and kesterite absorbers and avoid the formation of recombination active defects. However, a detailed analysis of defects and phase diagrams is rather complex for $CuIn_{1-x}Ga_xSe_2$ and other absorbers (Guillemoles, 2000). Incidentally, no indication has been found also for recombination of minority charge carriers at grain boundaries in thin-film solar cells based on CdTe (Scheer, 1999).

Doping by variation of the stoichiometry in $CuInSe_2$ has been demonstrated by Noufi *et al.* while the Cu to In and the Se to metal ratios are important (Noufi *et al.*, 1984). Well-controlled doping of chalcopyrite and kesterite thin-films is practically impossible. For example, epitaxial layers of $CuIn_{1-x}Ga_xSe_2$ are usually intrinsically p-type doped with densities of free holes (p_0) between $4 \cdot 10^{16}$ and $2 \cdot 10^{19}$ cm^{-3} where two different acceptors and compensating donors exist (Schroeder *et al.*, 1998).

Different defect configurations have been analyzed for $CuInSe_2$. It has been shown that the Cu vacancy is a shallow acceptor with very low formation energy, which can explain intrinsic p-type doping of $CuInSe_2$ thin-films (Zhang *et al.*, 1998).

$$V_{Cu} \leftrightarrow V_{Cu}^{-} + h^{+} \tag{9.21}$$

Low formation energies of defects cause self-compensation and prevent controlled doping. For example, the formation energy of the Cu vacancy becomes even negative for a Fermi-energy close to the conduction band edge of $CuInSe_2$ so that compensating Cu vacancies are formed together with incorporation of donor atoms (Zhang *et al.*, 1998). Self-compensation avoids the formation of pn-homo-junctions (see Chapter 4) as well as of ohmic tunneling contacts (see Chapter 5) in thin-film solar cells based on chalcopyrite or kesterite absorbers. On the other hand, self-compensation leads to an exceptional radiation hardness of thin-film solar cells based on $CuIn_{1-x}Ga_xSe_2$ (Jasenek and Rau, 2001).

Well-controlled n-type and p-type doping as well as high doping is also difficult or practically impossible for CdTe thin-films due to enhanced compensation and segregation effects at grain boundaries (Birkmire and Eser, 1997). CdTe thin-films are intrinsically p-type doped. Therefore pn-homo-junctions as charge-selective contacts and ohmic contacts cannot be formed by controlled doping of CdTe thin-films.

Despite the high probability for the formation of defects, the mobilities of electrons (μ_e) and of holes (μ_h) can be relatively large in chalcogenides. For example, μ_e and μ_h were up to $1100 \, cm^2/(Vs)$ in n-type doped (Triboulet *et al.*, 1974) or $80 \, cm^2/(Vs)$ in p-type doped (Yamada, 1960) CdTe crystals, respectively. In epitaxial layers of $CuIn_{1-x}Ga_xSe_2$, μ_e and μ_h were in the range of 20–$1100 \, cm^2/(Vs)$ (Schumann *et al.*, 1980) and 167–$311 \, cm^2/(Vs)$ (Schroeder *et al.*, 1998), respectively. In combination with response times of p-type doped $CuInSe_2/CdS$ detectors (5–150 ns, Wagner *et al.*, 1974) or photoluminescence lifetimes in $CuIn_{1-x}Ga_xSe_2$ (5–50 ns, Ohnesorge *et al.*, 1998), the relatively high values of μ_e and μ_h result in a diffusion length of the order of 1–2 μm.

9.4.5 *Contacts in chalcogenide solar cells*

CdTe thin-film solar cells are produced in super-strate configuration (Figure 9.26(a)). The TCO layer (SnO_2:F or Cd_2SnO_4) is deposited onto a glass substrate for high-efficiency CdTe solar cells. The electron mobility in Cd_2SnO_4 is relatively large so that a thickness of the TCO layer of only 0.2–0.3 μm is sufficient for a low series resistance (Wu, 2004). A layer of undoped $ZnSnO_x$ is deposited on the TCO layer.

Chalcopyrite and kesterite thin-film solar cells are produced in strate configuration (Figure 9.26(b)). Ohmic back contacts of chalcopyrite or kesterite absorbers are prepared on sputtered molybdenum layers (thickness of the order of 0.5–1 μm) where the ohmic contact with the absorber layer contact is formed with an interfacial $MoSe_2$ or MoS_2) layer (see, for example, Scheer *et al.*, 1993). Incidentally, $MoSe_2$ or MoS_2 are layered materials well known for use as lubricants. From this point of view, $MoSe_2$ or MoS_2 interface layers avoid stress caused by different thermal expansion of the substrate and the absorber layer.

Charge-selective contacts are usually realized with hetero-junctions (see Chapter 4) in thin-film solar cells based on chalcogenide absorbers. The layer forming the charge-selective contact in solar cells based on

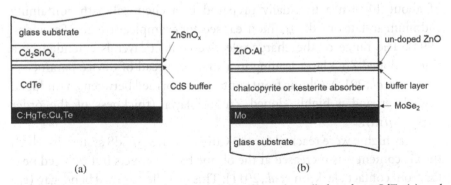

(a)   (b)

Figure 9.26. Schematic cross-sections of thin-film solar cells based on CdTe (a) and chalcopyrite or kesterite (b) absorbers.

chalcogenide absorbers is also called the buffer layer, which is a semi-conductor with a wide band gap. The buffer layer enables transfer of photogenerated electrons from the conduction band of the p-type doped absorber into the conduction band of the wide gap semiconductor whereas the transfer of photogenerated holes from the absorber into the buffer layer is blocked. Band gaps of materials that are used as buffer layers are 2.4 eV (CdS (Shin *et al.*, 1991)), 2.1 eV (In$_2$S$_3$ (Allsop *et al.*, 2006)) and between 2.6 eV (ZnO$_{0.5}$S$_{0.5}$) and 3.4 eV (ZnO$_{1-x}$S$_x$ (Meyer *et al.*, 2004)). The feasibility of CuInS$_2$/CdS charge-selective contacts has been demonstrated by Wagner *et al.* in their pioneering works (Wagner *et al.*, 1974; Shay *et al.*, 1975). In the production of PV modules with chalcogenide absorbers, CdS is usually used for the buffer layer. Incidentally, optical absorption in the buffer layer does not contribute to the I_{SC} of a chalcogenide thin-film solar cell. Therefore, the buffer layer has to be as thin as possible in order to minimize optical losses.

In CdTe solar cells, the CdS buffer layer is about 70 nm thick. The CdTe absorber layer is several-μm thick. A hetero-junction with a highly doped p-type semiconductor has to be prepared on top of the p-type doped CdTe thin-film in order to form an ohmic contact. Ohmic contacts with CdTe thin films can be formed with metal layers and alloys containing mercury (Hg) and/or Te such as ZnTe–Cu, HgTe or PbTe (Birkmire and Eser, 1997) or HgTe:Cu$_x$Te (Wu, 2004) thin films. A heat treatment at 150°C leads to the formation of a highly p-type doped surface region at the CdTe thin films and therefore to an ohmic contact (Birkmire and Eser, 1997).

In chalcopyrite or kesterite solar cells, a CdS buffer layer with a thickness of about 40–60 nm is usually prepared in a chemical bath containing cadmium acetate or CdSO$_4$, thiourea (see, for example, Britt and Ferekides, 1993). The range of the charge-selective contact layer is extended by a thin undoped ZnO layer leading to a smaller spread of device parameters (Scheer *et al.*, 2011). Ohmic front contacts are formed between an undoped ZnO layer and a highly doped ZnO:Al layer (thickness of the order of 0.5–1 μm).

Very high η were reached for the CuIn$_{1-x}$Ga$_x$Se$_2$/CdS system in which the Ga content was increased at the ohmic back contacts but reduced near the front contact (Jackson *et al.*, 2011). This so-called graded band-gap (see also Figure 9.28(a)) leads to a reduction of the density of minority charge carriers and therefore to a reduction of the recombination rate at the ohmic back contact.

Hetero-junctions between chalcogenide absorbers and CdS cannot be considered as metallurgically sharp contacts. The contact is extended to an intermixed region with a thickness of the order of 2 nm as demonstrated for a $CuInSe_2$/CdS interface (Cojacaru-Mirédin *et al.*, 2011). Metallurgical intermixing at interfaces makes a precise theoretical description of charge-selective contacts with chalcogenides rather complicated. Nevertheless, useful information can be obtained from models with sharp hetero-junctions (Klenk, 2001).

For a $CuIn_{1-x}Ga_xSe_2$ absorber with a CdS buffer layer, V_{OC} increases with increasing E_g as expected, but only within a range of E_g between 1 and about 1.2 eV. For larger E_g, V_{OC} increases much less than expected with increasing E_g and tends to saturate for larger E_g (Herberholz *et al.*, 1997). This phenomenon is also known as the V_{OC} problem of thin-film solar cells and is caused by the band-offsets at the charge-selective contact.

Figure 9.27 depicts simplified band diagrams of chalcopyrite solar cells based on $CuInSe_2$/CdS (a) and $CuInS_2$/CdS (b) charge-selective contacts. It is commonly agreed that a type I hetero-junction is formed at the

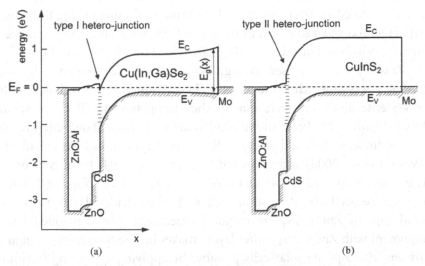

Figure 9.27. Simplified band diagrams of thin-film solar cells based on Cu(In,Ga)Se$_2$ (a) and CuInS$_2$ (b) with metallurgically sharp junctions and interface states. $E_g(x)$ denotes the graded band gap caused by an increase of the content of Ga towards the Mo contact.

$CuInSe_2/CdSe$ contact (Herberholz *et al.*, 1997; Klenk, 2001). Electrons crossing the $CdS/CuInSe_2$ interface accumulate in the interface region due to the positive conduction band offset of the order of $+0.31\,eV$ (theoretical analysis (Wei and Zunger, 1993)). Therefore, free electrons form an inversion layer at the $CuInSe_2$ side. As a consequence, the density of holes is strongly reduced at the charge-selective contact, and a solar cell based on $CuInSe_2/CdS$ is limited by bulk recombination in the $CuInSe_2$ absorber (Turcu *et al.*, 2002).

The conduction band offset at the $CdS/CuIn_{1-x}Ga_xSe_2$ interface reduces with increasing amount of Ga and a type II hetero-junction (see Chapter 8) is formed at the $CdS/CuInS_2$ contact (Herberholz *et al.*, 1997). Surface recombination becomes dominating in this case, and the limit of V_{OC} reduces to the difference between the E_C reduced by the conduction band offset at the $CdS/CuInS_2$ interface and E_V.

For engineering of the band-offsets at charge-selective contacts with chalcopyrites, In_2S_3 and $Zn(O,S)$ have been tested. In_2S_3 can be deposited by a spray ILGAR$^{®}$ (ion layer gas reaction) process (Fischer *et al.*, 2011). The spray ILGAR$^{®}$ process consists of cycles of spraying a chemisorbed layer of precursor metal ions, a purge, sulfurization step, a purge. Each cycle gives a thickness control of about 1 nm and additives as chlorine or oxygen can be incorporated via the precursor salt or water molecules during the purge step. High η comparable with those with a CdS buffer layer can be reached for solar cells based on $CuIn_{1-x}Ga_xSe_2/In_2S_3$ (Fischer *et al.*, 2011).

$ZnO_{1-x}S_x$ can be used to align both the valence (oxygen rich) and conduction (sulfur rich) band offsets at hetero-junctions with chalcopyrite or kesterite absorber layers due to the strong bowing (Persson *at al.*, 2006). Figure 9.28 shows the dependencies of the calculated energies of the conduction (E_V) and valence (E_C) band edges and of the calculated (Persson *at al.*, 2006) and measured (Meyer *et al.*, 2004) band gaps on x. The values of E_V and E_C can be varied over 0.9 and 1.1 eV for $x < 0.5$ and $x > 0.5$, respectively (Persson *at al.*, 2006). The calculated and measured band gaps of $ZnO_{1-x}S_x$ are in good agreement. The controlled band alignment with $ZnO_{1-x}S_x$ buffer layers makes in-line processing of highly efficient chalcopyrite solar cells possible by applying sputtering (Grimm *et al.*, 2012). Atomic layer deposition (ALD) by pulsing gaseous precursors allows as well for a controlled variation of the stoichiometry in $ZnO_{1-x}S_x$

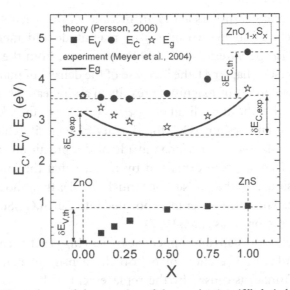

Figure 9.28. Dependence of the energies of the conduction (filled circles) and valence (filled squares) band edges and of the band gap (open stars and solid line) in $ZnO_{1-x}S_x$ on the stoichiometry. The symbols and the solid line correspond to theoretical analysis (Persson, 2006) and to experimental results (Meyer *et al.*, 2004), respectively. The arrows mark the theoretical and experimental variability of the energies of the conduction and valence band edges.

layers. High η has been demonstrated for chalcopyrite solar cells with a $ZnO_{1-x}S_x$ buffer layer deposited by ALD (Platzer-Björkman *et al.*, 2006) with a ZnS buffer layer prepared by chemical bath deposition (Nakada and Mizutani, 2002).

9.4.6 *Treatment of Cu(In,Ga)Se₂ absorbers with alkali elements*

As shown by preparation of chalcopyrite absorbers on different substrates, the values of V_{OC} and FF were higher for solar cells prepared on soda-lime glass than for those prepared on substrates which did not contain sodium (Na) (Hedström *et al.*, 1993). Therefore, in order to achieve high η, Na is added to the precursors before a Cu(In,Ga)Se₂ absorber is formed, for example, by depositing several tens of nm of NaF (Contreras *et al.*, 1997). Incidentally, fluorine is not incorporated into the chalcopyrite.

Without the incorporation of Na, Cu(In,Ga)Se₂ absorber layers are usually *p*-type doped and highly compensated. The incorporation of Na into $Cu(In,Ga)Se_2$ absorbers led to an increase of the density of majority charge carriers by about two orders of magnitude and therefore to an increase of V_{OC} (Contreras *et al.*, 1997). As known from the behavior of *pn*-junctions (see Chapter 4), the increase of the density of majority charge carriers by two orders of magnitude results in an increase of the diffusion potential, which is the upper limit of V_{OC}, by 0.12 V. For the alkali elements potassium (K) and cesium (Cs), the density of majority charge carriers increased by only one order of magnitude or did practically not increase, respectively. This has been explained by the fact that the ionic radius of Cu^+ is very similar to that of Na^+ but much smaller than those of K^+ and Cs^+, i.e. Na can be much better incorporated into $Cu(In,Ga)Se_2$ crystallites than K and Cs (Contreras *et al.*, 1997).

Indium atoms occupying copper vacancies (In_{Cu}^{2+}) act as compensating donors in $CuInSe_2$. The effect of Na on the density of majority charge carriers in $CuInSe_2$ is caused by the replacement of In_{Cu}^{2+} by the neutral Na_{Cu}, which does not have electronic states in the band gap of $CuInSe_2$ (Wei *et al.*, 1999).

Alkali elements can additionally be added to $Cu(In,Ga)Se_2$ absorbers after the formation of the chalcopyrite (post-deposition treatment). The post-deposition treatment of $Cu(In,Ga)Se_2$ absorber layers with KF led to a further increase of the V_{OC} and therefore of η (Chirilă *et al.*, 2013). The concentration of alkaline elements was increased at the Mo back- and at the charge-selective front-contacts. In contrast to the postdeposition treatment with NaF, the post-deposition treatment with KF resulted in a well passivated and stable surface of $Cu(In,Ga)Se_2$ without Cu and with a reduced amount of Ga (Chirilă *et al.*, 2013). Furthermore, the thickness of the charge-selective CdS layer could be reduced to 30 nm due to homogeneous deposition independent of the facets of crystallites (Friedlmeier *et al.*, 2015). Very high η can also be reached for solar cells with a $Zn(O,S)$ charge-selective contact layer prepared after post-deposition treatment of KF (Friedlmeier *et al.*, 2015).

An advantage, in addition to surface passivation, of post-deposition treatment with KF is that a relatively large amount of K can be incorporated into a chalcopyrite crystal without segregation of secondary phases and that E_g of $Cu_{1-x}K_xInSe_2$ is larger than E_g of $CuInSe_2$ (Muzillo *et al.*, 2016). Furthermore, the lifetime of photogenerated charge carriers was about 10 times larger in, for example, $Cu_{0.84}K_{0.14}InSe_2$ than in $CuInSe_2$ (Muzillo *et al.*, 2016).

The grading of the band gap is steeper in $Cu(In,Ga)Se_2$ absorber layers with post-deposition treatment with KF than without a post-deposition treatment (Friedlmeier *et al.*, 2015). The Ga/(Ga+In) ratio, for example, can be as high as 0.4 or 0.3 at the Mo back- and charge-selective front-contacts, respectively, and as low as 0.5 about 0.5 μm underneath the front-contact (Friedlmeier *et al.*, 2015). Steeper graded band gaps lead to lower densities of minority charge carriers at ohmic contacts and therefore to reduced recombination losses (see Chapter 3) and increased V_{OC} (see Chapter 4). A reduced Ga/(Ga+In) ratio or x causes a lower band gap of $CuIn_{1-x}Ga_xSe_2$ (see Figure 9.20(a)) and therefore a higher I_{SC} (see Chapter 2).

A post deposition treatment of $Cu(In,Ga)Se_2$ absorber layers with K has been compared with the treatment with rubidium and cesium (treatment with RbF and CsF, respectively) (Jackson *et al.*, 2016). Despite the fact that the highest V_{OC} has been obtained after treatment with KF, the highest η was reached after treatment with RbF (Jackson *et al.*, 2016). Ion exchange of the lighter by the heavier alkali elements seems to be important for reaching very a high η.

9.5 Hybrid Organic–Inorganic Metal Halide Perovskite Solar Cells

9.5.1 *The family of hybrid organic–inorganic metal halide perovskites*

A perovskite has the stoichiometry ABX_3 whereas A and B denote cations with the charge +1 and +2, respectively, and X denotes anions with the charge −1. One cation with the charge +2 and six anions form an octahedron (Figure 9.29). In a perovskite crystal, the octahedrons are connected with each other by sharing the anions between neighbored octahedrons. The free space between the octahedrons is occupied by the compensating cations with the charge +1.

Halides are salts in which the anions consist of elements of the seventh group of the periodic table, i.e. iodine, bromine and chlorine (I, Br and Cl, respectively). In metal halides, the cation belongs to a metal

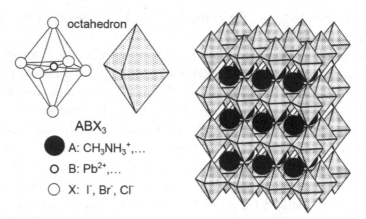

Figure 9.29. Structure of a hybrid organic–inorganic metal halide perovskite. The space between the rigid BX_6 octahedrons is filled with the organic cations.

ion such as the twofold positively charge lead ion (Pb^{2+}). Metal halides are inorganic. In hybrid organic–inorganic metal halide perovskites, the cations with the charge $+1$ are formed from small organic molecules such as the methylammonium ion ($CH_3NH_3^+$) or the formamidinium ion ($HC(NH_2)_2^+$). Methylammonium lead triiodide ($CH_3NH_3PbI_3$) is the most prominent hybrid organic–inorganic metal halide perovskite.

The family of hybrid organic–inorganic metal halide perovskites provides a large variability for the combination of different organic cations with metal ions and halide ions. Furthermore, the stoichiometry can be varied by partially replacing anions or cations by other metal ions with the charge $+2$ or other halide ions. Therefore, the variability is even huge in the family of hybrid organic–inorganic metal halide perovskites if taking into account ternary (ABX_3), quaternary ($A(B_{1-x}C_x)X_3$; $AB(X_{1-x}Y_x)_3$), $A_{1-x}D_yBX_3$), penternary (for example, $A_{1-x}D_xB(X_{1-y}Y_y)_3$)), etc. compounds (A and D denote two different organic cations or an organic cation and a metal ion with the charge $+1$; B and C denote two different metal ions with the charge $+2$; X and Y denote two different halide ions, the indices x and y denote the variability in the stoichiometry).

9.5.2 *Band gaps and optical absorption of hybrid organic–inorganic metal halide perovskites*

The band gaps of hybrid organic–inorganic metal halide perovskites can be tuned over a wide range. As examples, Figure 9.30 shows the dependence

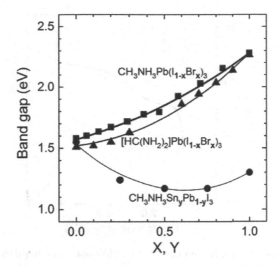

Figure 9.30. Dependence of the band gap of $CH_3NH_3Pb(I_{1-x}Br_x)_3$, $[HC(NH_2)_2]$ $Pb(I_{1-x}Br_x)_3$ and $CH_3NH_3Sn_yPb_{1-y}I_3$ on the stoichiometry (squares, triangles and circles, respectively). The data points were taken after Noh *et al.* (2013); McMeekin *et al.* (2016) and Zhu *et al.* (2016), respectively. The thick, medium and thin solid lines represent the bowing parameters of 0.3, 0.6 and 1.0, respectively.

of methylammonium lead triiodide bromide — $CH_3NH_3Pb(I_{1-x}Br_x)_3$, formamidinium lead triiodide bromide — $[HC(NH_2)_2]Pb(I_{1-x}Br_x)_3$ and methylammonium tin lead triiodide — $CH_3NH_3Sn_yPb_{1-y}I_3$ on the stoichiometry.

$CH_3NH_3PbI_3$ and $[HC(NH_2)_2]PbI_3$ and $CH_3NH_3PbBr_3$ and $[HC(NH_2)_2]PbBr_3$ practically have the same band gaps (1.52–1.58 and 2.27–2.29 eV, respectively). This means that the small organic cation has only minor influence on E_g, i.e. E_g of hybrid organic–inorganic metal halide perovskites is dominated by the $Pb(I,Br)_6$ octahedrons.

The iodides and bromides can be mixed over the whole range of stoichiometry so that E_g can be varied between the 1.55 ($CH_3NH_3PbI_3$) and 2.27 eV ($CH_3NH_3PbBr_3$). The bowing parameter (Equation (8.2)) depends on the organic cation and amounts to about 0.3 and 0.6 for $CH_3NH_3Pb(I_{1-x}Br_x)_3$ and $[HC(NH_2)_2]Pb(I_{1-x}Br_x)_3$, respectively. The tunability of E_g over a wide range and the excellent transparency of hybrid organic–inorganic metal halide perovskites at photon energies below E_g (see Figure 9.31 for $CH_3NH_3PbI_3$) make them very interesting for applications in multi-junction solar cells (see Chapter 6), for example,

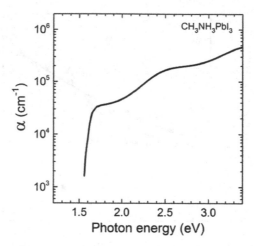

Figure 9.31. Absorption spectrum of CH₃NH₃PbI₃ obtained after data of Löper *et al.* (2015) and Dittrich *et al.* (2016).

as top cells in combination with crystalline silicon (c-Si) solar cells (see Chapter 7) (McMeekin *et al.*, 2016).

The metal ions Pb^{2+} and Sn^{2+} are mixed in $CH_3NH_3Sn_yPb_{1-y}I_3$. The band gaps of $CH_3NH_3SnI_3$ and $CH_3NH_3Sn_{0.5}Pb_{0.5}I_3$ are 1.3 and 1.15 eV, respectively (Zhu *et al.*, 2016). The approximated bowing parameter is of the order of 1.0 for $CH_3NH_3Sn_yPb_{1-y}I_3$.

The optical absorption of $CH_3NH_3PbI_3$ sharply sets on near E_g and the absorption coefficient amounts to about $3 \cdot 10^4 \, cm^{-1}$ already at a photon energy of 0.1 eV above E_g (Figure 9.31). Therefore, the thickness of thin-film absorbers (d_{abs}) based on hybrid organic–inorganic metal halide perovskites is of the order of several hundred nm. Incidentally, the E_g of the tetragonal phase of $CH_3NH_3PbI_3$, which is the relevant one for solar cells, increases with increasing temperature (see, for example, Dittrich *et al.*, 2015), in contrast to the conventional semiconductors such as c-Si or GaAs.

9.5.3 *Diffusion length and type of conductivity in $CH_3NH_3PbI_3$*

The diffusion length (L) of a photovoltaic absorber can be obtained from measurements of the diffusion coefficient (D) and lifetime (τ) of photogenerated charge carriers (Equation (3.5)) whereas the D can be obtained from the mobility (μ) by using the Einstein equation

(Equation 9.15)). However, measurements of μ and τ are usually performed under different experimental conditions that can have a rather strong influence on the result in the case of hybrid organic–inorganic metal halide perovskites. This is not the case if L is measured directly by using surface photovoltage (SPV) spectroscopy for which the SPV signals are kept constant and the dependence of the intensity is analyzed as a function of the absorption length (α^{-1}) (Goodman, 1961). In this method, the diffusion or transport length is obtained from the intersection point of the dependence of the light intensity on α^{-1}. In powders of $CH_3NH_3PbI_3$, L can be of the order of tens of μm (Figure 9.32). In layers of $CH_3NH_3PbI_3$ which are used for solar cells, the transport length is practically equal to the layer thickness and the transport length is related to the drift length. Furthermore, the transport length is limited by the grain size in $CH_3NH_3PbI_3$ layers.

In contrast to inorganic semiconductor crystals, hybrid organic–inorganic metal halide perovskites are not very stable and even the control of the type of conductivity is challenging. The change of the sign of SPV signals can serve as an indication of the type of the conductivity due to the

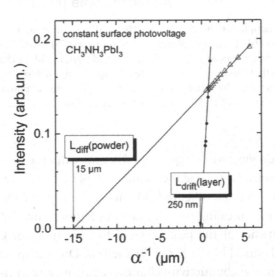

Figure 9.32. Measurement of the diffusion length of a $CH_3NH_3PbI_3$ powder (triangles) and of the drift length of a $CH_3NH_3PbI_3$ layer (circles) following the analysis after Goodman (1961). In this method, the light intensity is plotted as a function of the absorption length whereas the surface photovoltage signal is kept constant (see also Dittrich *et al.*, 2016).

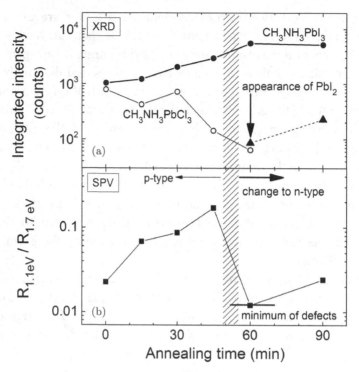

Figure 9.33.　Example for the change of $CH_3NH_3PbI_3$ (filled circles), $CH_3NH_3PbCl_3$ (open circles) and PbI_2 (triangles) phases (a) and of defects (b) in a $CH_3NH_3PbI_3$ layer as a function of annealing time at 100°C. Between 45 and 60 min, PbI_2, a change from *p*-type to *n*-type conductivity and a minimum of defects appeared (analysis of X-ray diffraction and of the ratio between the surface photovoltage signals excited at photon energies below and above the band gap of $CH_3NH_3PbI_3$, see for details Naikeaw *et al.*, 2015).

change from downward (*p*-type doped semiconductor depleted near the surface) to upward (*n*-type semiconductor depleted near the surface) band bending. As an example, Figure 9.33 shows the type of the conductivity as a function of annealing time (annealing temperature 100°C) together with the intensities of the peaks related to the phases of $CH_3NH_3PbI_3$, $CH_3NH_3PbCl_3$ and PbI_2 in X-ray diffraction. During annealing between 45 and 60 min, the conductivity changed from *p*-type to *n*-type whereas the $CH_3NH_3Cl_3$ phase disappeared nearly completely and PbI_2 phase appeared. The appearance of the PbI_2 phase was caused by a deficit of CH_3NH_3I due to evaporation. The change of the type of the conductivity was caused by a change from an excess to a deficit of iodine ions in the perovskite lattice.

Figure 9.33 also shows the ratio of the SPV signals measured at 1.1 and 1.7 eV, i.e. measured in the ranges of excitation of mobile charge carriers from defect states in the band gap and from states in the valence band into the conduction band, respectively. This ratio has a maximum at 45 and a minimum at 60 min, i.e. the SPV signals related to defects in the band gap has a maximum just before and a minimum just after switching from the excess to the deficit of iodine ions in the perovskite lattice. As a conclusion, precipitates of PbI_2 getter impurities from $CH_3NH_3PbI_3$, which leads to a reduction of defect states (passivation, see Chapter 3). Incidentally, a maximum of η is achieved for annealing at 100°C for 60 min. However, Figure 9.33 demonstrates that the realization of a charge-selective contact with a *pn*-homo-junction is impossible for hybrid organic–inorganic metal halide perovskites. On the other hand, the formation of a $CH_3NH_3PbI_3/PbI_2$ interface leads to additional passivation (Supasai *et al.*, 2013).

With regard to Figure 9.33, the electronic properties are not stable under ongoing annealing at 100°C, which is not that much higher than the maximum temperature of solar cells under operation (see also task T1.5). The stabilization of hybrid organic–inorganic metal halide perovskites over a long time is still an issue and a well-defined adjustment of stabilized electronic properties of, for example, $CH_3NH_3PbI_3$ will demand additional measures in the preparation.

9.5.4 *Preparation of hybrid organic–inorganic metal halide perovskites*

Halides of organic molecules and metal halides are soluble in organic solvents. Therefore, hybrid organic–inorganic metal halide perovskites can be easily prepared from a solution by evaporating the solvent and by crystallizing the perovskite phase. Amazingly, the formation of layers of hybrid organic–inorganic metal halide perovskites from a solution at low temperatures can be applied for solar cells with high and even very high solar energy conversion efficiency. Therefore, the crystalliza-tion of the perovskite phase is extremely robust against the formation of recombination-active defect states in the band gap by incorporation of residual solvent and/or by incorporation of impurity molecules and atoms from precursor materials and solvents.

The robustness against the formation of defect states in the band gap is an exclusive property of hybrid organic–inorganic metal halide perovskites due to the ionic character of bonds. Positively and negatively charged ions repel holes and electrons, respectively, in different regions and reduce therefore the probability for recombination. Furthermore, the robustness against the formation of defect states in the band gap enabled researchers in numerous laboratories to prepare solar cells of high η. In addition, the robustness against the formation of defect states in the band gap is an enormous driving force for research and development of high-efficiency low-cost solar cells and PV modules based on hybrid organic–inorganic metal halide perovskites.

For the preparation of solar cells with very high η, it is important to avoid losses of the halide due to out-diffusion during crystallization of the hybrid organic–inorganic metal halide perovskites. Furthermore, the grains of perovskite crystallites should be larger than d_{abs} since the transport length is limited by grain boundaries. The volume of uncontrolled recombination can be additionally reduced if the lateral grain size is larger than d_{abs} by many times.

Figure 9.34 gives an idea about some principle preparation routes for hybrid organic–inorganic metal halide perovskites. In the simplest case, a metal halide such as PbI_2 and an organic halide such as CH_3NH_3I are dissolved, for example, in γ-butyrolactone, spin-coated onto a substrate and slightly heated. The temperature of crystallization can be increased up to 200°C (so-called hot-casting) if using solvents with high boiling temperature and the lateral size of $CH_3NH_3PbI_3$ crystallites can be increased if mixing PbI_2 and CH_3NH_3Cl precursors (Nie *et al.*, 2015). In contrast to $CH_3NH_3Pb(Br_{1-x}I_x)_3$, $CH_3NH_3Pb(I_{1-x}Cl_x)_3$ can exist only in narrow range of stoichiometry. Therefore, the chloride can form seeds for the crystallization of large $CH_3NH_3PbI_3$ grains.

The solvent has strong influence on the crystallization of hybrid organic–inorganic metal halide perovskites, for example, due to a specific influence on wetting, formation of complexes and maximum reaction temperature. Solvent engineering combines several solvents with each other in the precursor solution, for example, γ-butyrolactone and

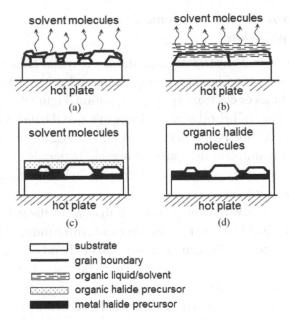

Figure 9.34. Some schematics for principle preparation routes of hybrid organic–inorganic metal halide perovskites. Evaporation of solvents and recrystallization from a solution (a), controlled recrystallization by solvent engineering with an additional liquid organic layer (b), reaction from metal halide and organic halide precursor layers under controlled recrystallization with solvent molecules (c), reaction from a metal halide precursor layer and organic halide molecules (d).

dimethylsulfoxide, with an additional organic liquid layer deposited in a second step (so-called dripping, for example, of toluene) (Jeon *et al.*, 2014). This technique makes use of the formation of intermediate phases before annealing.

The formation of large grains of hybrid organic–inorganic metal halide perovskites is possible from a double precursor layer of the metal halide and of the organic halide in a controlled atmosphere of solvent molecules (Xiao *et al.*, 2014; Zhu *et al.*, 2016).

As an alternative to solution-based processing, thin-films of hybrid organic–inorganic metal halide perovskites can be also obtained in a vacuum process. In such case, a precursor layer of a metal halide such as $PbCl_2$ is deposited in vacuum and converted in an atmosphere containing organic halide molecules such as CH_3NH_3I (Liu *et al.*, 2013).

9.5.5 *Solar cells with hybrid organic–inorganic metal halide perovskites*

The absorber layer of hybrid organic–inorganic metal halide perovskites is placed between two layers forming charge-selective contacts for the separation of holes or electrons (pin configuration, Figure 9.35). The deposition temperature of hybrid organic–inorganic metal halide perovskites is low and the electronic properties are very robust against most substrates. Therefore, many different inorganic or organic materials can be used for the formation of charge-selective contacts with hybrid organic–inorganic metal halide perovskites.

Semiconductors with band gaps larger than E_g of the hybrid organic–inorganic metal halide perovskite absorber and with conduction or valence band edges aligned with the conduction or valence band edges, respectively, are used for electron- or hole-selective contacts, respectively. The offsets between the valence bands or between the conduction bands of the hybrid organic–inorganic metal halide perovskite and the electron- or hole-selective contacts serve as barriers for photogenerated holes or electrons, respectively.

For the formation of charge-selective electron contacts with hybrid organic–inorganic metal halide perovskites, metal oxides with wide band

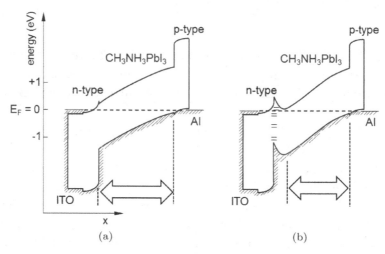

Figure 9.35. Idealized band diagrams of thin-film solar cells with a pin configuration based on $CH_3NH_3PbI_3$ absorber without (a) and with (b) electron traps at the charge-selective electron contact. The arrows mark the drift length of photogenerated charge carriers.

gaps and a relatively large electron affinity such as TiO_2 (Zhou *et al.*, 2014) or ZnO nanoparticles (You *et al.*, 2016) or organic electron acceptors such as phenyl-C_{61}-butyric acid methyl ester (PCBM, see also Chapter 10) (Malinkiewicz *et al.*, 2014) have been applied. In contrast, metal oxides with wide band gaps and a relatively low electron affinity such as NiO_x nanoparticles (You *et al.*, 2016) or organic hole conductors such as 2.2′,7,7′-tetrakis-(N,N-di-4-methoxyphenylamino)-9,9′-spirobifluorene (spiro-OMeTAD) (Liu and Kelly, 2014), polytriarylamine (Jeon *et al.*, 2014) or poly(4-butylphenyl-diphenyl-amine) (Malinkiewicz *et al.*, 2014) have been used hole-selective contacts in thin-film solar cells based on hybrid organic–inorganic metal halide perovskites.

Light-induced degradation of solar cells based on hybrid organic–inorganic metal halide perovskites is possible under operation. Trapping, for example, of electrons, near the charge-selective contact can cause chemical reactions in the absorber layer and therefore to the formation of additional trapped or fixed negative charge (Figure 9.35(b)). This charge can cause a reduction of the transport length of photogenerated charge carriers and/or an increase of surface recombination. In general, a competitive long-term stability has not been reached yet for solar cells based on hybrid organic–inorganic metal halide perovskites and degradation mechanisms are under investigation. Incidentally, it has been shown, for example, that the ammonium group in $CH_3NH_3PbI_3$ decomposed under illumination (Nickel *et al.*, 2017). Furthermore, a partial replacement of organic cations by Cs^+ resulted in an increased stability of so-called triple cation perovskite solar cells (Saliba *et al.*, 2016a), whereas Cs^+ and Rb^+ ions can be additionally combined (Saliba *et al.*, 2016b).

9.6 Summary

PV thin-film absorbers are formed on foreign substrates, which do not provide a certain structure for the growth of crystals with a low density of defects. Therefore, inherent passivation mechanisms of defects are required in order to achieve lifetimes which are long enough for collecting the photogenerated charge carriers before they recombine.

Thin-film PV makes use of the integrated series connection of solar cells in PV modules. For this purpose, thin layers of transparent conductive oxides (TCOs) are applied and techniques for the deposition of homogeneous thin films on large areas have been developed. TCO layers cause optical losses due to reflection, fundamental absorption and free carrier absorption. The resistance of TCO layers is a major reason for resistive losses in thin-film solar cells. In thin-film PV modules, the areas of the back contacts, the PV absorber and the front contact are separated by the P1, P2 and P3 cuts, respectively, in order to separate the thin-film solar cells from each other.

Inherent passivation mechanisms are related to the introduction of atoms passivating defects by chemical reactions during the deposition process or to the formation of crystallites with low densities of defects during phase transitions (see Figure 9.36). Defects are directly passivated by a large amount of hydrogen in hydrogenated amorphous silicon and micro-crystalline silicon (a-Si:H and μc-Si:H, respectively). Crystallites with low densities of defects are obtained for chalcogenides (CdTe, chalcopyrites — $CuIn_{1-x}Ga_x(Se_{1-y}S_y)_2$ and kesterites — $Cu_2ZnSn(Se_{1-x}S_x)_4$) and hybrid organic–inorganic metal halide perovskites, for example, $CH_3NH_3PbI_3$. Thin-film silicon, CdTe and chalcopyrite PV modules are established on the market, whereas thin-film solar cells based on kesterites and hybrid organic–inorganic metal halide perovskites are still on the research level.

To date, there are four main preparation routes for PV thin-film absorbers (see the overview in Figure 9.36).

Silicon thin-film absorbers are directly deposited from silane and hydrogen by PECVD (plasma enhanced chemical vapor deposition). Layers of a-Si:H can be n-type and p-type doped whereas the density of dangling bonds increases with increasing density of dopants. The realization of pin structures enables large drift lengths of photogenerated charge carriers in the undoped (intrinsic — i) layer of the pin structure.

Thin films of CdTe can be deposited directly due to congruent evaporation (close space sublimation — CSS). A recrystallization of CdTe thin films in the presence of $CdCl_2$ is required in order to achieve very high solar energy conversion efficiency (η).

Metal precursor layers are transformed into thin films of chalcopyrites and kesterites during selenization or sulfurization whereas several phase transitions usually follow each other during the processing. The presence

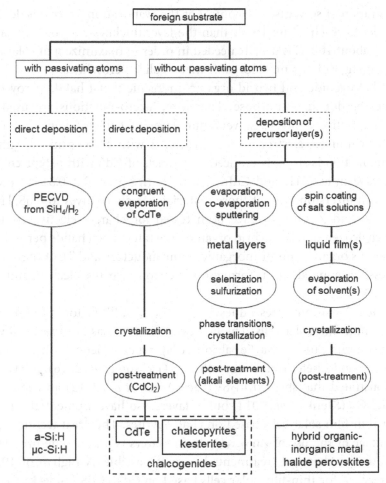

Figure 9.36. Overview about main principles and routes for deposition of homogeneous PV absorbers for thin-film solar cells. Incidentally, the material-specific hydrazine-based deposition of kesterites has not been considered in this overview.

of secondary phases in chalcopyrite and kesterite absorber layers has to be avoided by applying well-controlled annealing regimes. A post-deposition-treatment with alkali metals or their fluorides such as potassium fluoride (KF) or rubidium fluoride (RbF) are required in order to reach very high η for chalcopyrite solar cells (Jackson *et al.*, 2016).

Liquid precursor layers of salt solutions are transformed into solid layers of hybrid organic–inorganic metal halide perovskites during the evaporation of solvents. The control of the evaporation process is decisive in order to reach grains larger than the layer thickness. A post-annealing step at about 100°C is usually needed in order to maximize η of solar cells based on hybrid organic–inorganic metal halide perovskites.

Chalcogenide and hybrid organic–inorganic metal halide perovskites cannot be doped by purpose. Therefore, hetero-junctions are used for charge-selective contacts. A very thin layer of CdS is mainly used for the formation of electron-selective contacts with chalcogenides. Surface recombination is strongly reduced at contacts of CdS with p-type chalcogenides such as CdTe and $CuIn_{1-x}Ga_xSe_2$ (for $x < 0.2$–0.3) since CdS forms an inversion layer at those materials. As alternatives to CdS, layers of In_2S_3 and $ZnO_{1-x}S_x$ have been tested for charge-selective contacts with chalcogenides. For hybrid organic–inorganic metal halide perovskites, numerous organic and/or inorganic semiconductors and TCOs have been successfully applied for the formation of charge-selective electron and hole contacts.

Table 9.1 summarizes values of E_g, I_{SC}, V_{OC}, fill factor (FF) and η of record thin-film solar cells. The highest η (22.6%) has been reached with a chalcopyrite $CuIn_{1-x}Ga_xSe_2$ absorber layer, a post-deposition-treatment with RbF and a hetero-junction with CdS (Jackson *et al.*, 2016). It has been demonstrated that sprayed In_2S_3 (Sáez-Araoz *et al.*, 2012) and sputtered $ZnS_{1-x}O_x$ (Klenk *et al.*, 2014) buffer layers also have a potential for high η for thin-film solar cells based on chalcopyrites. The great technological advantage of the use of $ZnS_{1-x}O_x$ buffer layers is that in-line processing without wet chemical treatments becomes possible. A high η of 19.6% was reached for thin-film solar cells based on CdTe/CdS (Gloeckler *et al.*, 2013).

For wide gap thin-film solar cells based on a-Si:H, $CuInS_2$ or $CuGaSe_2$, η is reduced due to insufficient values of V_{OC}, i.e. the diode saturation current densities are quite high. Here, an improved understanding of chemical processes at charge-selective and ohmic contacts and their relation to defect formation is required.

With regard to the variability of E_g of chalcopyrite, multi-junction solar cells based on, for example, $CuInSe_2$ and $CuGaSe_2$ bottom and top cells, respectively, seem promising. However, it is still challenging to realize (i) a top cell with a high transparency for photons with

Table 9.1. Values of the band gap (E_g) of thin-film absorbers, short circuit current density (I_{SC}), open circuit voltage (V_{OC}), fill factor (FF) and solar energy conversion efficiency (η) for different thin-film record solar cells. References are given. MA, FA and FDT denote methylammonium — $CH_3NH_3^+$, formamidinium — $CH[NH_2]_2^+$ and a fluorene-dithiophene derivative (Saliba *et al.*, 2016), respectively.

	E_g (eV)	I_{SC} (mA/cm^2)	V_{OC} (V)	FF (%)	η (%)	Reference
a-Si:H	1.7	16.75	0.886	67.8	10.1	Benagli *et al.* (2009)
a-Si:H/μc-Si	1.7/1.1	14.4	1.41	72.8	14.7	Konagai (2011)
CdTe/CdS	1.4	20.25	0.876	79.4	21.0	First Solar (2014)
$CuIn_{1-x}Ga_xSe_2$/CdS	1.15	37.8	0.741	80.6	22.6	Jackson *et al.* (2016)
$CuIn_{1-x}Ga_xSe_2$/In_2S_3	1.2	35.5	0.630	72.0	16.1	Sáez-Araoz *et al.* (2012)
$CuIn_{1-x}Ga_xSe_2$/ $ZnS_{1-x}O_x$	1.13	37.2	0.717	78.6	21.0	Friedlmeier *et al.* (2016)
Cu(In,Ga)S_2/CdS	1.53	18.9	0.960	68.0	12.3	Hiroi *et al.* (2016a)
Cu(In,Ga)S_2/ZnMgO	1.53	23.4	0.920	72.2	15.5	Hiroi *et al.* (2016b)
$Cu_2ZnSn(Se,S)_4$/ $CdTe$/In_2S_3	1.07	38.9	0.466	69.8	12.7	Kim *et al.* (2014)
TiO_2/(MA,FA)$Pb(I,Br)_3$/ FDT	1.60	22.7	1.148	76	20.2	Saliba *et al.* (2016c)

energy below E_g of the top cell, (ii) a top cell with a very high V_{OC} and, (iii) transparent recombination contacts for the integrated series connection. The achievement of a V_{OC} of 0.96 V for a chalcopyrite solar cell with E_g of 1.53 eV (Hiroi *et al.*, 2016a) might be first step towards efficient chalcopyrite tandem solar cells.

The reserves of indium and tellurium are rather limited in relation to global application of thin-film solar cells based on $CuIn_{1-x}Ga_xSe_2$ or CdTe absorbers (Andersson, 2000). Kesterite-based solar cells contain only earth abundant elements and kesterite PV modules can be produced with similar technologies as chalcopyrite solar cells. It seems that kesterite absorber layers have the potential for high η. Different technological approaches are under development to ascertain materials limitations in kesterite-based solar cells (Delbos, 2012).

Thin-film solar cells have the highest potential for sustainable PV energy production due to short energy payback times (0.7 years for CdTe, 1.1 years for Si thin-film, 1.2 years for $CuIn_{1-x}Ga_xSe_2$ for PV systems installed in south of Italy; for comparison: 1.5 years for crystalline silicon PV systems (Wild-Scholten *et al.*, 2010)). The degradation rate of PV modules is also very important for sustainable PV energy production. Chalcopyrite thin-film PV modules installed after 2010 have degradation rates even below 0.5%/year, whereas the degradation rates of CdTe thin-film PV modules were about 2–3%/year (Jordan *et al.*, 2016).

Thin-film PV modules based on a-Si:H have a high shunt resistance so that η is also high at reduced light intensity. Thin-film PV modules based on a-Si:H have also the lowest decrease of η with increasing temperature so that they provide the highest amount of energy produced per kW_p installed in tropical regions, for example. Furthermore, the development of a-Si:H was key for the realization of charge-selective a-Si:H/c-Si hetero-junctions for solar cells with very high η based on c-Si (see Chapter 7).

Thin-film absorbers based on hybrid organic–inorganic metal halide perovskites have an excellent transparency for photon energies below the E_g. Furthermore, both the electron- and hole-selective contacts can be well passivated in solar cells based on hybrid organic–inorganic metal halide perovskites so that the absorber layers can be really thin. As a main advantage, η of solar cells based on hybrid organic–inorganic metal halide perovskites can exceed 20% and the solar cells can be fabricated at low temperatures. This makes solar cells based on hybrid organic–inorganic metal halide perovskites very interesting for applications in tandem solar cells with a high-efficiency c-Si solar cell as the bottom cell (Albrecht *et al.*, 2016). However, the long-term stability of solar cells based on hybrid organic–inorganic metal halide perovskites is not yet sufficient for their practical application in PV.

9.7 Tasks

T9.1: Bulk and tunneling resistances at grain boundaries in a TCO layer

Compare the bulk and tunneling resistances at grain boundaries in highly doped TCO layers ($n_0 = 5 \cdot 10^{20}$ cm^{-3} cm^{-3}, $\mu_n = 20$ cm^2/Vs) with an average grain size of $d_g = 200$ nm and a surface band bending at grain boundaries of (a) $\phi_0 = 1.2\,\text{V}$ and (b) $\phi_0 = 0.3\,\text{V}$. The effective electron mass (m_e^*) is 0.26 times the free electron mass (m_e).

T9.2: TCO contact stripe of thin film solar cells

Find the optimal width of a TCO ($R_\square = 5\,\Omega\square$) contact stripe for a high efficiency pin-pin a-Si:H solar cell.

T9.3: Optical losses in TCO layers

Calculate I_{SC} as a function of the wavelength corresponding to the band gap of photovoltaic absorbers on basis of the AM1.5d spectrum. The absorber layer has a surface ZnO:Al layer ($n_0 = 10^{21}\,\text{cm}^{-3}$) with a thickness of 0.5 or $1\,\mu$ m. The refractive index of the absorber layer is approximately 3. Compare the maximum I_{SC} with I_{SC} obtained for both surface layers in the cases of a-Si:H and CuInSe$_2$ absorber layers.

T9.4: Burstein shift in ZnO:Al

Calculate the increase of the optical band gap in highly doped ZnO:Al for densities of free electrons of 10^{20} and $10^{21}\,\text{cm}^{-3}$ by taking into account the energy dependence of the density of states (Chapter 2) and an effective electron mass of $0.265 \cdot m_e$.

T9.5: Open-circuit voltage in an a-Si:H solar cell

Estimate the open-circuit voltage in an a-Si:H solar cell by taking into account the densities of majority charge carriers in the p- and n-type doped regions and the density of photo-generated charge carriers in the un-doped a-Si:H absorber layer. Discuss the importance of light trapping for thin-film solar cells with un-doped a-Si:H absorber layers.

T9.6: Chalcopyrite tandem solar cell

Estimate the maximum possible energy conversion efficiency of a chalcopyrite tandem solar cell by taking into account data given in Table 9.1 and compare the value with the world record chalcopyrite solar cell. Discuss the importance of the V_{OC} problem.

T9.7: Advantages of CdTe thin-film technology

Explain why the thin-film technology of CdTe solar cells is so stable and robust with respect to temperature regimes and purity standards of materials and ambience.

T9.8: Strategic potential of CdTe, chalcopyrite and kesterite solar cells

Estimate the mass of Te, In and Sn in solar cells based on CdTe, $CuIn_{0.75}Ga_{0.25}Se_2$ and $Cu_2ZnSn(Se_{1-x}S_x)_4$ absorbers needed (a) for $1\,W_p$ of installed power and (b) for a supply of $2\,kW$ per capita for all people of mankind by only one type of those solar cells. Compare with the annual production of Te (80 t in 2012; George (2013)), In (640 t in 2011; Tolcin (2012)) and Sn (240000 t in 2012; Carlin (2013)).

10

Nanocomposite Solar Cells

Materials with different properties are combined in nanocomposites to achieve functionalities needed in solar cells but which cannot be realized with only one of these materials. Nanocomposite absorbers offered the opportunity for application of highly absorptive materials with very short diffusion or drift lengths of photogenerated charge carriers in solar cells. The concept of nanocomposite materials is explained and an example of folded absorbers is shown. Next, the quantum confinement effect, the concept of quantum dot absorbers and measures of surface conditioning are introduced. The principles of organic photovoltaic absorbers and the realization of local charge separation with donor–acceptor hetero-junctions in organic solar cells are illuminated, the exceptional role of C_{60} molecules as acceptors is highlighted, and examples for conjugated polymers and oligomers are given. The principles of dye-sensitized solar cells and their realization with sintered networks of TiO_2 nanoparticles sensitized with dye molecules and contacted with redox couples of an electrolyte are given in detail. Passivation concepts and the inherent stability of dye-sensitized solar cells are mentioned. The concept of nanocomposite absorbers with hybrid organic–inorganic metal halide perovskites and nanoporous metal oxides is given. The basic idea of nanophotonic concepts for solar cells is briefly explained and examples are given. The potential of nanocomposite solar cells for sustainable solar energy conversion is discussed.

10.1 The Concept of Nanocomposite Materials for Photovoltaics

10.1.1 *New quality of properties in nanocomposite materials*

A nanocomposite is a material in which two or more materials interpenetrate each other on a scale of 1–500 nm without changing the phase of each individual material (Figure 10.1) such that the individual properties of each component are conserved. The combination of very different properties of individual materials results in qualitatively new properties of nanocomposite materials. For example, materials with very different mechanical and thermal properties can interpenetrate each other and form a new material such as aerogels, which can be used for ultralight and heat-insulating structures. Actually, microcrystalline hydrogenated silicon (μc-Si:H; see Chapter 9) is a thin-film absorber consisting of silicon nanocrystals embedded in amorphous hydrogenated silicon (a-Si:H) where silicon nanocrystals provide delocalization of photogenerated charge carriers and a-Si:H provides a high absorption coefficient and hydrogen for passivation of dangling bonds.

Each component contributes with a specific functionality to the new quality of a nanocomposite. Functionalities important for solar cells are, amongst others, light absorption, photogeneration, charge separation, charge transport, surface passivation and series connection of nanoparticles. A new quality can be, for example, an extremely low thermal conductivity (in the case of aerogels) or a stable silicon thin-film absorber with a band gap of 1.1 eV based on silicon (in the case of μc-Si:H).

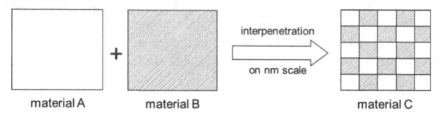

material A   material B   material C

Figure 10.1.   Schematic of a nanocomposite formed by interpenetration of two materials A and B with different properties on a scale of about 1–500 nm resulting in qualitatively new properties of material C.

A qualitatively new property of a nanocomposite cannot be achieved with the homogeneous individual materials.

In nanocomposites for photovoltaics, (i) materials transparent to most of sun spectrum such as surfactant molecules are combined with materials absorbing light very well such as semiconductor quantum dots (QDs), (ii) materials accepting electrons very well such as C_{60} molecules are combined with materials donating electrons such as phthalocyanine molecules, (iii) materials with high conductivity of charge carriers such as TiO_2 or electrolytes are combined with materials not conducting charge carriers such as dye molecules and (iv) transparent materials with a high-refractive index are combined with materials having a low refractive index. The new quality of these material combinations is related to (i) a tunable band gap of a photovoltaic nanocomposite absorber based on semiconductor QDs, (ii) charge separation at donor–acceptor bulk hetero-junctions in organic solar cells, (iii) application of dye molecules for light absorption in dye-sensitized solar cells and (iv) possible incorporation of photonic structures into nanocomposite solar cells.

A certain individual material can have several functionalities in a nanocomposite solar cell. For example, the structure of a nanocomposite is often defined by the morphology of a nanostructured skeleton of a transparent metal oxide such as nanoporous TiO_2 or ZnO nanorods, which also form charge-selective contacts and connect individual solar cells with a locally very thin absorber layer in parallel with each other. Furthermore, the morphology and the refractive index of nanostructured transparent metal oxides can contribute to light scattering and light trapping.

The optimization of a certain functionality of a given material in a nanocomposite absorber is, in general, a complex task since the variation of any local parameter in a nanocomposite absorber can decisively influence other functionalities. The challenge of photovoltaic nanocomposite absorbers is to combine materials in such a way that light absorption, local charge separation and transport of photogenerated charge carriers are optimized at a minimum of optical, recombination and resistive losses limiting the short-circuit current (I_{SC}) density, the diode saturation current density (I_0) and the fill factor (FF) of a solar cell.

10.1.2 *Photovoltaic absorbers with very short diffusion or drift lengths*

Several concepts of nanocomposite absorbers aim for the incorporation of absorber materials with very short or even extremely short diffusion or drift lengths of photogenerated charge carriers into solar cells. Related absorber materials are, for example, dye molecules that do not form conducting or semiconducting layers. A nanocomposite absorber can be treated in general as a photovoltaic absorber with an effective absorption coefficient and an effective diffusion or drift length of photogenerated charge carriers.

The thickness of a photovoltaic absorber should be three times larger than the absorption length in order to absorb more than 95% of incoming light (see Chapter 2). Figure 10.2(a) shows the schematic cross section of a solar cell with a photovoltaic absorber material having a diffusion or drift lengths of photogenerated charge carriers (L) much shorter than the absorption length. Such an absorber is called an absorber with very short L. It is obvious that efficient solar cells cannot be prepared from homogeneous absorbers with very short L since the vast majority of photogenerated charge carriers recombine before reaching the charge-selective contact.

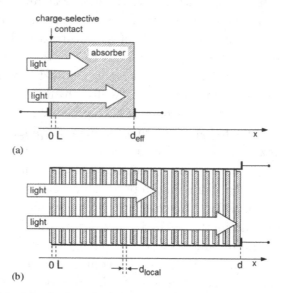

Figure 10.2.   Principle of solar cells with an absorber material having a diffusion or drift length (L) much shorter than the absorption length of the compact absorber layer (a) and for locally thin absorber layers with local charge-selective contacts (b). The individual solar cells in the stack of many solar cells are connected in parallel.

Absorber materials with very short L can only be applied in efficient solar cells if the local thickness of the absorber (d_{local}) is shorter than L. In this case, most of the photogenerated charge carriers can reach the charge-selective contact. However, only a small fraction of photons in the sun spectrum is absorbed in an optically thin absorber layer. Therefore, a stack of many optically thin absorber layers is needed for complete absorption of sunlight with photon energies above the optical band gap of the absorber material (Figure 10.2(b)). Each of the thin absorber layers in such a stack should be connected individually with a charge-selective contact so that each of the absorber layers is part of a complete solar cell.

Stacked solar cells with very thin absorber layers have to be connected in parallel because it is impossible to fulfill the current-matching condition for a large number of solar cells with very short L. The point is that an integrated series connection of many stacked solar cells with very short L would require a large variability of individual absorber layer thicknesses with respect to the absorption length, which is not possible in principle due to the very short L. The sum or the integral of all local absorber thicknesses results in the effective absorber layer thickness (d_{eff}), which should be three times larger than the effective absorption length. Incidentally, there is the hope that implementing nanophotonic concepts will help to drastically reduce d_{eff}.

The values of d_{local} and the geometry of individual solar cells with very thin absorber layers can vary depending on the kind of materials used and technologies applied. A solar cell consisting of many individual solar cells with locally very thin absorbers and which are connected in parallel can be treated as one solar cell with one effective nanocomposite absorber.

An absorber with a very short L can be folded by using, for example, the surface of ZnO nanorods where the ZnO nanorods additionally serve as the electron conductor. A related system has been demonstrated for a solar cell based on ZnO nanorod/In_2S_3:Cu/CuSCN (see Dittrich *et al.*, 2011 and references therein) where the In_2S_3:Cu absorber has a band gap of about 1.5 eV and CuSCN serves as the hole conductor.

An absorber layer folded between the electron and hole conductors is illustrated in Figure 10.3. The I_{SC} density increases with increasing d_{eff} and increasing surface area of the supporting matrix. For example, the I_{SC} density increased with increasing length of ZnO nanorods in the ZnO nanorod/In_2S_3:Cu/CuSCN system (Kieven *et al.*, 2008). However, the increase of the contact area causes an increase of surface recombination

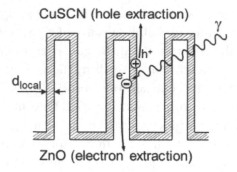

Figure 10.3. Schematic nanocomposite absorber based on ZnO nanorod/In$_2$S$_3$:Cu/ CuSCN.

and therefore of I_0. The increase of I_0 leads to a decrease of the open-circuit voltage (V_{OC}). Furthermore, the decrease of d_{local} to about 10 nm also led to an increase of I_{SC} in the ZnO nanorod/In$_2$S$_3$:Cu/CuSCN system due increasing drift of photogenerated charge carriers in the increased electric field between the electron and hole conductors. However, V_{OC} and FF decreased with decreasing d_{local} due to an increase of local shunts (Belaidi *et al.*, 2008). This example shows that very high solar energy conversion efficiencies (η) cannot be reached with related systems without a drastic reduction of surface recombination and/or without a drastic improvement of light trapping.

10.2 Quantum Dot-Based Nanocomposite Solar Cells

10.2.1 *Quantum confinement in semiconductor nanocrystals*

Electrons and holes influence each other due to the Coulomb attractive force. The attraction between a free electron and a free hole leads to the formation of a bond state. The bond state between a free electron and a free hole is called exciton (Figure 10.4). Excitons (see, e.g., Scholes and Rumbles, 2006) are treated as electrically neutral quasi-particles, which can diffuse in photovoltaic absorbers before they recombine.

Excitons are characterized by the energy and by the radius of the electron–hole orbit. In an exciton, electrons and holes move together, similarly to orbitals of electrons in atoms. In contrast to electrons and nuclei in atoms, the masses of electrons and holes in excitons are rather similar such that the center of mass is between the electron and the hole.

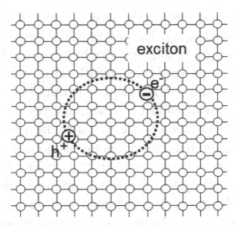

Figure 10.4. Schematic of an exciton.

The effective mass of the exciton (μ_{eh}) is given by the effective masses of the free electron (m_e^*) and of the free hole (m_h^*).

$$\frac{1}{\mu_{eh}} = \frac{1}{m_e^*} + \frac{1}{m_h^*} \tag{10.1}$$

The energy and the radius of the electron–hole orbit with the lowest energy of an exciton, E_{e1h1} and a_{e1h1}, respectively, can be estimated by taking into account μ_{eh}, the relative dielectric constant of the photovoltaic absorber (ε_r) and the ionization energy and the Bohr radius of the hydrogen atom ($E_{ion}^H = 13.56\,\text{eV}$, $a_0 = 0.051\,\text{nm}$) The radius $a_{e1h1,b}$ is called the exciton radius of the bulk semiconductor.

$$E_{e1h1,b} = \frac{\mu_{eh}}{m_e} \cdot \frac{1}{\varepsilon_r^2} \cdot E_{ion}^H \tag{10.2}$$

$$a_{e1h1,b} = \frac{m_e}{\mu_{eh}} \cdot \varepsilon_r \cdot a_0 \tag{10.3}$$

Effective electron and hole masses have been measured practically for all inorganic crystalline semiconductors. For PbS, for example, m_e^* and m_h^* are about 0.2·me (Barton, 1971) and 0.1·me (Cuff *et al.*, 1964), me is the mass of the free electron ($9.1 \cdot 10^{-31}$ kg). The exciton radius is therefore 18 nm in PbS crystals. The values of E_{e1h1} are usually much lower than the thermal energy in inorganic photovoltaic (PV) absorbers at room temperature due to the low μ_{eh} and due to the large ε_r. Therefore, excitons can be observed in most semiconductor crystals only at low temperatures.

In bulk semiconductors, the dimension of the crystal with a volume V is much larger than $a_{e1h1,b}$. The energy of the first exciton transition is lower than the band gap by a few meV in bulk semiconductors and is not important for solar cells based on conventional semiconductor absorbers such as crystalline silicon (c-Si) or gallium arsenide (GaAs).

The radius of a semiconductor nanocrystal (r_{QD}) is less than $a_{e1h1,b}$ and therefore limits the radius of the exciton in the nanocrystal. The wave function of an exciton in a semiconductor nanocrystal can be compared with standing waves vibrating between fixed points like a string on a guitar. The half wavelength of the first harmonic of the string is equal to the length of the string. The frequency of a wave is proportional to the reciprocal wavelength. The energy of a wave is proportional to its frequency. Therefore, the energy of a standing wave of a string in string in a guitar increases with decreasing length of the string. The same happens in a semiconductor nanocrystal. The energy of an exciton increases with decreasing r_{QD}. The exciton binding energy becomes much larger that the thermal energy for very low values of r_{QD} so that excitons dominate electronic properties in quantum dots (QDs). The increase of the energy of an exciton with decreasing r_{QD} is called quantum confinement. A QD is a semiconductor nanocrystal that shows quantum confinement.

The quantum confinement has consequences for optical absorption of QDs. An electron–hole pair photogenerated in a QD has to form an exciton since the interaction between the electron and the hole cannot be avoided. Therefore, optical absorption in QDs is only possible if the photon energy is larger than the sum of the band gap of the bulk semiconductor from which the QD is made and the binding energy of the exciton. In other words, photons with an energy a little bit larger than the band gap of the bulk semiconductor cannot be absorbed in a QD due to the absence of available electronic states for free holes and for free electrons in the QD. Therefore, the absorption edge (E_{abs}) of semiconductor QDs shifts to the energy of the first exciton transition in the semiconductor QDs. As a result, the absorption edge or optical band gap depends on the size of semiconductor nanocrystals.

Figure 10.5 compares the exciton radius in a bulk semiconductor and in a QD (a), the extension of the corresponding wave functions (b) and optical absorption in a bulk semiconductor and in a QD (c).

The increase of E_{abs} due to quantum confinement opens the opportunity to use semiconductors with very low band gaps (E_g) such as PbS ($E_g = 0.37$ eV for PbS; Scanlon, 1958) in solar cells and to tune E_g for optimum absorption.

For application in solar cells, QDs have to be embedded into a surrounding matrix consisting of another material. The relative dielectric constant of the surrounding material (ε_{out}) is usually different than ε_r so

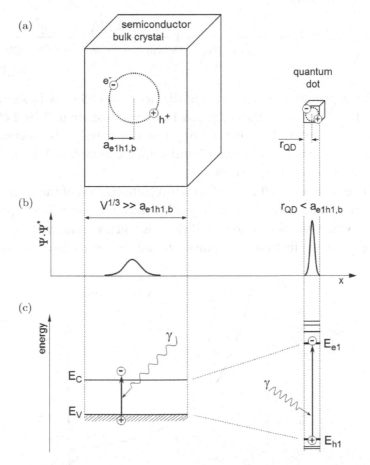

Figure 10.5. Schematic of a bulk semiconductor crystal (bulk) and a semiconductor QD (a), of the extension of an exciton in the bulk and in the QD (b) and of optical absorption in the bulk and in the QD (c). $a_{e1h1,b}$, r_{QD}, E_{h1} and E_{e1} denote the exciton radius in the bulk, the radius of the QD and the energies of the hole and of the electron for absorption in the first exciton transition, respectively.

that a correlation with respect to the change of the Coulomb potential is needed. Both the quantum confinement effect and the correction term of the Coulomb energy depend not only on the size of the semiconductor nanocrystal but also on the shape, the given electron and hole wave functions, the band structure of the semiconductor crystal, the surface conditioning of the QDs and the distance between neighboring QDs in a layer of QDs. In a first approximation, E_{abs} can be calculated as a function of r_{QD} (Brus, 1984):

$$E_{abs}(r_{QD}) \approx E_g + \frac{\hbar^2 \cdot \pi^2}{2 \cdot r_{QD}^2 \cdot \mu_{eh}} - \left(\frac{1}{\varepsilon_{out}} + \frac{0.79}{\varepsilon_r} \right) \cdot \frac{q^2}{4\pi \cdot \varepsilon_0 \cdot r_{QD}}$$

(10.4)

The dependence of E_{abs} on the diameter of PbS-QDs is shown in Figure 10.6. The absorption edge reaches a value of about 1 or 2 eV for a diameter of PbS-QDs of about 4 or 3 nm, respectively. This means that PbS-QDs with diameters between 3 and 4 nm are suitable as PV absorbers with respect to the sun spectrum.

The absorption coefficient of semiconductor QDs is of the same order as the absorption coefficient of the bulk semiconductor at larger photon energies where the absorption coefficient is usually somewhat higher in the region of the first exciton transition and somewhat lower at photon

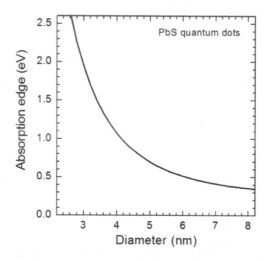

Figure 10.6. Dependence of the absorption edge on the diameter of PbS-QDs.

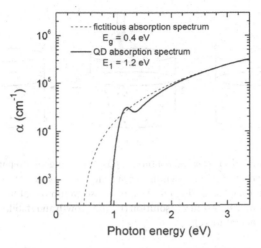

Figure 10.7. Absorption spectrum of a fictitious PbS-QD. The dotted line gives the approximate absorption spectrum of PbS.

energies a little bit above the first exciton transition (see, e.g., Moreels *et al.*, 2009 for optical absorption in PbS-QDs). This is schematically demonstrated for PbS-QDs in Figure 10.7. The absorption coefficients of bulk PbS and PbS-QDs are equal at high photon energies.

High-absorption coefficients above 10^4 cm^{-1} are reached close to the absorption edge of QDs. This makes semiconductor QD layers very attractive for solar cells with very thin absorber layer thicknesses.

PbS-QDs are of great interest due to the abundance of lead and sulfur in the earth's crust and due to the high absorption coefficient of PbS at photon energies above 1 eV.

10.2.2 *Solution-based fabrication of quantum dot absorbers*

Colloidal semiconductor QDs are grown in solution at low temperatures between room temperature and 100–300°C within short time periods of between seconds and several minutes. Solutions with stabilized QDs can be stored for long time and further processed by casting, layer-by-layer growth or printing. These technological advantages and the excellent absorption behavior make colloidal semiconductor QDs very interesting for future large-scale applications in solar cells.

Nearly perfect colloidal QDs of metal sulfide or metal selenide are be grown from organometallic precursors dissolved in organic compounds

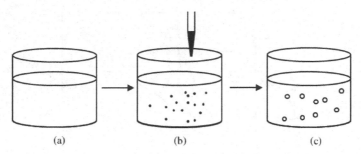

Figure 10.8. Fabrication steps of colloidal QDs consisting of preparation of separate organometallic and sulfur- or selenium-containing solutions (a), fast nucleation of semiconductor nanocrystals during injection of the selenium or sulfur precursor solution into the hot organometallic precursor solution (b) and growth and stabilization and colloidal QDs at reduced temperature (c).

(see, e.g., Brumer *et al.*, 2005). Incidentally, colloidal QDs have to be grown in an atmosphere free from oxygen and water in order to avoid oxidation of precursors.

Preparation of colloidal QDs consists of three main steps (Figure 10.8). In the first step, two separate solutions with organometallic and sulfur or selenium precursors are prepared, for example, by dissolving lead (II) acetate or selenium powder in oleic acid (OA) and/or trioctylphosphine (TOP). In the second step, the solution with the sulfur or selenium precursors is injected into the hot solution with the organometallic precursors and fast nucleation of colloidal QDs takes place. Incidentally, instead of solutions with sulfur or selenium precursors, H_2S or H_2Se can also be used for providing sulfur or selenium. During the third step, seeds grow to colloidal QDs with the desired diameter.

It is important to stabilize and passivate colloidal QDs in the suspension in order to avoid aggregation of colloidal QDs and continuation of chemical reactions at the surface of colloidal QDs. Stabilization and passivation of colloidal QDs is achieved with organic shells around colloidal QDs. The organic shell consists of, for example, TOP molecules (Figure 10.9(a)). Molecules stabilizing QDs are also called surfactant molecules.

Semiconductor QDs are extracted from the colloidal solution by centrifugation. A so-called stock solution contains stabilized colloidal QDs at high concentration. Layers of colloidal QDs can be cast directly from stock solutions. However, long surfactant molecules form barriers

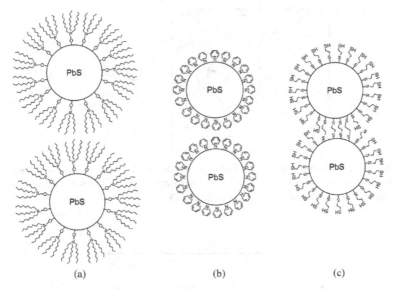

Figure 10.9. Examples for PbS QDs with stabilizing surfactant molecules of TOP in stock solution (a), with short pyridine molecules as intermediate surfactants exchanged in solution (b) and with short ethane–dithiol molecules as linking surfactants exchanged on the deposited QD layer (c).

for the transfer of excitons and/or charge carriers between neighboring semiconductor QDs. Therefore, long surfactant molecules have to be exchanged by short surfactant molecules so that charge transport becomes possible in photovoltaic absorbers based on semiconductor QDs. For example, TOP molecules can be exchanged in solution by much smaller pyridine molecules (Figure 10.9(b)). Semiconductor QD absorber layers are mechanically stabilized by linking neighboring nanocrystals with very short so-called linker molecules, such as ethane-dithiol molecules (Luther *et al.*, 2008b) (Figure 10.9(c)).

A PV absorber based on colloidal QDs is a nanocomposite absorber consisting of semiconductor nanocrystals embedded in an organic matrix of linker molecules (Figure 10.10). The nanocomposite absorber has a band gap given by the quantum confinement of the semiconductor QDs and an absorption length of the order of several 100 nm.

The dielectric constant of the nanocomposite absorber with colloidal QDs depends on the dielectric constants of the inorganic semiconductor and of the organic molecules, on the size of the semiconductor

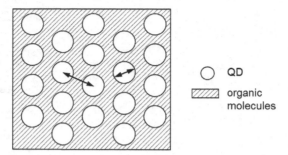

Figure 10.10. Schematic of a photovoltaic nanocomposite absorber containing semiconductor QDs and organic linker molecules. The arrows mark the diameter of a given colloidal QD and the distance between two given colloidal QDs.

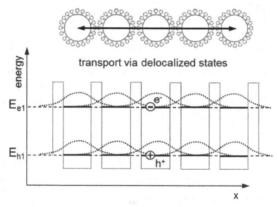

Figure 10.11. Chain of semiconductor quantum dots with ideally overlapping wave functions of free electrons and holes. The arrow marks charge transport in mini bands.

nanocrystals and on the distance between neighboring semiconductor nanocrystals. The calculation of the dielectric function of nanocomposites as an effective medium with colloidal QDs remains challenging.

10.2.3 Charge transport and separation in quantum dot solar cells

The wave functions of electrons, holes and excitons can partially overlap between neighboring QDs. The overlap of wave functions is also called coupling. Figure 10.11 shows an idealized picture of a chain of QDs. Ideally coupled wave functions can form so-called mini bands in which free electrons and holes may be transported at high mobility. However, numerous sources of disorder (fluctuations of distances, diameters and shapes of QDs, local variations of surface coverage with surfactant molecules, etc.) make a band-like transport of free charge carriers in colloidal QD layers unlikely (Guyot-Sionnest, 2012).

The surface of QDs plays a decisive role for the charge transport in photovoltaic absorber layers made from colloidal QDs (see also tasks T10.4 in Section 10.7). The quenching of the photoluminescence of colloidal QDs in solutions is an indication for the density of surface defects. The photoluminescence efficiency of colloidal QDs in stock solutions is of the order of 10% (see, e.g., Talapin *et al.*, 2001), i.e. only one out of ten QDs has no recombination active defects. The photoluminescence efficiency of colloidal QDs has been increased to 70% by growing a very thin layer of a semiconductor with a larger band gap around the semiconductor nanocrystal, for example, CdS around CdSe-QD forming a passivating type I hetero-junction (Talapin *et al.*, 2001). However, this kind of surface passivation is not suitable for charge transport.

The density of surface defects increases with each surface treatment od colloidal QDs. For example, an average density of surface hole traps at CdSe-QDs increased from about 2 traps per QD after washing to 7 or 12 traps per QD after subsequent exchange of surfactants by pyridine or benzenedithiol, respectively, as estimated by analysis of surface photovoltage transients (Zillner *et al.*, 2012). Furthermore, the distribution of surface states is very broad at semiconductor QDs (Fengler *et al.*, 2013) due to the large variability of defect configurations at QD surfaces.

An exciton dissociates rapidly in a QD containing surface defects due to trapping at defect states. Charge transport in QD layers can be considered as a superposition of tunneling between neighboring defect states or neighboring QDs depending on the density and distribution of surface states at colloidal QDs. Figure 10.12 shows a chain of colloidal QDs with surface defects. Hopping between localized states dominates the charge transport. The density of defect states in colloidal QD layers is related to the volume occupied by one QD (about $100 \, nm^3$) and is therefore of the order of 10^{19}–$10^{20} \, cm^{-3}$, where only a small fraction of these defects are usually recombination active.

Electron and hole mobilities in colloidal QD layers can be of the order of 0.9 and $0.2 \, cm^2/(Vs)$, as shown with field-effect transistors

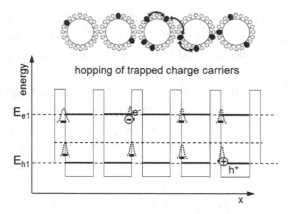

Figure 10.12. Chain of semiconductor QDs with surface states and localized wave functions of trapped electrons and holes. The arrows mark hopping events of a charge carrier between neighboring surface states. Incidentally, excitation of trapped charge carriers into delocalized states is also possible (trap-limited transport). Further, trapped charge carriers can lead to local variations in E_{h1} and E_{e1} (energetic disorder).

with channels consisting of PbSe-QD layers (Talapin and Murray, 2005). The electron and hole mobilities in colloidal QD layers are significantly lower than for undoped a-Si:H in which disorder and localized defect states also play a dominant role (see Chapter 9). The diffusion length of photogenerated charge carriers (L_{diff}) can be calculated by using Equation (9.16). In semiconductor QD layers, L_{diff} is of the order of 10 nm, which is much shorter than the absorption length. Therefore, drift is important for transport of photogenerated charge carriers in solar cells based on colloidal QDs.

The drift length of photogenerated charge carriers in colloidal QD layers (L_{drift}) can be estimated by using Equation (9.17) and by making a reasonable assumption about a built-in electric field. A value of L_{drift} of the order of 100 nm is obtained for an electric field of about 10^4 V/cm. This demonstrates that the realization of very efficient solar cells with QD absorber layers is challenging.

A QD-based solar cell with drift-controlled transport of photogenerated charge carriers was demonstrated Schottky contacts with PbSe-QD layers (Luther *et al.*, 2008a). The external quantum efficiency of Schottky contact solar cells with PbSe-QD layers was increased to 60% by optimizing the layer structure for optical absorption (Law *et al.*, 2008).

The importance of depletion regions for solar cells with PbS-QDs has been demonstrated based on a TiO$_2$/PbS-QD depletion region (Pattantyus-Abraham *et al.*, 2010).

Colloidal QD layers can be *p*-type doped by adding silver atoms into PbS-QDs (Liu *et al.*, 2012). This can help to adjust regions of built-in electric fields within a pin concept (Chapter 9). The concept of depleted hetero-junction colloidal QD solar cells (Pattantyus-Abraham *et al.*, 2010) combined with a charge-selective pin structure seems most promising for reaching high η with semiconductor QD absorbers. Figure 10.13 shows an idealized band diagram of a solar cell with a thin *p*-type doped QD layer and a moderately *n*-type doped QD layer deposited on a wide-gap

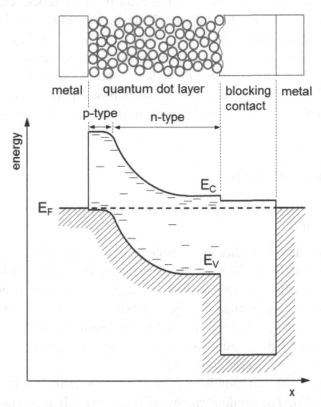

Figure 10.13. Idealized band diagram of a solar cell with a semiconductor QD absorber placed between a very thin highly *p*-type doped layer and a depleted *n*-type doped layer on top of a wide band gap buffer layer. The conduction and valence band edges (E_C and E_V) correspond to the E_{h1} and E_{e1} states in the QDs, respectively.

semiconductor accepting electrons from the n-type doped QD layer. The valence and conduction band edges of the QD absorber layer are obtained by averaging over the distribution of diameters of QDs. The η of solar cells with a PbS-QD absorber has been increased to more than 6% by increasing the width of the depletion region in a junction between a highly doped p-type layer of PbS-QDs and a moderately n-type doped PbS-QD (Liu *et al.*, 2012).

The values of V_{OC} are quite low in comparison to the band gap of colloidal QD absorber layers. For example, values of V_{OC} of 0.51 and 0.24 V were obtained for E_g about 1.3 eV (PbS-QDs, depleted, on TiO$_2$ (Pattantyus-Abraham *et al.*, 2010)) and 1.1 eV (PbSe-QDs, Schottky barrier (Luther *et al.*, 2008a)), respectively. The value of V_{OC} has been increased to 0.45 V for a band gap of E_g of 1.03 eV in the case of ternary PbS$_x$Se$_{1-x}$ QDs (Schottky barrier, Ma *et al.*, 2009). The value of V_{OC} has been further increased to 0.63 V by implementing a ligand exchange based on the treatment in a solution with methylammonium iodide (CH$_3$NH$_3$PbI$_3$) (Lan *et al.*, 2016). This treatment resulted in a higher degree of ligand exchange and in an improved passivation of defects by iodine (Lan *et al.*, 2016).

10.3 Organic Solar Cells

10.3.1 *Organic semiconductors with conjugated π -electron systems*

Different kinds of hybridization of electron in s- and p-orbitals are possible for the formation of chemical bonds between carbon atoms in organic molecules. Carbon bonds can exist in single, double and triple bonds. Double bonds are formed between carbon atoms in the case of sp^2 hybridization. A double bond with sp^2 hybridization is called a conjugated bond. The sp^2 hybridization is the origin for electrical conductivity in organic polymers (Heeger *et al.*, 1988).

The chemical C=C bonds are formed by the σ-orbitals in the xy-plane (Figure 10.14). The binding energy of the σ-orbitals is of the order of 4 eV. The p_z-orbitals along the z-axis do not participate in the chemical bond. The p_z-orbitals are quite large so that p_z-orbitals can overlap between neighboring carbon atoms with a double bond.

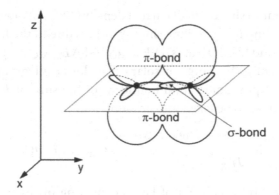

Figure 10.14. Configuration of a conjugated bond.

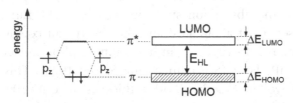

Figure 10.15. Formation of the HOMO and LUMO bands due to overlapping p_z orbitals in conjugated bonds and HOMO–LUMO gap.

The overlap of p_z-orbitals leads to the formation of weaker bonding and of anti-bonding states. These bonds are also called π-bonds. Electrons participating in π-bonds are called π-electrons. Bonding and anti-bonding states of overlapping p_z-orbitals form the so-called HOMO (highest occupied molecular orbital) and LUMO (lowest un-occupied molecular orbital) bands in molecules with a conjugated π-electron system (Figure 10.15). Electrons and missing electrons can be transferred in the LUMO and HOMO bands similarly to the transport of free electrons and holes in the conduction and valence bands of an inorganic semiconductor. Therefore, organic semiconductors can be formed from organic molecules with a conjugated π-electron system.

The difference between the energies at the highest value of the HOMO and the lowest value of the LUMO band is called the HOMO–LUMO gap (Figure 10.15) of an organic semiconductor (E_{HL}). The values of E_{HL} range between 1.5 and 2.5 eV for numerous organic semiconductors, i.e. related organic semiconductors are well suited as photovoltaic absorbers.

The maximum short-circuit current density (I_{SC}^{max}) of organic solar cells is reduced in comparison to inorganic absorbers due to the limited widths of the HOMO and LUMO bands (ΔE_{HOMO} and ΔE_{LUMO}, respectively). For the calculation of I_{SC}^{max} of organic solar cells, a lower integration boundary of E_{HL} and an upper integration boundary of the sum of E_{HL}, ΔE_{HOMO} and ΔE_{LUMO} have to be taken into account.

$$I_{SC}^{max} = q \cdot \int_{E_{HL}}^{E_{HL}+\Delta E_{HOMO}+\Delta E_{LUMO}} \Phi_{sun}(h\nu) \cdot d(h\nu) \qquad (10.5)$$

The absorption coefficient of layers consisting of organic conjugated molecules can be very high in the absorption maximum. As an example, Figure 10.16 shows the configuration of a CuPc (copper phthalocyanine) molecule and the absorption spectrum of a layer of CuPc molecules. Absorption sets in at about 1 eV and the absorption coefficient reaches a value of about $2.4 \cdot 10^5$ cm^{-1} in maximum at photon energies around 2 eV for a CuPc layer (Schechtman and Spicer, 1970). This means that organic photovoltaic absorbers with a thickness of only 100–200 nm are

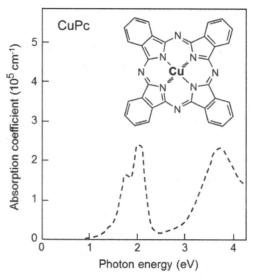

Figure 10.16. Configuration and absorption spectrum of CuPc (copper phthalocyanine) in the range of the air mass (AM) 1.5 spectrum. Data points after (Schechtman and Spicer, 1970).

already sufficient for complete absorption of light with photon energies around the absorption maximum.

There is practically an unlimited variability for the design of conjugated organic molecules. A first classification makes a difference between small conjugated organic molecules (oligomers) and large conjugated organic molecules (conjugated polymers). For example, phthalocyanines such as CuPc or ZnPc and porphyrins belong to the class of oligomers. Oligomers can be packed in dense layers in which p_z-orbitals of neighboring molecules overlap so that charge transport is possible across the complete layer of the organic semiconductor.

The family of poly-(1,4-phenylene-vinylene) (PPV) played an important role for the development of organic polymer solar cells (see, for example, Brabec *et al.*, 2001a). PPV consists of a phenyl ring and two carbon atoms with a double bond. The semiconducting polymer is formed by repeating the basic unit (Figure 10.17(a)).

Organic molecules can be modified by replacing, for example, a hydrogen atom by a side group. In this way, the structure or diameter of molecules can be changed and properties needed for technological processing such as solubility in certain organic solvents can be modified. MEH-PPV (see, for example, Yu *et al.*, 1995) is given as an example in Figure 10.17(a). Other families of conjugated polymers used for organic solar cells belong, for example, to poly-thiophenes (see, for example, Günes *et al.*, 2007). Figure 10.17(a) shows a rather complex poly-thiophene (PBDB-T), which has been applied in polymer solar cells with high efficiency (Zhao *et al.*, 2016a).

The C_{60} molecule (Kroto *et al.*, 1985) played a decisive role for the development of organic solar cells (Figure 10.17(b)). A C_{60} molecule contains 60 carbon atoms arranged in a football-like sphere and belongs to the group of fullerenes. C_{60} molecules are stable and strong acceptor molecules (Haddon *et al.*, 1986), and can accept several electrons. C_{60} molecules are so stable that they can be deposited by thermal evaporation. The phenyl-C_{61}-butyric acid methyl ester (PCBM) molecule (Hummelen *et al.*, 1995) is a soluble derivative of a C_{60} molecule which can be deposited by casting or spin coating from solutions.

Alternative, fullerene-free, acceptor molecules can also be applied in organic solar cells. For example, the rather complex ITIC molecule

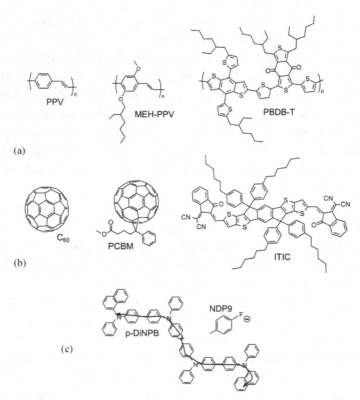

Figure 10.17. Examples of hole-conducting conjugated polymers of the poly-vinylene-phenylene (PPV and MEH-PPV) and of the poly-thiophene (PBDB-T) families (a), of C_{60}, PCBM and fullerene-free (ITIC) acceptor molecules (b) and of a p-type doped organic semiconductor based on p-DiNPB molecules and NDP9 ionic dopants (c).

(Figure 10.17(b), Zhao *et al.*, 2016a) contains a conjugated core with dithiophenes and phenyl rings, cyano groups, which provide the quality of electron acceptors, and hexylphenyl side groups linked to the core.

Organic semiconductors can be n-type or p-type doped. For example, p-type doping of organic semiconductors can be achieved by adding some small electron acceptor molecules. The small electron acceptor molecules capture electrons from conjugated organic molecules. The charge at the small immobile anions is compensated by mobile positive charge or holes in the π-electron system of the conjugated organic molecule. As an example, a p-DiNPB molecule p-type doped with a NDP9 molecule (Schüppel *et al.*, 2010) is shown in Figure 10.17(c).

Organic semiconductors can be n-type doped by adding electron donors, for example, metal ions or electron-donor molecules. Organic semiconductors can be also highly doped. The most prominent highly p-type doped organic semiconductor is PEDOT:PSS in which poly(3,4-ethylenedioxythiophene) is doped with poly-styrenesulfonate (Groenendaal *et al.*, 2000).

The mobility of charge carriers and therefore the diffusion coefficient can vary over many orders of magnitude in organic semiconductors depending on ordering and architecture of molecules (Coropceanu *et al.*, 2007). For example, disordered chains in layers of conjugated polymers result in a hole mobility between 10^{-7} and 10^{-4} cm^2/(Vs) (see, for example, Lebedev *et al.*, 1997). Electron and hole mobilities of 0.4 and 0.8 cm^2/(Vs), respectively, were shown in pure organic crystals such as anthracene (Kepler and Hoestery, 1974). A high intrachain hole mobility of 600 cm^2/(Vs) was found for local rigid conjugated ladder elements (Prins *et al.*, 2006). For comparison, the electron mobility in a graphene layer can be as high as $2 \cdot 10^5$ cm^2/Vs) (Bolotin *et al.*, 2008). These examples give an impression about the importance of molecular design for optimization of charge transport in organic solar cells.

Organic molecules have the advantage that all chemical bonds are saturated. Therefore, organic semiconductors do not have electronic states caused by unsaturated dangling bonds or surface states in the HOMO-LUMO gap. This opens broad opportunities for the design of interfaces with organic semiconductors.

10.3.2 *Excitons in organic semiconductors*

After photogeneration of an electron hole pair in an organic semiconductor, an exciton (see, for example, Scholes and Rumbles, 2006) is formed within an extremely short period of time. The energy of the electron and of the hole in the exciton is significantly lower than E_{HL} due to the low dielectric constant of organic semiconductors (Figure 10.18).

The exciton binding energy in organic semiconductors can be estimated using Equation (10.2) and taking into account an ε_r of 2–4 in organic semiconductors. Effective masses of electrons and holes of the order of m_e can be assumed in organic semiconductors. The resulting exciton binding energy is of the order of 0.4–0.8 eV in organic semiconductors.

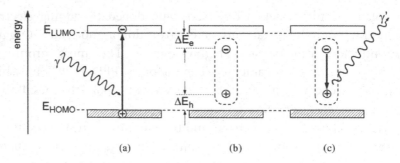

Figure 10.18. Schematic of light absorption in organic semiconductors (a), an exciton (b) and radiative recombination of an exciton (c).

The exciton binding energy corresponds to a loss in the maximum chemical potential of photogenerated electrons and holes and cannot be regained after dissociation of the exciton. Therefore, the high binding energy of excitons is an important fundamental loss mechanism limiting V_{OC} in organic solar cells.

The Bohr radius of the electron–hole orbit at the lowest energy of the exciton can be estimated using Equation (10.3). As a result, excitons are extended only over several atoms in inorganic semiconductors. This strong localization causes low diffusion constants and very short diffusion lengths of excitons in organic semiconductors.

The lifetime of excitons is of the order of 1 ns in organic semiconductors (see, for example, Markov *et al.*, 2005). The diffusion length of excitons is of the order of 5 nm in PPV derivatives (Markov *et al.*, 2005) or 14 nm in ladder-type poly(*p*-phenylene) (Haugeneder *et al.*, 1999), i.e. the exciton diffusion length is much shorter than the absorption length in organic PV absorbers. Therefore, small aggregates of organic semiconductors and local dissociation of excitons at the boundaries are required for an organic nanocomposite absorber.

10.3.3 *Donor–acceptor hetero-junction and organic nanocomposite absorber*

Excitons are neutral quasi particles. They have to dissociate within the photovoltaic absorber before photogenerated electrons and holes can be separated at charge-selective contacts. Organic molecules have the ability to attract or accept electrons from other molecules (electron-acceptor molecule) or to give away or donate an electron to another molecule (electron-donor molecule). The ability of molecules to accept or to donate

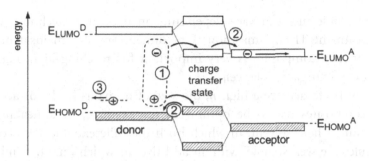

Figure 10.19. Principle of charge separation at a donor–acceptor hetero-junction including the steps of dissociation of an exciton via a charge transfer state (1), transfer of the electron and hole into the LUMO and HOMO bands of the acceptor and donor molecules, respectively (2) and polaronic transport in the donor molecules (3).

electrons is used to dissociate excitons at molecular donor–acceptor hetero-junctions (Yu *et al.*, 1995).

Excitons diffuse to a donor–acceptor hetero-junction. If the attractive force of an acceptor molecule to an electron that is part of an exciton is strong enough, then the exciton forms an intermediate state from which the electron can be transferred into the acceptor molecule (Sariciftci *et al.*, 1992) (dissociation of the exciton via a charge transfer state, see Figure 10.19). The lifetime of the charge separated configuration is of the order of milliseconds at a temperature of 80 K, i.e. several orders of magnitude longer than the exciton lifetime (Smilowitz *et al.*, 1993).

The first donor–acceptor junction has been demonstrated for potential applications in solar cells with a double layer of CuPc and a perylene tetracarboxylic derivative (Tang, 1986). The ultrafast photoinduced electron transfer from a conducting polymer into C_{60} (Sariciftci *et al.*, 1992; Brabec *et al.*, 2001b) and its application for solar cells has been shown by Sariciftci *et al.* (1993). C_{60} molecules or derivatives of C_{60} molecules and other fullerenes such as C_{70} are applied as acceptor molecules in most kinds of efficient organic solar cells.

The diffusion length of excitons is much shorter than the absorption length in organic semiconductors. Therefore, the donor–acceptor hetero-junction is dispersed or folded in organic photovoltaic absorbers (blended). Domains of donor and acceptor molecules interpenetrate each other in organic photovoltaic absorbers on a scale of the order of 1–10 nm, i.e. organic photovoltaic absorbers are nanocomposites, which are also called organic blends. The local dimensions of domains of donor or acceptor molecules can vary depending on deposition technology and post-treatment. The optimization of the dimensions of domains in donor–acceptor nanocomposites is very important for reaching high quantum efficiencies in organic solar cells.

In order to achieve a high η, the size of domains in donor-acceptor nano-composites has to be optimized, for example, by annealing and combining a host solvent, in which both the fullerene and the acceptor molecules are well soluble, with an additive, in which only the fullerene molecules are soluble but not the acceptor molecules (Zhao *et al.*, 2016).

From the point of view of charge transport, organic photovoltaic absorbers can be treated as nanocomposite semiconductors, the valence and conduction bands of which are formed by the HOMO band of donor molecules and by the LUMO band of acceptor molecules, respectively. However, in a donor–acceptor nanocomposite, a certain density of each of the components is required for getting separate transport paths for electrons and holes connected between two external contacts (Figure 10.20). The connectivity of conducting units in a system is called percolation. The percolation threshold is about 16% for polymer/PCBM nanocomposites (Hotta *et al.*, 1987).

The diffusion length of photogenerated and locally separated electrons and holes in donor–acceptor nanocomposites can be obtained by using Equation (9.16). The electron and hole mobilities sensitively depend on the composition of donor–acceptor nanocomposites and amount, for example, to the order of 10^{-5}–10^{-4} cm^2/(Vs) for blended polyfluorene/PCBM layers (Pacios *et al.*, 2003). The lifetimes of charge-separated states are

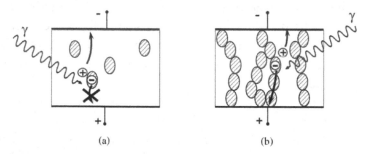

(a)   (b)

Figure 10.20. Schematic donor–acceptor nanocomposite with light absorption, exciton dissociation and local charge separation without percolation (a) and with percolation (b) of conduction paths for photogenerated charge carriers.

of the order of ms at low temperature (Smilowitz *et al.*, 1993). However, photocurrents can decrease within only 1–10 ns, for example, in MEH-PPV/C$_{60}$ nanocomposites (Lee *et al.*, 1993). Therefore, the resulting diffusion length of photogenerated electrons and holes can be of the order of 100 nm or even much less depending on disorder and/or the density of defects in donor–acceptor nanocomposites.

Donor–acceptor nanocomposites have two band gaps. The transport gap ($E_{HL}^{transport}$) is the difference between the lowest energy of the LUMO band of the acceptor molecules ($E_{LUMO}^{acceptor}$) and the highest energy of the HOMO band of the donor molecules (E_{HOMO}^{donor}).

$$E_{HL}^{transport} = E_{LUMO}^{acceptor} - E_{HOMO}^{donor} \tag{10.6}$$

The absorption gap of a donor–acceptor nanocomposite ($E_{HL}^{absorption}$) is related to the lowest of the HOMO–LUMO gaps of the acceptor or donor molecules, respectively.

$$E_{HL}^{absorption} = E_{LUMO}^{donor} - E_{HOMO}^{donor} \tag{10.7'}$$

or

$$E_{HL}^{absorption} = E_{LUMO}^{acceptor} - E_{HOMO}^{acceptor} \tag{10.7''}$$

The absorption and transport band gaps of organic photovoltaic nanocomposite absorbers limit I_{SC} and V_{OC}, respectively, of organic solar cells.

10.3.4 *Charge-selective and ohmic contacts in organic solar cells*

The potential energy of electrons decreases and the potential energy of holes increases at the charge-selective electron contact and *vice versa* at the charge-selective hole contact. The change of the potential energy at charge-selective contacts can be realized, for example, with a *pn*-junction where minority photogenerated charge carriers become majority charge carriers.

The ionization energy of the HOMO band of donor molecules and the electron affinity of the LUMO band of acceptor molecules are important as well for charge-selective and ohmic contacts in organic solar cells from the point of view of losses in potential energy. As an example, Figure 10.21

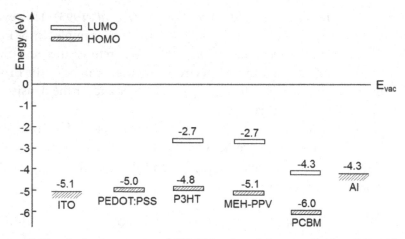

Figure 10.21. Ionization energies of HOMO bands and electron affinities of LUMO bands for P3HT, MEH:PPV, PEDOT:PSS and PCBM and work functions of Al and ITO.

gives the energies of HOMO and LUMO bands of P3HT and MEH-PPV as donor molecules and of PCBM as an acceptor molecule. The work functions of indium tin oxide (ITO) and Al contacts are given as well. PEDOT:PSS is used as a highly doped organic semiconductor at the ohmic contact with ITO.

The ionization energies of the HOMO bands of P3HT and MEH-PPV are well aligned with the ionization energy of highly doped PEDOT:PSS so that holes can be transferred from P3HT or MEH-PPV into PEDOT:PSS without additional losses in potential energy. Therefore, p-type doped PEDOT:PSS is a hole-collecting contact for P3HT. The ionization energy of the highly doped PEDOT:PSS is well aligned with the work function of ITO. Therefore ITO and PEDOT:PSS form an ohmic contact. The electron affinity of PCBM is well aligned with the work function of Al so that electrons can be transferred from PCBM into Al without additional losses in potential energy.

The ionization energies of HOMO bands can be engineered over a relatively wide range by varying side groups at a conjugated polymer. Figure 10.22 shows examples for polythiophenes with different side groups and corresponding ionization energies of the HOMO bands changing between -4.83 and -5.45 eV for the given examples (Scharber *et al.*, 2006). The values of V_{OC} increased with increasing ionization energy of the

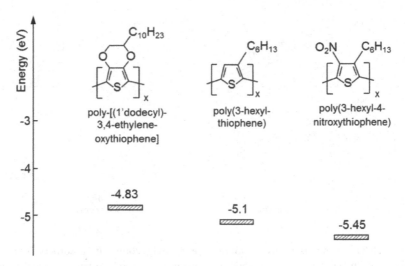

Figure 10.22. Ionization energy of HOMO bands for polythiophenes with different side groups (values taken from Scharber *et al.*, 2006).

HOMO band (Scharber *et al.*, 2006), i.e. potential energy losses at contacts can be minimized by molecular engineering.

Donor and acceptor molecules electrically compensate each other in organic photovoltaic nanocomposite absorbers and as such a donor–acceptor nanocomposite absorber can be treated like an undoped or intrinsic semiconductor. Furthermore, metal/organic semiconductor interfaces can often be treated as contacts close to the Schottky limit (Equation (5.5)) since all chemical bonds are saturated in organic molecules. It has been shown, for example, that the barriers for hole and electron injection are 0.2 and 0.1 eV, respectively, for the ITO/MEH-PPV/Ca system (Parker, 1994), which corresponds closely to the Schottky limit. However, one has to keep in mind that there are also organic semiconductors forming dipole barriers at metal contacts (Hill *et al.*, 1998).

In organic solar cells, barriers at metal/organic semiconductor interfaces usually correlate well with the metal work function (Φ_m), which is applied in metal/insulator/metal contact systems. The property that many organic semiconductors do not form dipole layers at metal contacts is used for realizing charge-selective contacts in organic polymer solar cells, which are processed from organic solutions. The potential energy is lower for electrons at the metal contact with a low Φ_m (Figure 10.23) so that

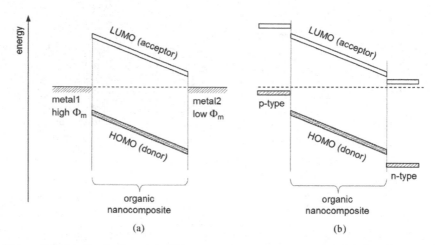

Figure 10.23. Band diagrams of donor–acceptor nanocomposite absorbers contacted between two metals with different work function (Φ_m) (a) and between p-type and n-type doped organic semiconductors (b).

photogenerated electrons can be collected. The potential energy is lower for holes at the metal contact with the high Φ_m so that photogenerated holes can be collected.

Work functions or potential energies of free charge carriers can also be engineered by doping of organic semiconductors. Offsets or alignment of HOMO and LUMO bands at organic semiconductor/organic semiconductor hetero-junctions follow from the differences of the ionization energies and electron affinities, respectively.

Figure 10.23(b) shows a schematic band diagram of a pin organic solar cell. In this type of solar cells, a donor–acceptor nanocomposite absorber is sandwiched between a p-type doped, for example, p-DiNPB doped with NDP9 or ZnO, and an n-type doped, for example, n-type doped C_{60} or MoO_3, organic or inorganic semiconductors, respectively. The pin concept is applied for organic solar cells based on small organic molecules, which can be deposited by thermal evaporation (see, for example, Schüppel *et al.*, 2010) and based on polymers (see, for example, Zhao *et al.*, 2016a).

Additional layers of organic molecules can reduce recombination at ohmic contacts by reducing diffusion of excitons to ohmic contacts. For example, a layer of bathophenanthroline (Bphen) molecules blocks diffusion of excitons exciton blocker from a C_{60} layer to an Al contact

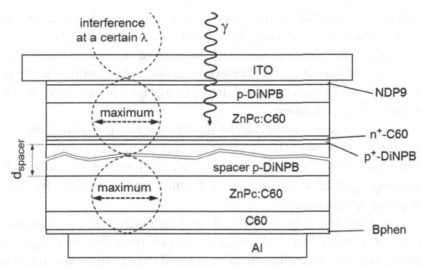

Figure 10.24. Schematic cross-section of an organic pin–pin solar cell with a *p*-DiNPB spacer for optimization of absorption. The layer thicknesses are adjusted so that the maximum light intensity is reached in the donor–acceptor nanocomposite absorber layers.

(Figure 10.24) but enables transfer of electrons from C_{60} to the ohmic contact (see, for example, Schueppel *et al.*, 2010).

The V_{OC} of organic solar cells operated under illumination of AM1.5 is given approximately by the difference of $E_{LUMO}^{acceptor}$ and E_{HOMO}^{donor} and an energy of the order of about 0.3 eV (Scharber *et al.*, 2006).

$$q \cdot V_{OC} \approx E_{LUMO}^{acceptor} - E_{HOMO}^{donor} - 0.3\,\text{eV} \qquad (10.8)$$

The energy of 0.3 eV in Equation (10.8) is related to the difference between the quasi-Fermi levels of electrons and holes and the edges of the LUMO and HOMO bands of acceptor and donor molecules, respectively, in the donor–acceptor nanocomposite absorber. The value of 0.3 eV is quite similar, for example, to the difference of the band gaps of c-Si (see Chapter 7) or GaAs (see Chapter 8) and V_{OC} in record c-Si or GaAs solar cells, respectively, under illumination at AM1.5.

10.3.5 *Optimization routes of organic solar cells*

Organic semiconductors offer some unique opportunities for optimization of solar cells. The absorption gap of organic semiconductors and the energies of the HOMO bands can be tuned over a wide range. In this way, the discrepancy between the absorption and transport gaps can be minimized so that, in turn, V_{OC} can be maximized (see, for example, Wilke *et al.*, 2012) at maximum I_{SC}. However, it is not really clear how far losses due to exciton binding energy and due to energy needed for dissociation of excitons can be minimized.

Organic semiconductors can be highly doped so that ohmic contacts can be formed between layers of organic semiconductors with different absorption gaps. Therefore, stacked multijunction organic solar cells absorbing light in a broader spectral range can be realized by using integrated series connection.

In the spectral range near the absorption maximum, the absorption length of organic semiconductors is shorter than the wavelength of exciting light divided by the refractive index of the organic semiconductor. Furthermore, layers of organic semiconductors with absorption maxima in distinct optical ranges can be combined. The combination of different organic semiconductors with a very short absorption length at certain wavelengths and with negligible absorption at other wavelengths gives the opportunity for sophisticated photon management, which leads to a maximum of light intensity in regions of maximum absorption in layer systems of organic semiconductors. Losses due to very short diffusion length of excitons and transmission losses at wavelengths not very close to the absorption maximum can be reduced in this way. Figure 10.24 shows a schematic cross section of an optimized organic solar cell including a transparent spacer for optimum photon management between two stacked organic solar cells in pin-spacer-pin configuration (Drechsel *et al.*, 2005). The solar energy conversion efficiency can exceed 12% for such an organic solar cell at an overall thickness of the $ZnPc:C_{60}$ nanocomposite absorber less than 100 nm.

The mobility of electrons and holes usually increases with increasing temperature in organic nanocomposite absorbers. This effect can even lead to an increase of η with increasing temperature, in contrast to conventional solar cells based on inorganic semiconductors for which V_{OC} and η decrease with increasing temperature. Therefore, η of organic solar cells can be optimized for an expected average temperature range at which an organic solar cell is operated.

10.4 Dye-Sensitized Solar Cells

10.4.1 *Light absorption in dye molecules*

The absorption of a photon by a dye molecule leads to the excitation of an electron from the ground state into an excited state of the dye molecule (Figure 10.25). The absorption spectra of dye molecules are usually rather narrow. There is a huge variability of natural and synthetic dye molecules absorbing light in different spectral regions, which are very interesting for PV solar energy conversion.

Dye molecules strongly absorb light in their absorption maximum. The absorption of dye molecules is described by the molar extinction coefficient (ε_{mol}). The molar extinction coefficient is defined by the decadic decrease of the intensity of light transmitted through a dye solution, in contrast to the absorption coefficient, which is defined by the exponential decrease of the intensity of transmitted light through a layer. The number of photons passing a cuvette with transparent windows and a thickness d_K of the solution is given by the following equation (reflection losses are considered in N_{ph0}):

$$N_{ph}(\lambda) = N_{ph0}(\lambda) \cdot 10^{-\varepsilon_{mol} \cdot d_K \cdot C_{dye}} \tag{10.9}$$

The concentration of dye molecules in the solution (C_{dye}) is given in the unit mol/L or M. Therefore the molar extinction coefficient is given in the unit of $M^{-1} cm^{-1}$. For example, 90% of photons are absorbed by the dye molecules if the product $\varepsilon_{mol} \cdot d_K \cdot C_{dye}$ is equal to one.

Ruthenium-based dye molecules played a very important role in the development of dye-sensitized solar cells. As an example, Figure 10.26

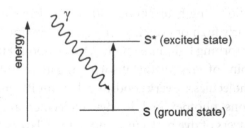

Figure 10.25. Photon absorption by a dye molecule.

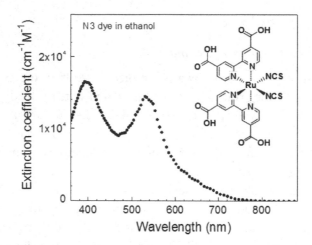

Figure 10.26. Configuration and extinction spectrum of N3 dye molecules in ethanol.

shows the configuration and the spectrum of ε_{mol} of the so-called N3 (N stands for Nazeeruddin who developed the given dye molecule and a whole family of related dyes (Nazeeruddin *et al.*, 1993)) molecules dissolved in ethanol. The N3 dye molecule contains one Ru atom which is surrounded by two bi-pyridine, and two thiocyanate moieties. Two carboxylic groups are connected with each bi-pyridine moiety. The geometric size of such a molecule is of the order of 1 nm. The N3 dye molecule has a relatively broad extinction spectrum and a large ε_{mol} of about 14,000 $M^{-1}cm^{-1}$ in the maximum at 532 nm. Absorption of N3 dye molecules sets in at wavelengths around 750 nm. A second absorption maximum appears for N3 dye molecules at about 400 nm.

A density of about $4 \cdot 10^{16}$ N3/cm³ is necessary for the absorption of 90% of incident photons with energies close to the absorption maximum. The density of $4 \cdot 10^{16}$ N3/cm³ would correspond to about 500 monolayers or to an absorption length of about 250 nm of densely packed N3 dye molecules. Similarly high or even higher ε_{mol} are obtained for many other dye molecules absorbing light in a range between about 800 and 400 nm.

From the point of view of the absorption gap and of the absorption coefficient, dye molecules are very good absorbers for PV energy conversion. However, electrons photoexcited in dye molecules are not mobile in densely packed layers of dye molecules and so thick layers of dye molecules cannot be applied for solar cells.

10.4.2 *Dye sensitization of a sintered network of TiO₂ nanoparticles*

For application in solar cells, electrons photoexcited in dye molecules have to be transferred from excited dye molecules into a material in which electrons are mobile. Metal oxides with large E_g are suitable materials for the transport of electrons. Dye molecules can be adsorbed on the surface of metal oxides with large E_g.

The transfer of mobile charge carriers into an insulator is called injection. Ultrafast electron injection from excited dye molecules into semiconductor electrodes has been shown (Eichberger and Willig, 1990).

Electrons photoexcited in a dye molecule can be injected into the conduction band of a metal oxide if the energy of the excited electron is equal or larger than the conduction band edge of the metal oxide. The transfer of electrons photoexcited in a dye molecule at a photon energy lower than E_g of the metal oxide is called sensitization of the metal oxide. Sensitization means that a metal oxide with a large E_g becomes sensitive in terms of a photocurrent induced by photons with energy less than E_g of the metal oxide.

The anatase phase of TiO₂ is usually applied for electron transport in dye-sensitized solar cells (O'Regan and Grätzel, 1991). Figure 10.27 depicts the unit cell of anatase and the absorption spectrum of an anatase layer

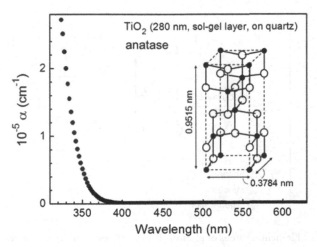

Figure 10.27. Unit cell of TiO₂ (anatase) and absorption spectrum of a sol–gel TiO₂ layer.

deposited by a sol–gel process. Titanium is sixfold coordinated in TiO_2. The unit cell of anatase is quite large due to the distorted positions of four oxygen atoms from a plane in the central TiO_6-octahedron of the unit cell (for details see Diebold, 2003 and references therein). The band gap of anatase varies between 3.2 and 3.4 eV depending on preparation. Optical absorption sets in for anatase layers or nanoparticles at wavelengths shorter than 370–380 nm.

Dye molecules can be adsorbed at the surface of anatase by linking with carboxylic groups resulting in a stable $\cdots$–O–Ti–O–C–$\cdots$ surface bond configuration. Therefore, carboxylic groups are usually attached to dye molecules used for dye-sensitized solar cells (Nazeeruddin *et al.*, 1993).

The injection of an excited electron from a dye molecule into the conduction band of TiO_2 (see also Figure 10.28) can be described by the following expression:

$$S^* \rightarrow S^+ + e^-_{TiO_2} \qquad (10.10)$$

S^* and S^+ denote the excited and charged states of a dye molecule. The mobility of electrons is about $1\,cm^2/(Vs)$ in TiO_2 crystals (Tang *et al.*, 1994), i.e. injected electrons can be transported well in TiO_2.

The dielectric constant of TiO_2 is larger than the dielectric constant of the surrounding ambience. Therefore, the potential energy of an electron injected into a TiO_2 nanoparticle increases towards the surface of the

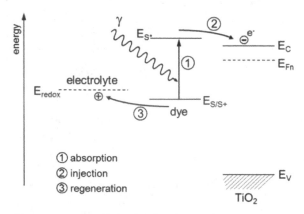

Figure 10.28. Elementary steps of photon absorption, electron injection from the excited dye molecule into the conduction band of TiO_2 and of regeneration of the dye molecule.

TiO_2 nanoparticle. This so-called dielectric screening (Keldysh, 1979) is important for low surface recombination in TiO_2 nanoparticles.

The diameter of a dye-sensitized TiO_2 nanoparticle should not be too small in order to avoid tunneling recombination of an electron injected into the TiO_2 nanoparticle with a charged dye molecule at the surface. The optimum diameter of TiO_2 nanoparticles is of the order of 25 nm for dye-sensitized solar cells.

The internal surface area of dye-sensitized solar cells has to be folded by 500–1000 times for optimum light absorption by adsorbed dyes. Large internal surface areas can be reached with porous or nanoporous structures. TiO_2 nanoparticles can be well prepared in large quantity (production of white pigments) and sintered to networks of interconnected nanoparticles at moderate temperatures of about 450°C.

A nanoporous electrode of interconnected TiO_2 nanoparticles can be produced, for example, by screen-printing of a paste containing TiO_2 nanoparticles on a transparent contact such as SnO_2:F followed by firing in air (O'Regan and Grätzel, 1991). TiO_2 nanoparticles are mixed with an organic binder to form a printable paste. Organic molecules of the binder are burned to CO_2 and sintering necks are formed between neighboring TiO_2 nanoparticles during firing in air for about 30 min. The internal surface area of a nanoporous TiO_2 electrode (np-TiO_2) is about 1000 times larger than the geometric area for a thickness of the np-TiO_2 electrode of about 30 μm.

Dye molecules are deposited on np-TiO_2 electrodes from a dye solution within 6–12 h. The deposition of dye molecules into np-TiO_2 electrodes is the most time-consuming step for the preparation of dye-sensitized solar cells. The deposition time of dye molecules into np-TiO_2 electrodes is limited by the low diffusion coefficient of dye molecules in solution. Incidentally, only dye molecules that do not form clusters or aggregates during their deposition into np-TiO_2 electrodes can be used for dye-sensitized solar cells.

10.4.3 Local charge separation and charge transport in dye-sensitized solar cells

Local charge separation takes place in dye-sensitized solar cells due to the injection of excited electrons from dye molecules into TiO_2. The remaining charged dye molecules have to be neutralized or regenerated for subsequent absorption events. An electric contact covering the whole internal surface area of the dyed np-TiO$_2$ electrode is required for regeneration of charged dye molecules. Related electric contacts can be realized with electrolytes completely filling the pores between interconnected TiO$_2$ nanoparticles (O'Regan and Grätzel, 1991). The positive charge is transferred from the charged dye molecule into the electrolyte during the elementary step of regeneration (Figure 10.28).

An electrolyte is characterized by the redox potential (E_{redox}) of the charge-transporting ionic species. The redox potential of an electrolyte is equivalent to the Fermi-energy of a semiconductor. The density of ions in electrolytes is huge in comparison to the density of photogenerated charge carriers in PV absorbers. Therefore, E_{redox} is fixed in dye-sensitized solar cells. The difference between the Fermi-energy of injected electrons in the interconnected TiO$_2$ nanoparticles (E_{Fn}) and E_{redox} corresponds to the Fermi-level splitting in illuminated bulk semiconductors.

The np-TiO$_2$ electrode with adsorbed dye molecules surrounded by a redox electrolyte can be considered as a complex photovoltaic nanocomposite absorber (Figure 10.29). The maximum energy of the dye in the ground state and the minimum energy of the dye in the excited state correspond to the absorption gap of the nanocomposite absorber. The E_{redox} and E_{Fn} correspond to the Fermi-energies of the majority and minority charge carriers, respectively, in the nanocomposite absorber.

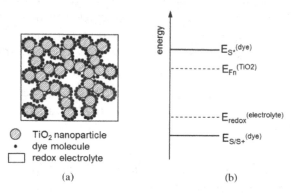

Figure 10.29. Nanocomposite absorber of a dye-sensitized solar cell consisting of interconnected TiO$_2$ nanoparticles, dye molecules and electrolyte (a) and energy scheme of the nanocomposite absorber in a dye-sensitized solar cell (b).

The regeneration of charged dye molecules requires a redox reaction during which an electron is transferred from an ionic species in the electrolyte to the charged dye molecule. The extraction of an electron from an ionic species in the electrolyte increases the oxidation state of the ionic species in the electrolyte, i.e. the ionic species is further oxidized (Figure 10.30). The transfer of an electron from an ionic species in the electrolyte to the charged dye molecule is equivalent to the transfer of a hole from the charged molecule into the electrolyte.

The iodide (I^-/I_3^-) redox electrolyte played a decisive role in the development of dye-sensitized solar cells (O'Regan and Grätzel, 1991). The first great advantage of the I^-/I_3^- electrolyte is that the negatively charged ionic species are repulsed from the negatively charged TiO$_2$ nanoparticles. In this way, the recombination of electrons from TiO$_2$ nanoparticles into the electrolyte is strongly suppressed. The second great advantage of the I^-/I_3^- electrolyte is that a catalytic activation is required for electron

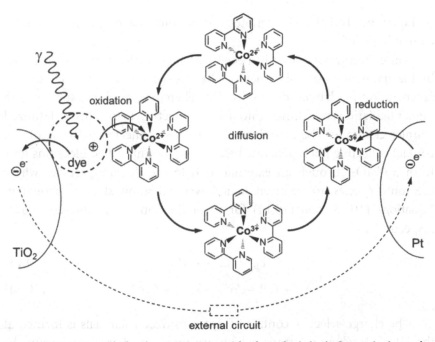

Figure 10.30. Example for charge transport with a hypothetic ionic Co species in a redox electrolyte of a dye-sensitized solar cell.

transfer at conducting electrodes. This is the prerequisite for realizing a very simple charge-selective contact with the np-TiO$_2$/dye/electrolyte nanocomposite absorber and a SnO$_2$:F electrode into which only electrons can be transferred. The regeneration of charged dye molecules can be described by the following expression:

$$2S^+ + 3I^- \rightarrow 2S + I_3^- \tag{10.11}$$

Equation (10.11) shows that two charged dye molecules and three I^- ions are involved in one elementary regeneration step. This complex redox reaction limits the regeneration rate.

An alternative redox reaction is based on ionic molecular complexes with a Co^{2+}/Co^{3+} redox couple (Nusbaumer *et al.*, 2001, 2003). Figure 10.30 illustrates the charge transport in an electrolyte via hypothetic OR$-$Co^{2+}/OR$-$Co^{3+} species.

$$S^+ + OR-Co^{2+} \rightarrow S + OR-Co^{3+} \tag{10.12}$$

In Equation (10.12), OR$-$ denotes the organic rest of the ionic species containing Co^{2+} or Co^{3+}.

The concentration of ionic species with higher oxidation state increases in the nanocomposite absorber of the dye-sensitized solar cell due to regeneration of charged dye molecules. Therefore, the ionic species with higher oxidation state diffuse into the bulk electrolyte towards a platinized counter electrode. The platinized electrode is connected with the ohmic contact collecting photogenerated electrons from np-TiO$_2$. Electrons flow from np-TiO$_2$ through an external load to the Pt nanoparticles where the ionic species are reduced to a lower oxidation state as shown in Equations (10.13) and (10.14) for the reduction of I_3^- and OR$-$Co^{3+}, respectively.

$$e_{Pt}^- + I_3^- \rightarrow 3I^- \tag{10.13}$$

$$e_{Pt}^- + OR-Co^{3+} \rightarrow OR-Co^{2+} \tag{10.14}$$

The charge-selective contact of dye-sensitized solar cells is formed at the interface where electrons, which are minority charge carriers in the nanocomposite absorber, become majority charge carriers. At the same

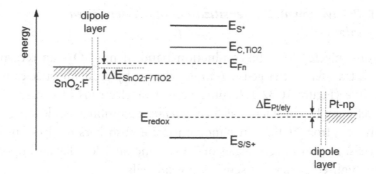

Figure 10.31. Energy diagram of a dye-sensitized solar cell under open-circuit voltage.

time, charge transfer from the redox electrolyte is blocked at the charge-selective contact. The charge-selective contact is at the np-TiO$_2$/SnO$_2$:F interface in dye-sensitized solar cells (Figure 10.31).

Permanent polarization takes place at heterocontacts and metal–electrolyte contacts due to the formation of interface dipoles. Interface dipoles introduce additional drops of the electrostatic potential reducing the potential at external leads.

The difference between E_{Fn} and E_{redox} in the nanocomposite absorber of a dye-sensitized solar cell is reduced at the contacts by the voltage drop across the dipole at the SnO$_2$:F/np-TiO$_2$ interface ($\Delta E_{SnO_2:F/TiO_2}$) and by the dipole at the electrolyte/Pt interface ($\Delta E_{Pt/ely}$). The value of V_{OC} of a dye-sensitized solar cell becomes

$$q \cdot U_{ph} = E_{Fn} - E_{redox} - \Delta E_{SnO_2:F/TiO_2} - \Delta E_{Pt/ely} \qquad (10.15)$$

Dipoles at SnO$_2$:F/np-TiO$_2$ and at electrolyte/Pt interfaces should be minimized for getting a maximum photovoltage at the external leads of a dye-sensitized solar cell. Furthermore, the energy of the excited state of the dye should be as close as possible to the conduction band edge of np-TiO$_2$ (minimization of E_{S^*}-E_{C,TiO_2}) and the redox potential of the electrolyte should be as close as possible to the ground state of the dye molecule (minimization of E_{redox}-$E_{S/S+}$) for maximizing η of dye-sensitized solar cells. This can be achieved by designing dye molecules and by taking the optimal redox couple in the electrolyte. Incidentally, the liquid electrolyte can be replaced by organic hole-transporting materials such as spiro-OMeTAD (Bach *et al.*, 1998).

10.4.4 *Passivation and co-sensitization of dye-sensitized solar cells*

Electrons injected into the conduction band of np-TiO$_2$ can recombine by back transfer to charged dye molecules or by reducing species in the electrolyte (Figure 10.32). Recombination of electrons in dye-sensitized solar cells can be considered as local surface recombination leading to an effective electron lifetime in nanocomposite absorbers of dye-sensitized solar cells. Passivation strategies of TiO$_2$ nanoparticles play an important role for high-efficiency dye-sensitized solar cells.

The recombination rates to charged dye molecules (R_{S+}) and to species in the electrolyte (R_{ely}) are proportional to the density of free electrons in np-TiO$_2$ (n_{TiO_2}). The density of available species in the electrolyte is much higher than n_{TiO_2} while the density of charged dye molecules is of the order of n_{TiO_2} and depends on the regeneration rate (R_{reg}). The rate constants are not well known for np-TiO$_2$/dye/electrolyte systems (Peter, 2009). The point is that the rate constants depend in a complex way on species in the electrolyte and their concentrations, on geometry, moieties and side groups of dye molecules as well as on bond configurations of dye molecules at TiO$_2$.

Elementary recombination events of electrons and electron diffusion are retarded by fast trapping of electrons within ps into exponentially distributed trap states near the conduction band edge (E_C) of np-TiO$_2$. Exponentially distributed trap states are caused by disorder in the relatively large unit cell of anatase in TiO$_2$ nanoparticles. The energy and the density of trap states characterizing the exponential distribution are of the order of 100 meV and 10^{19}–10^{20} cm^{-3}, respectively (Dittrich, 2000).

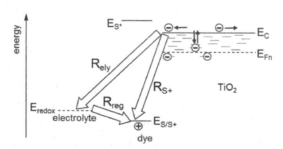

Figure 10.32.   Trapping and transport of electrons in TiO$_2$ and recombination of electrons with charged dye molecules or ions or molecules in the electrolyte.

De-trapping times are longer for electrons trapped in deeper trap states. The average de-trapping time decreases with increasing n_{TiO_2} since more of the deepest traps are occupied when E_{Fn} shifts towards E_C. The effective electron lifetime (τ_n) decreases and the effective electron diffusion coefficient increases with increasing n_{TiO_2} so that their product and hence the effective electron diffusion length remain constant (Fisher *et al.*, 2000). The intensity-dependent effective electron lifetime can be described by the change of the density of trapped electrons (n_{t,TiO_2}) and the change of n_{TiO_2} (Bisquert and Vikhrenko, 2004).

$$\tau_n = \tau_0 \cdot \left(1 + \frac{\partial n_{t,TiO_2}}{\partial n_{TiO_2}} \right) \qquad (10.16)$$

High values of I_{SC} are obtained for efficient dye-sensitized solar cells, for example, 21 mA/cm^2 for a solar cell with a Ru-based dye (Han *et al.*, 2012). A high value of I_{SC} gives evidence that the effective electron diffusion length is about 3 times larger than the thickness of the nanocomposite absorber in efficient dye-sensitized solar cells. The value of n_{TiO_2} can be estimated by using Equation (3.9) if assuming a reasonable lifetime of 1–10 ms and an absorption length of 30 μm in a nanocomposite absorber of a dye-sensitized solar cell. The resulting value of n_{TiO_2} is of the order of 10^{16}–10^{17} cm^{-3}.

The V_{OC} in dye-sensitized solar cells is related to the change of the Fermi-energy of electrons in TiO$_2$. A value of V_{OC} of 0.743 V (Han *et al.*, 2012) corresponds to a change of n_{TiO_2} under illumination by more than 12 orders of magnitude. The density of free electrons in the dark (n_{0,TiO_2}) is equilibrated by the redox potential of the electrolyte.

The value of n_{0,TiO_2} can be reduced and therefore V_{OC} can be increased if the difference between E_C and E_{redox} is increased. For a given electrolyte, an increase of the difference between E_C and E_{redox} can be achieved if the electron affinity of np-TiO$_2$ is reduced, for example, by implementing a surface dipole layer where the surface dipole layer should be dense enough to avoid penetration of screening ions from the electrolyte to the TiO$_2$ surface. Dipole layers can be realized by diade molecules consisting of a donor and an acceptor moiety linked with a bridge with a conjugated π-electron system (see, for example, Macor *et al.*, 2012). The acceptor moiety of the diade points to the TiO$_2$ surface. An additional advantage of a dense passivating diade layer is that the distance between the TiO_2 surface and the electrolyte is increased so that the recombination rate of electrons from np-TiO_2 into the electrolyte is reduced.

The spectral absorption range and therefore I_{SC} of a dye-sensitized solar cell can be increased by co-sensitization with dye molecules having complementary extinction spectra. Figure 10.33 shows an example of a well-passivated TiO_2 surface with diade molecules also providing cosensitization by light-absorbing donor moieties.

Figure 10.34 shows the configuration of two diade molecules which have the functions of co-sensitization, TiO_2 surface passivation and minimization of n_{0,TiO_2} (Yella *et al.*, 2011). The molecules contain long chain side groups filling additional space between neighboring adsorbed molecules. The application of related diade molecules allowed the use of a Co^{2+}/Co^{3+}-based electrolyte giving the opportunity for further increase of the difference between E_C and E_{redox}. A value of V_{OC} as high as 0.935 V

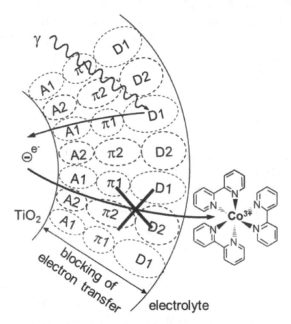

Figure 10.33. Principle of passivation and co-sensitization of TiO_2 nanoparticles in dye-sensitized solar cells with densely packed diade molecules containing small acceptor moieties (A1 and A2, pointed to the TiO_2 surface), π conjugated bridges (π1 and π2) and large light-absorbing donor moieties (D1 and D2, pointed to the electrolyte).

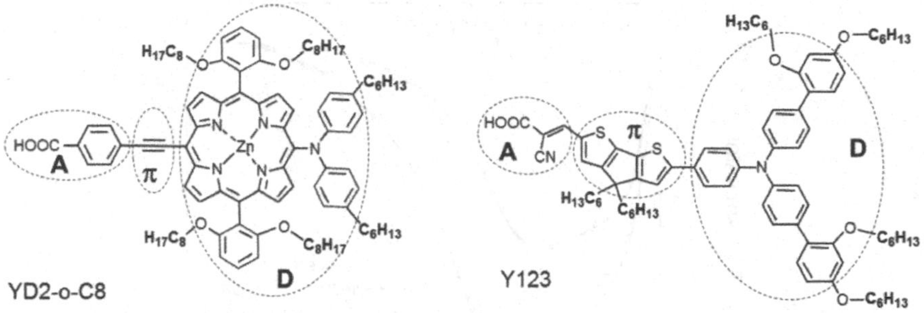

Figure 10.34. Diade dye molecules with acceptor, π-conjugated bridge and donor moieties used in co-sensitized solar cells with a Co^{2+}/Co^{3+}-based redox electrolyte by Yella *et al.* (2011). The molecules contain long C_6H_{13} and/or C_8H_{17} side groups for additional passivation.

and an η of 12.3% were reached for a dye-sensitized solar cell with the given diade molecules (Yella *et al.*, 2011). However, the values of V_{OC} remain still quite low in comparison to the onset of optical absorption in dye-sensitized solar cells. A decrease of dye aggregation and additional passivation is reached by enveloping porphyrin molecules with alkoxyl and/or alkyl chains (Wang *et al.*, 2012).

10.4.5 *About the inherent stability of dye-sensitized solar cells*

Anatase is an efficient photocatalyst, which can be used, for example, for the oxidation of organic molecules or for water splitting (Fujishima and Honda, 1972). Free holes in the valence band of anatase contribute to the decomposition of organic molecules, which is useful for water cleaning (see, for example, Vautier *et al.*, 2001) but leads to degradation of dye-sensitized solar cells. Therefore, the illumination of dye-sensitized anatase nanoparticles with ultraviolet light must be avoided, i.e. optical filters transmitting visible light but absorbing ultraviolet light have to be placed in front of a dye-sensitized solar cell.

Oxygen or water molecules can be activated at titania surfaces so that oxidation of organic molecules becomes possible. This effect is used, for example, for cleaning of TiO_2 surfaces during firing in air at 450°C. The degradation of N3 dye molecules in air already starts at much lower temperatures and is demonstrated in Figure 10.35 for heating in air at 200°C for 30 min. Therefore, oxygen and water molecules have to be avoided in electrolytes of dye-sensitized solar cells.

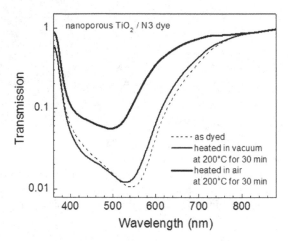

Figure 10.35. Transmission spectra of N3 dye molecules deposited onto nanoporous TiO_2 before (dashed line) and after heating in vacuum (thin solid line) or air (thick solid line) at 200°C for 30 min.

Free holes and reactive surface defects can be thermally generated in nanoporous anatase. An example of decomposition of N3 dye molecules by less than 10% is given in Figure 10.35 after heating in vacuum at 200°C for 30 min. the band gap of anatase is about 3.2 eV. Therefore, the generation rate of thermally generated charge carriers at room temperature is reduced by about 10 orders of magnitude in comparison to 200°C if considering the density of intrinsic charge carriers (Equation (3.32)). This shows that degradation can be neglected for dye molecules deposited on anatase and stored at room temperature in vacuum or in ambience free of oxygen or water molecules.

Free radicals leading to a degradation of dye molecules in dye-sensitized solar cells may be formed during photoexcitation of dye molecules. The number of excitation and neutralization cycles which have to sustain a dye molecule during the lifetime of a dye-sensitized solar cell can be estimated if taking into account a lifetime of the solar cell of 25 years and a density of dye molecules of about $5 \cdot 10^{16}$ cm^{-2} corresponding to about 500 monolayers of dye molecules. The integrated flux of photons absorbed at AM1.5 is about $1 \cdot 10^{17}$ cm^{-2}s^{-1} for a dye-sensitized solar cell. Therefore, each dye molecule in a dye-sensitized solar cell can absorb about one or two photons within one second. It has been demonstrated that dye molecules sustain more than 10^9 cycles of excitation and neutralization, which is more than enough for the lifetime of a dye-sensitized solar cell. The stability of dye-sensitized solar cells has been tested under thermal aging under AM1.5 at 60°C and no degradation has been observed after 1000 h (Wang *et al.*, 2005).

The formation of free radicals in dye-sensitized solar cells is suppressed due to ultrafast transfer of excited electrons from dye molecules into the conduction band of TiO_2 (Eichberger and Willig, 1990) and due to the extremely low thermal generation rate of free charge carriers in TiO_2. This provides an inherent stability of excited dye molecules. Therefore, degradation of dye-sensitized solar cells is related to components such as the sealing of contacts, which can be technologically optimized.

10.4.6 *Solar cells sensitized with methylammonium lead triiodide*

Methylammonium lead triiodide ($CH_3NH_3PbI_3$) belongs to the family of hybrid organic–inorganic metal halide perovskites (see Chapter 9) and can also be applied for sensitization of np-TiO_2 (see, for example, Burschka *et al.*, 2013). Electrons photogenerated in $CH_3NH_3PbI_3$ are injected into TiO_2. $CH_3NH_3PbI_3$ and/or its precursors are salts that can be dissolved. Therefore, $CH_3NH_3PbI_3$ can be infiltrated into a porous network of interconnected TiO_2 nanoparticles.

A nanocomposite of np-TiO_2/$CH_3NH_3PbI_3$ is a PV absorber, in which electrons are transported through interconnected TiO_2 nanoparticles. The absorption band gap of the np-TiO_2/$CH_3NH_3PbI_3$ nanocomposite corresponds to the band gap of $CH_3NH_3PbI_3$. The Fermi-level splitting in an np-TiO_2/$CH_3NH_3PbI_3$ nanocomposite is given by the difference between the Fermi-energies of electrons in np-TiO_2 and of holes in $CH_3NH_3PbI_3$.

Figure 10.36 shows a schematic band diagram of a complex nanocomposite solar cell based on nanocomposite layers. The charge-selective contact for electrons is formed at the junction between a compact TiO_2 layer and the np-TiO_2 in the np-TiO_2/$CH_3NH_3PbI_3$ nanocomposite (region A).

The transport of electrons in np-TiO_2 gives an alternative opportunity for the formation of local charge-selective contacts for holes. For this purpose, a barrier for electrons has been applied at local TiO_2/ZrO_2 hetero-junctions between the layers of np-TiO_2/$CH_3NH_3PbI_3$ (region A in Figure 10.36) and np-ZrO_2/$CH_3NH_3PbI_3$ (region B in Figure 10.36)

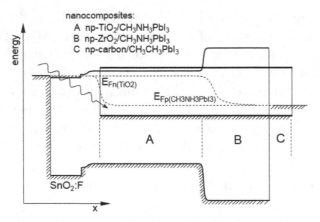

Figure 10.36. Schematic band diagram of a complex nanocomposite solar cell sensitized with methylamine lead iodide. The nanocomposite solar cell contains a compact-TiO_2/np-TiO_2 charge-selective contact for electrons, an np-TiO_2/$CH_3NH_3PbI_3$ absorber (A), a barrier for electrons based on local TiO_2/ZrO_2 hetero-junctions, an np-ZrO_2/$CH_3NH_3PbI_3$ absorber (B) and an ohmic contact between the np-ZrO_2/$CH_3NH_3PbI_3$ layer (B) and an np-carbon/$CH_3NH_3PbI_3$ layer (C).

nanocomposites (Mei *et al.*, 2014). Incidentally, the band gap of ZrO_2 (between about 6 and 7 eV depending on the phase (see, for example, Králik *et al.*, 1998) is larger than that of TiO_2 and a type I hetero-junction is formed at the TiO_2/ZrO_2 interface.

A part of the sunlight is absorbed in the np-ZrO_2/$CH_3NH_3PbI_3$ nanocomposite. Electrons or holes photogenerated in $CH_3NH_3PbI_3$ are not injected into ZrO_2 due to its large band. Therefore, the nanoparticles of ZrO_2 play the role of a scaffold in the np-ZrO_2/$CH_3NH_3PbI_3$ nanocomposite (Mei *et al.*, 2014). In general, nanoparticles of any metal oxide with a very large band gap, such as Al_2O_3, can be used as scaffolds for nanocomposite absorbers with $CH_3NH_3PbI_3$ (Lee *et al.*, 2012).

The Fermi-energy of the holes in $CH_3NH_3PbI_3$ can be contacted with an np-carbon/$CH_3NH_3PbI_3$ layer (Mei *et al.*, 2014). The $CH_3NH_3PbI_3$ is not active in the np-carbon/$CH_3NH_3PbI_3$ nanocomposite (region C in Figure 10.36), i.e. the interface between the np-ZrO_2/$CH_3NH_3PbI_3$ and np-carbon/$CH_3NH_3PbI_3$ layers serves as an ohmic or recombination contact. Incidentally, the surface recombination velocity is very high at ohmic contacts (see Chapter 4). Therefore, the thickness of the np-ZrO_2/$CH_3NH_3PbI_3$ layer should be larger than the diffusion length of electrons in order to avoid surface recombination at the ohmic back contact.

The infiltration of nanoporous layers and layer systems of metal oxides with $CH_3NH_3PbI_3$ has some technological advantages. On one side, nanoporous metal oxide layers can be printed and formed by sintering. On the other side, the infiltration of nanopores with $CH_3NH_3PbI_3$ (and related compounds) can be strongly enhanced by adding molecules containing a carboxylic and an ammonium side groups, for example, 5-ammoniumvaleric acid, which can bind to the surface of metal oxides or which can be incorporated into the structure of $CH_3NH_3PbI_3$, respectively (Mei *et al.*, 2014). Related solar cells are stable over a relatively long time due to passivation (Mei *et al.*, 2014).

The composition of the PbI_6 octahedrons in $CH_3NH_3PbI_3$ (Figure 9.29) can be varied if combining small organic cations, i.e. methylammonium, with large organic cations, for example, butylammonium ($C_4H_9NH_3^+$). Double layers of the large organic cations act as spacers between disconnected layers of PbI_6 octahedrons. The combination of large and small organic cations creates a nanocomposite of 3D perovskite domains arranged in 2D layer structures with the stoichiometry $(C_4H_9NH_3)_2(CH_3NH_3)_{n-1}Pb_nI_{3n+1}$. It has been shown that the stability of solar cells based on perovskite domains in 2D layer structures can be strongly increased in comparison solar cells with homogeneous $CH_3NH_3PbI_3$ (Tsai *et al.*, 2016).

10.5 Light Concentration by Nanophotonic Concepts for Solar Cells

10.5.1 *Optical confinement for solar cells*

Scattering changes the direction of light. Light can be homogeneously scattered at nanoparticles in a homogeneous medium (Figure 10.37(a)). If the nanoparticle is placed at the boundary between two media with different refractive indices, the light is scattered preferentially into the material with the higher refractive index (Figure 10.37(b)). If scattered light meets lateral periodic structures then constructive or destructive interference may occur depending on wavelengths of light, distances, heights and shapes of structures and optical constants of scattering and

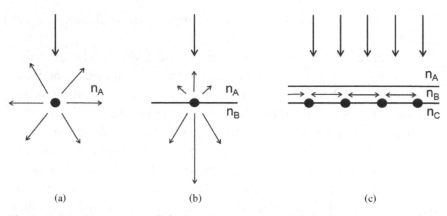

Figure 10.37. Homogeneous light scattering at a nanoparticle in a homogeneous medium (a), preferential forward scattering of light at a nanoparticle placed at the boundary between two materials with different refractive indices (b) and trapping in a thin layer of light scattered at an array of nanoparticles displaced at the boundary between the thin layer and a material (c). n_A, n_B and n_C denote the refractive indices in the different media ($n_B > n_A, n_C$).

wave-guiding materials. Light of a given wavelength can be trapped in thin films by combining scattering and constructive interference (Figure 10.37(c)) (Atwater and Polman, 2010).

For trapping, light can be scattered at nanostructures by interaction with free charge carriers in metal nanoparticles (Basch *et al.*, 2012) or by Mie scattering in dielectric nanoparticles (Spinelli *et al.*, 2012). In ideal case, scattered light is trapped and propagates only in the PV absorber. For example, the reflectivity of a silicon wafer has been reduced to 1.3% over a broad spectral range by forming an array of silicon nanoparticles coated with Si_3N_4 on top of the wafer (Spinelli *et al.*, 2012).

Figure 10.38 shows an example for interference of two lateral concentric wave sources. Related wave sources may arise, for example, by scattering of incoming light at two neighboring metal nanoparticles. Constructive interference appears when a maximum of one wave meets the maximum of the other wave. The intensity is increased in the regions of constructive interference but disappears in regions of destructive interference. There is no need to deposit absorber material in regions with destructive interference, i.e. the application of nanophotonic structures can help to drastically reduce the amount of absorber materials for solar cells.

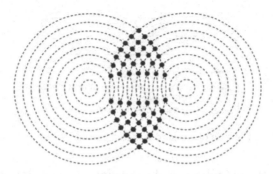

Figure 10.38. Example for two interfering concentric wave sources with a constructive interference pattern (black dots).

Furthermore, the extension of regions with constructive interference can be much smaller than the wavelength of the light, i.e. light can be concentrated in nanostructures so that the density of photogenerated charge carriers can be strongly increased in comparison to bulk PV absorbers. This effect gives the opportunity of increasing the Fermi-level splitting in PV absorbers and therefore also increasing V_{OC}.

Interference patterns depend on the geometry, the wavelength and other parameters of a structure at which interference takes place. A given interference pattern is also called mode. Localized and propagating modes are distinguished.

The planning of nanophotonic structures demands sophisticated simulation tools which take into account the Maxwell equations, the complete geometry of the system and all relevant material parameters and processes (see, for example, Ferry *et al.*, 2011a). The realization of theoretically optimized nanophotonic structures in solar cells requires sophisticated technologies for duplicating nanophotonic structures on large areas. Typical lateral structure sizes of nanophotonic structures are of the order of 50–500 nm. For comparison, a spatial resolution of 10 nm can be reached by soft imprinting over a range of more than 150 mm (Polman and Atwater, 2012).

10.5.2 *Enhancement of optical absorption in very thin absorbers*

The photon flux can be confined within a very thin absorber layer on top of a nanoplasmonic rear reflector. The excitation of plasmons in a well conducting metal such as silver is directed by a strip or by a flat dot with

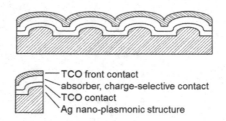

Figure 10.39.   Cross-section of a solar cell with a possible nanoplasmonic rear reflector.

height of tens of nm and a lateral extension of the order of 100–500 nm (Figure 10.39). Incoming light that is not absorbed by the very thin absorber layer is scattered at the nanoplasmonic rear reflector. The distribution of the photon flux in the very thin absorber layer follows from scattering at plasmons, Mie scattering and interference.

The structure shown in Figure 10.39 is very similar to the structure of thin-film solar cells but with one important difference: conventional thin-film solar cells do not have a nanoplasmonic rear reflector. The feasibility of a solar cell with an ultrathin a-Si:H solar cell and a nanoplasmonic rear reflector has been demonstrated (Ferry *et al.*, 2011b). An η of 9.6% has been reached for an optimized structure and an intrinsic a-Si:H layer with a thickness of only 90 nm where I_{SC} was even higher for the solar cell with the ultrathin a-Si:H absorber than for the present record cell based on a standard a-Si:H absorber (Benagli *et al.*, 2009). In addition, degradation of a-Si:H thin-film solar cells caused by the Stabler–Wronski effect (Staebler and Wronski, 1977) is strongly reduced in solar cells with an ultrathin a-Si:H absorber due to the increased electric field in the undoped a-Si:H layer.

Nano-photonic rear reflectors open the opportunity of realizing solar cells with ultrathin absorbers without folding of the internal surface area. For example, the optimal local thickness of the absorber layer in a solar cell based on ZnO/In$_2$S$_3$:Cu/CuSCN is of the order of 30 nm (Belaidi *et al.*, 2008). Wang *et al.* showed theoretically that about 90% of the incident light can be absorbed by a related solar cell in superstrate configuration with an absorber layer thickness of only 5 nm if taking into account a refractive index of 3.5 for the substrate (Wang *et al.*, 2013). However, a proof of concept with, for example, QD absorber remains challenging.

10.5.3 *Nanowire array solar cells*

The amount of absorber material can be reduced in solar cells in which the nanophotonic structure is formed by a nanocomposite consisting of light-absorbing nanowires and a dielectric matrix. For example, a photocurrent of more than 90% of the theoretical maximum can be achieved for solar cells based on an array of nanowires with a length of 2 μm, a diameter around 200 nm and a geometric FF of about 0.2, as shown by theoretical analysis (Kupec *et al.*, 2010).

The catalytic growth of semiconductor nanowires is possible by using liquid metal/semiconductor interfaces (Figure 10.40). In this vapor–liquid–solid (VLS) deposition technique, first, an array of metallic nanoparticles can be deposited on a substrate by using nanoimprint lithography (Mårtensson *et al.*, 2004).

The array of the metal nanoparticles defines the distance between the growing semiconductor nanowires. In the second step, the crystal growth is activated by increasing the temperature above the melting threshold of the metal nanoparticle. In third step, atoms from the ambience are dissolved in the liquid metal droplet and diffuse to the liquid metal/semiconductor interface. The crystal growth is catalytically activated at the liquid metal/semiconductor interface so that incoming atoms are incorporated into the growing semiconductor nanowire. Sidewalls have to be passivated during the growth by sophisticated measures. For example, sidewalls of InP nanowires can be passivated by adding traces of HCl to the reaction gas (Wallentin *et al.*, 2013). The temperature is reduced to terminate the growth of semiconductor nanowires.

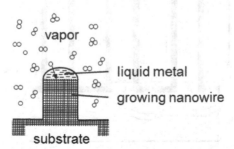

Figure 10.40. Schematic of catalytic growth of a semiconductor nanowire underneath of a metallic droplet.

There are strong advantages for the catalytic growth of semiconductor nanowires based on $A^{III}B^{V}$ semiconductors (see also Chapter 8). First, epitaxial growth of perfect semiconductor nanowires is possible. Second, doping of semiconductor nanowires can be performed with the usual dopants so that charge-selective and ohmic contacts can be formed. Third, strain can better relax in nanowires so that semiconductors with larger differences in the lattice constants can be grown on each other.

The space between semiconductor nanowires has to be filled with a dielectric such as SiO_2. Obviously, the density of interface states at the semiconductor/SiO_2 interface will be high so that the surface recombination velocity will be high as well. There is not enough space in semiconductor nanowires for additional measures of surface passivation. Therefore, the density of photogenerated free charge carriers has to be minimized by implementing a drift field for free charge carriers. This has been realized in InP nanowire array solar cells (Wallentin *et al.*, 2013). Figure 10.41 shows a schematic cross section of a nanocomposite solar cell based on an InP nanowire array with a drift field in an un-doped region of InP. The drift length is about 1.3 μm in an InP nanowire. The

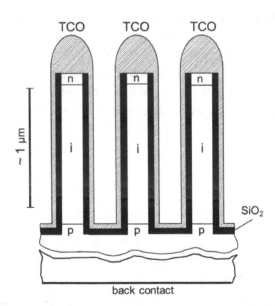

Figure 10.41.   Schematic cross-section of a solar cell based on an InP nanowire array.

drift time of photogenerated charge carriers is of the order of 10–100 ps, which is much shorter than the non-radiative recombination lifetime (see also task T10.5 in Section 10.7). The implementation of a drift field as a measure for surface passivation of semiconductor nanowires is not possible for conventional solar cells and opens the opportunity of reaching very η with nanostructured solar cells.

Despite the fact that the InP nanowires cover only 12% of the surface area of the solar cell, a value of I_{SC} (24.6 mA/cm^2) close to that of the planar record InP solar cell (29.5 mA/cm^2) has been achieved (Wallentin *et al.*, 2013). Further values of V_{OC} close and even larger (0.906 V) than that of the planar record InP solar cell (0.878 V) have been reached with solar cells based on InP nanowire arrays (Wallentin *et al.*, 2013). This shows that nanowire structures are well suited for solar cells from the point of view of optimization of I_{SC} and reducing recombination losses.

10.6 Summary

Lead selenide, lead sulfide, conjugated organic molecules and polymers and dye molecules absorb light very well but cannot be directly applied as absorbers in solar cells due to very low band gaps (PbSe, PbS), very short diffusion length of excitons (layers of conjugated organic molecules and polymers) and missing charge transport (layers of dye molecules). However, the band gaps of PbSe and PbS can be increased by quantum confinement, excitons dissociate at donor–acceptor hetero-junctions and electrons can be injected from excited dye molecules into the conduction band of a metal oxide. The combination of PbSe and PbS quantum dots with separating and passivating shells of organic molecules, the implementation of local dissociation of excitons and local charge separation brought up new classes of nanocomposite absorbers for solar cells: quantum dot QD solar cells (QD-SC), organic solar cells (OSC) and dye-sensitized solar cells (DSSC), respectively. (see also Table 10.1).

In solid-state nanocomposite solar cells, the absorbers are embedded between two charge-selective contacts with materials which have a low and a high work function in order to extract electrons and holes, respectively. The charge-selective contacts can be realized with metals, p-type and n-type doped semiconductors or metal oxides such as TiO$_2$,

Table 10.1. Functionality and materials of nanocomposite absorbers in QD-based solar cells (QD-SCs), organic solar cells (OSCs), dye-sensitized solar cells (DSSCs), solar cells with very thin absorber (vtaSCs) and in nanowire solar cells (nwSCs).

Type of solar cell	Functionality of the nanocomposite	Materials
QD-SC	Tunable band gap	$PbSe_xS_{1-x}$ QD core, passivating organic molecule shell
OSC	Local dissociation of excitons	Conjugated donor and acceptor molecules and polymers, C_{60} and PCBM as acceptor molecules
DSSC	Local charge separation	Dye molecules, np-TiO_2, redox electrolyte
SSC	Local charge separation	Composites of $CH_3NH_3PbI_3$ with np-TiO_2, np-ZrO_2 and np-carbon
2D–3D	Stabilization and passivation	Perovskite domains in 2D layer structures
vtaSC	Light confinement in a very thin absorber layer	Structured metal or refraction array below the absorber layer
nwSC	Light confinement in semiconductor nanowires	Structured array of semiconductor nanowires

ZnO, In_2O_3:Sn or MoO_3. For achieving a high V_{OC}, the mismatch between the corresponding bands (optimization of band alignment) and the density of defect states in the band gap have to be minimized at charge-selective hetero-junctions. Contacts with liquid electrolytes are based on electron transfer into (reduction) or from (oxidation) mobile ionic species.

Scattering and interference of light at nanoparticles, nanostripes or nanowires are combined in photonic nanocomposites with materials having different refractive indexes. Light can be confined in very thin layers or in nanowires of nanowire arrays by exploring nanophotonic concepts (see also Table 10.1). The thickness of the absorber layer is much thinner than the absorption length in solar cells with very thin absorber (vtaSCs), for example, less than 100 nm in a-Si:H vtaSCs. Semiconductor nanowires such as InP nanowires are a part of the photonic nanocomposite in nanowire solar cells (nwSCs). Therefore, solar cells with high η can be produced whereas the amount of absorber material can be strongly reduced. QD-SCs, OSCs and DSSCs can be combined with nanophotonic concepts as well.

The values of I_{SC}, V_{OC}, FF and η are summarized in Table 10.2 for some kinds of record solar cells related to QD-SCs, OSCs, DSSCs, vtaSCs and nwSCs. Related material combinations are given as well.

The highest stable η has been reached with InP-nw solar cells (13.8%) for which the role of surface recombination has been strongly reduced (Wallentin *et al.*, 2013). A high potential for further improvement of η can be expected for nwSCs.

The η of vtaSCs based on a-Si:H (9.6%) (Ferry *et al.*, 2011b) has potential for further increase and for implementation of nanophotonic structures in large scale production. Limitations are given by the increase of shunts with decreasing thickness of undoped a-Si:H towards 30 nm, similarly to ZnO/In_2S_3:Cu/CuSCN-based solar cells (Belaidi *et al.*, 2008).

Dye-sensitized solar cells played a decisive role for the development of the concept of nanocomposite solar cells. Even at the beginning of the 90ies, DSSCs showed relatively large η of 7.1–7.9% (O'Regan and Grätzel, 1991) and an optimum was reached soon for Ru-based dye molecules and iodide electrolyte. The optimization potential for this material combination seems quite narrow despite the fact that an increase of η to 11.28% has been reached by co-sensitization (Han *et al.*, 2012). The value of V_{OC} has been increased to 0.94 V by replacing the iodide electrolyte by a Co-based electrolyte and an η of 12.9% has been reached by co-sensitization (Grätzel, 2012). However, a broad implementation of DSSCs is challenging for cost reasons and its limited outdoor application due to the liquid electrolyte. For example, large-scale DSSC modules degraded by more than 10% after only half a year of outdoor operation (Toyoda *et al.*, 2004). Furthermore, the replacement of the liquid electrolyte by OMeTAD led to a drastic decrease of η to 5% in DSSCs based on Ru-dyes (Wang *et al.*, 2010).

The replacement of the dye by hybrid organic–inorganic metal lad halide perovskite $(CH_3NH_3)PbI_3$ allowed a reduction of the thickness of the nanocomposite absorber to about 0.3–0.4 μm and an increase of η to almost 15% (Burschka *et al.*, 2013). This solid-state semiconductor-sensitized solar cell seems to have a great potential for further improvement of solution-based production. Long term stability of solar cells with $CH_3NH_3PbI_3$ can be increased by introducing passivating molecules (Mei *et al.*, 2014) or by separating layers with a perovskite structure by double

Table 10.2. Values of short-circuit current density (I_{SC}), open-circuit voltage (V_{OC}), fill factor (FF) and solar energy conversion efficiency (η) for different types of nanocomposite solar cells. For all DSSCs the electrode material is nanoporous TiO_2.

Type of solar cell	Materials	I_{SC} (mA/cm²)	V_{OC} (V)	FF	η (%)	Reference
QD-SC	PbS-QDs	24.3	0.61	0.71	10.6	Lan et al. (2016)
OSC	Small molecules	5.38	3.31	0.74	13.2	Heliatek (2016); Uhrich (2017)
OSC	Polymer, fullerene-free	16.26	1.08	0.636	11.2	Chen et al. (2017)
OSC	Polymer, with fullerene	19.7	0.791	0.735	11.5	Zhao et al. (2016b)
DSSC	Co^{2+}/Co^{3+}, YD2-o-C8/Y123	17.75	0.94	0.76	12.9	Grätzel (2012)
DSSC	I^-/I_3^-, Ru-dye/Y1	20.88	0.743	0.727	11.28	Han et al. (2012)
DSSC	OMeTAD, Ru-dye	8.27	0.848	0.71	5.0	Wang et al. (2010)
SSC	$CH_3NH_3PbI_3$	20.0	0.993	0.73	15.0	Burschka et al. (2013)
2D–3D	$(C_4H_9NH_3)_2(CH_3NH_3)_3Pb_3I_{13}$	16.76	1.01	74.1	12.5	Tsai et al. (2016)
vtaSC	a-Si:H, 90 nm	16.94	0.864	0.66	9.6	Ferry et al. (2011b)
nwSC	InP-nw	24.6	0.779	0.724	13.8	Wallentin et al. (2013)

layers of large organic cations (Tsai *et al.*, 2016). The detailed mechanisms of degradation of $CH_3NH_3PbI_3$ and related compounds are under investigation.

Solar energy conversion efficiencies above 10% have been reached in several laboratories for organic solar cells. So far, the highest η has been reached for polymer organic solar cells containing fullerene (11.7%; Zhao *et al.*, 2016b) and for multijunction organic solar cells based on small molecules (13.2%; Heliatek, 2016; Uhrich, 2017). For organic solar cells without fullerene, an η above 11% was also shown (Chen *et al.*, 2017) and η remained above 11% even after long-term annealing at 100°C (Zhao *et al.*, 2016a). Polymer solar cells can be further optimized via engineering of side groups, which have a strong influence on the dimension of aggregates in the nanocomposite and on the alignment of HOMO and LUMO energies. A potential for η above 15% is expected for organic solar cells.

Low values of V_{OC} and FF limited η of QD-SCs in the past (Pattantyus-Abraham *et al.*, 2010). The replacement of the charge-selective hetero-junction with TiO_2 by a *pn*-junction led to a strong increase of I_{SC} (Liu *et al.*, 2012). Passivation concepts with halide ions led to an increase of η of PbS-QD-SCs above 10% (Lan *et al.*, 2016).

Nanocomposite absorbers can be produced from earth-abundant elements, for example, from lead and sulfur (PbS-QDs). If assuming that stable PV modules can be fabricated from PbS-QD absorbers, only 10% of the yearly produced amount of Pb and S will be sufficient for electricity production for mankind (see also task T10.5). This means that in terms of a material extraction cost index and annual electricity potential index (Wadia *et al.*, 2009), the strategic potential of PbS and $CH_3NH_3PbI_3$ is larger than the strategic potential of c-Si photovoltaics by five to ten orders of magnitude! Similar is true for vtaSCs based on a-Si:H or for OSCs.

The application of design rules for nanocomposite absorbers and the consequent implementation of nanophotonic concepts allow for a drastic reduction in absorber material and processing temperatures. This leads to a reduction of the energy payback time and therefore to an improvement of the sustainability of PV solar energy production. However, competitive long-term stability of PV modules operated outdoors is a prerequisite for sustainability as well.

10.7 Tasks

T10.1: Solar cell with extremely thin absorber

Estimate the factor by which the surface area of an absorber with very low diffusion of drift length has to be increased (roughness factor). The values of the absorption length, of the mobility and lifetime of photogenerated charge carriers are 2 μm, 1 cm^2/(Vs) and 1 ns, respectively.

T10.2: Surface atoms at colloidal quantum dots

Compare the amount of surface atoms with the number of bulk atoms of colloidal PbS-QDs with diameters of 2, 4 and 6 nm. The density of PbS is 7.6 g/cm^2. The bond length is 0.297 nm in PbS.

T10.3: QD absorber layer

Ascertain I_{SC}^{max} for absorber layers with colloidal PbS-QDs with diameters of 2, 4 and 6 nm. The effective electron and hole masses, the relative dielectric constant of PbS, the relative dielectric constant of the shell and the band gap of PbS are about 0.2 and 0.1 times the free electron mass, 14, 3 and 0.37 eV, respectively. Discuss the importance of accurate measurements of effective masses and uncertainties of dielectric properties in QD absorbers.

T10.4: Strategic potential of PbS-QD solar cells and solar cells sensitized with $CH_3NH_3PbI_3$

Estimate the amount of lead which would be needed to completely support the electricity consumption of mankind produced only with solar cells based on PbS-QDs as compared with solar cells sensitized only with $CH_3NH_3PbI_3$. Make reasonable assumptions for the effective thickness of the absorber layer, for the electricity demand and for the η of related solar cells. The density of PbS is 7.6 g/cm^3. Compare your estimated amount with the annual production of lead (5.2 t/a in 2012 (Guberman, 2013)).

T10.5: Absorption and transport gaps in nanocomposite absorbers

What role do the different absorption and transport gaps play for the maximum η of nanocomposite solar cells? Compare with the Shockley–Queisser limit.

T10.6: InP-nw solar cell

The implementation of a drift field in pin InP-nw solar cell gives the opportunity of reducing the role of surface recombination despite poor surface passivation. Estimate the effective lifetime in an InP-nw solar cell and compare it with the drift time. Make a reasonable assumption about the surface recombination velocity. The drift mobility of photogenerated holes is $250 \, \text{cm}^2/(\text{Vs})$ in InP.

Appendix
Solutions to Tasks

A.1 Solutions to Chapter 1

T1.1: Power installed and energy produced

In Germany and in the south of Spain, the average power of the sun is 0.12 and 0.20 kW/m^2, respectively. The time of one year corresponds to 8760 h. With regard to Equation (1.11) one can write:

$$W_{el} = \frac{\langle P_{sun} \rangle}{P_{sun}(AM1.5)} \cdot t \cdot P_{inst} \qquad (T1.1')$$

$$W_{el} = \frac{0.12(0.2)kW/m^2}{1\,kW/m^2} \cdot 8760\,h \cdot 1000\,kW \qquad (T1.1'')$$

In total, $1.05 \cdot$ GWh and $1.75 \cdot$ GWh can be produced in Germany and in the south of Spain, respectively, during the first year of operation of a PV power plant with 1 MW$_p$.

T1.2: Degradation time of PV modules

The time-dependent degradation of η as following:

$$\eta(t) = \eta_0 \cdot (1 - k_{deg})^t \qquad (T1.2)$$

The initial η at time 0 is denoted by η_0. The degradation time of PV modules (Δt_η) is derived from Equation (T1.2):

$$\eta \Delta t = \log_{(1-k_{deg})} \eta = \frac{\log \eta}{\log(1 - k_{deg})} \qquad (T1.3)$$

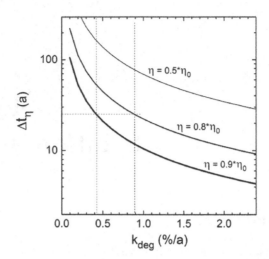

Figure T1.1. Dependence of the degradation time of PV modules on k_{deg} for reaching 90%, 80% and 50% of the initial value (thick, medium and thin lines, respectively).

The dependence of Δt_η on k_{deg} is plotted in Figure T1.1 after which η reduced to 90%, 80% and 50% of η_0. In order to reach 90% or 80% of η_0 after 25 years, k_{deg} is about 0.4% or 0.9%/a, respectively, which is of the order of k_{deg} for most of the worldwide tested PV modules (Jordan *et al.*, 2016).

Half of η_0 is reached for k_{deg} equal to 0.35%/a after about 200 years, which is of the order of the lifetime of a building. PV modules, for which k_{deg} is above 2%/a, are not really useful for building-integrated PV.

T1.3: Energy payback factor of PV modules

The EPBF reached after a certain time is defined as:

$$\text{EPBF}(t) = \frac{W_{el}(t)}{W_{el}(t_{EPB})} \tag{T1.4}$$

$W_{el}(t)$ and $W_{el}(t_{EPB})$ can be obtained by integrating the power of the PV modules over up to t and t_{EPB}, respectively. It can be assumed that the average power of the sun is independent of the operation time of PV modules. Then, EPBF only depends on the ratio of η integrated over time

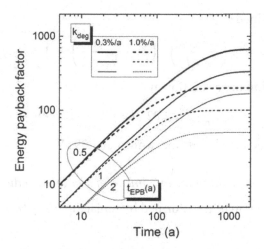

Figure T1.2. Dependence of the energy payback factor on operation time of PV modules for k_{deg} equal to 0.3%/a and 1%/a (solid and dashed lines, respectively) and for t_{EPB} equal to 0.5, 1 and 2 years (thick, medium and thin lines, respectively).

up to t and t_{EPB}, respectively.

$$\text{EPBF}(t) = \frac{\int_0^t (1 - k_{\text{deg}})^{t'} dt'}{\int_0^{t_{\text{EPB}}} (1 - k_{\text{deg}})^{t'} dt'} = \frac{(1 - k_{\text{deg}})^t - 1}{(1 - k_{\text{deg}})^{t_{\text{EPB}}} - 1} \quad \text{(T1.5)}$$

The dependencies of EPBF on operation time of PV modules are shown in Figure T1.2 for k_{deg} and t_{EPB} equal to 0.3% and 1%/a and to 0.5, 1 and 2 years, respectively. For time periods of the order of 10 years, EPBF is dominated by t_{EPB}. For example, EPBF is about 5 or 20 after 10 years of operation of a PV module with t_{EPB} of 2 (current status in Germany) or 0.5 (target for Spain in 2020) years, respectively. The maximum EPBF is reached after times longer than 100 years. As a possible limit, the maximum EPBF can be as high as 300 for k_{deg} and t_{EPB} equal to 0.3%/a and 1 year, respectively. However, one has to keep in mind that k_{deg} can non-linearly increase in time due to fatigue of materials and contacts.

T1.4: Photon energy in the maximum of blackbody radiation at T_S

The first derivative of Equation (1.5) has to be set to zero for getting the maximum of the sun spectrum. The following equation for the non-trivial solution is obtained

$$\exp\left(\frac{E_{ph}}{k_B \cdot T_s}\right) \cdot \left(3 - \frac{E_{ph}}{k_B \cdot T_s}\right) - 3 = 0 \tag{T1.6}$$

This equation has a solution at:

$$E_{ph}\big|_{max} = 2.82 \cdot k_B \cdot T_s = 1.39\,\text{eV} \tag{T1.7}$$

The sun spectrum has its maximum at 1.39 eV.

T1.5: Wavelength of light in the maximum of blackbody radiation at T_S

Equation (1.5) has to be transformed by replacing υ by c/λ and by replacing dE_{ph} by

$$dE_{ph} = -\frac{hc}{\lambda^2} \cdot d\lambda \tag{T1.8}$$

The sun spectrum as a function of wavelength is obtained:

$$\frac{dJ_S(\lambda)}{d(\lambda)} = \frac{8\pi \cdot c^2 \cdot h}{\lambda^5} \cdot \frac{1}{\exp\left(\frac{hc}{\lambda \cdot k_B \cdot T_S}\right) - 1} \tag{T1.9}$$

The first derivative of Equation (1.9) has to be set to zero for getting the wavelength in the maximum of the sun spectrum. The following equation for the non-trivial solution is obtained.

$$\exp\left(\frac{hc}{\lambda \cdot k_B \cdot T_s}\right) \cdot \left(5 - \frac{hc}{\lambda \cdot k_B \cdot T_s}\right) - 5 = 0 \tag{T1.10}$$

Equation (T1.10) has a solution at:

$$\lambda\big|_{max} = \frac{hc}{4.96 \cdot k_B \cdot T_s} = 502\,\text{nm} \tag{T1.11}$$

The sun spectrum has its maximum at 502 nm.

T1.6: Maximum temperature of a solar cell under operation

A solar cell is cooled only by thermal emission if convection is absent. The solar cell receives power from the solar radiation via the front surface and from the homogeneous blackbody radiation of the surrounding environment (at temperature T_0) via the front and back surfaces (sidewalls neglected). A part of the power received from sun is converted into electric power, which does not contribute to heating the solar cell. The solar cell is heated to the temperature T_{SC} and emits blackbody radiation via the front and back surfaces (sidewalls neglected). The area of the solar cell, η and the Stefan–Boltzmann constant are denoted by A, η and σ_S, respectively. The balance of energy fluxes gives the following equation:

$$A \cdot P_{sun} + 2 \cdot A \cdot \sigma_s \cdot T_0^4 = \eta \cdot A \cdot P_{sun} + 2 \cdot A \cdot \sigma_s \cdot T_{SC}^4 \qquad (\text{T1.12})$$

The temperature of the solar cell can be calculated after transforming Equation (T1.12):

$$T_{SC} = \left[T_0^4 + \frac{1 - \eta}{2 \cdot \sigma_s} \cdot P_{sun} \right]^{1/4} \qquad (\text{T1.13})$$

The temperature on earth is about 300 K and a temperature of the environment around a satellite of 10 K is assumed. Then the temperature of the solar cell is 356, 350 and 340 K for η of 10%, 20% and 40%, respectively, i.e. solar cells can reach about 70°C under operation on earth at AM1.5. The temperature of solar cells on a satellite is 298, 289 and 270 K for η of 10%, 20% and 40%, respectively, i.e. the temperature conditions are comparable to those on earth.

T1.7: Air mass

The angle of incidence (θ) is measured to the surface normal. The air mass is ratio between the optical path of light and the thickness of the atmosphere, i.e. the reciprocal of the cosine of θ.

$$AM(\theta) = \frac{1}{\cos \theta} \qquad (\text{T1.14})$$

At high angles of incidence, i.e. during the morning and evening hours, the air mass is overestimated due to the curvature of the earth and a correction term has to be implemented into Equation (T1.14).

T1.8: Tolerable series and parallel resistances

The combination of Equations (1.19) and (1.29′) and (1.29″) gives the expressions for calculating the tolerable series and parallel resistances

$$R_{s,tol} < \frac{1}{100} \cdot \frac{\frac{k_B \cdot T}{q} \cdot \ln\left(\frac{I_{SC}}{I_0} + 1\right)}{I_{SC}} \qquad (\text{T1.15}')$$

$$R_{p,tol} > 100 \cdot \frac{\frac{k_B \cdot T}{q} \cdot \ln\left(\frac{I_{SC}}{I_0} + 1\right)}{I_{SC}} \qquad (T1.15'')$$

Room temperature and a linear response of the short circuit current density are assumed. The tolerable series and parallel resistances of the solar cell with $I_0 = 10^{-19}$ A/cm^2 and $I_{SC} = 0.025$ A/cm^2 are equal to 0.4 and 4000 Ωcm^2, respectively. The tolerable series and parallel resistances are 1 mΩcm^2 and 10 Ωcm^2, respectively, under operation at concentration factor of 400.

T1.9: Current–voltage characteristics

The current–voltage characteristics are simulated by using the following equation:

$$I = I_{SC} - I_0 \cdot \left[\exp\left(\frac{q \cdot U}{k_B \cdot T}\right) - 1\right] - \frac{U}{R_P} \qquad (T1.16)$$

The current is multiplied with the potential for getting the power–voltage characteristics. The values of V_{OC} and FF are 1.017 V and 88.4% ($I_{SC} = 100$ mA/cm^2, $R_P = 100$ kΩcm^2), 0.957 V and 87.8% ($I_{SC} = 10$ mA/cm^2, $R_P = 100$ kΩcm^2), 1.017 V and 87.5% ($I_{SC} = 100$ mA/cm^2, $R_P = 1$ kΩcm^2) and 0.956 V and 80% ($I_{SC} = 10$ mA/cm^2, $R_P = 1$ kΩcm^2).

A.2 Solutions to Chapter 2

T2.1: I_{SC} current density

The short circuit current density can be calculated by taking into account that photons with energy below E_g and above $E_g + \Delta E$ ($E_g = 1.6$ cV, $\Delta E = 0.3$ eV) do not contribute to photo current generation.

$$I_{SC} = q \cdot \int_{E_g}^{\infty} \Phi_{sun}(E_{ph}) \cdot dE_{ph} - q \cdot \int_{E_g + \Delta E}^{\infty} \Phi_{sun}(E_{ph}) \cdot dE_{ph} \qquad (T2.1)$$

The values of I_{SC} are about 24 and 15 mA/cm^2 for solar cells with band gaps of 1.6 and 1.9 eV, respectively. Therefore I_{SC} is about 9 mA/cm^2 for a solar cell with a band gap of 1.6 eV and a width of the absorption band of 0.3 eV.

T2.2: Temperature-dependent I_{SC} density

The I_{SC} density increases by about 1% if the temperature of a c-Si solar cell increases from 25 to 60°C.

T2.3: Lambertian limit of light trapping

For Lambertian light scattering the intensity of scattered light is distributed within a sphere as shown in Figure 2.13. The surface area of this sphere (A_0) corresponds to the total intensity of scattered light. The area of the sphere segment around the critical angle for total reflection (A_{loss}) corresponds to the intensity of the light lost due to scattering into a cone of angle less than the critical angle for total internal reflection (θ_C). The maximum intensity of scattered light is denoted by $2 \cdot R$ (R is the radius of the sphere).

The increase of the optical path is proportional to the ratio between the total intensity of scattered light and the intensity lost by scattering at angles less than θ_C. With respect to the geometric situation shown in Figure T2.1 it can be written:

$$X_{path} = \frac{A_0}{A_{loss}} = \frac{4\pi \cdot R^2}{2\pi \cdot R \cdot h} = \frac{2R}{h} \tag{T2.2}$$

$$\tan \theta_C = \frac{h}{a} = \frac{a}{2R - h} \tag{T2.3}$$

Equation (T2.3) can be transformed to:

$$\frac{2R}{h} = \frac{1}{\tan^2 \theta_C} + 1 = \frac{1}{\sin^2 \theta_C} \tag{T2.4}$$

The limit of Lambertian light trapping is obtained by using Equations (T2.4), (T2.2), the definition of θ_C and the fact that scattered

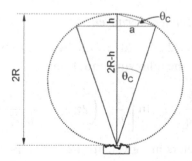

Figure T2.1. Geometric situation for Lambertian light scattering. The maximum intensity of scattered light is denoted by $2R$.

light passes the absorber for a second time after total internal reflection.

$$X_{\text{path}} = 2 \cdot \left(\frac{n_S}{n_{AR}} \right)^2 \tag{T2.5}$$

T2.4: Temperature-dependent effective density of states

Equation (2.27) is used for calculating the effective density of states at the conduction band edge. The values of k_B, m_e and h are $1.38 \cdot 10^{-23}$ J/K, $9.1 \cdot 10^{-31}$ kg and $6.626 \cdot 10^{-34}$ Js, respectively. The temperatures are 273.14 and 343.14 K.

$$N_C = 2 \cdot \left(2\pi \cdot k_B \cdot T \cdot \frac{m_e^*}{h^2} \right)^{\frac{3}{2}} \tag{T2.6}$$

The values of N_C are $2.2 \cdot 10^{19}$ cm^{-3} at 0°C and $3.1 \cdot 10^{19}$ cm^{-3} at 70°C.

T2.5: Temperature-dependent density of minority charge carriers

Regarding to Equations (2.29) and (2.27) the density of minority charge carriers (n_0) can be calculated using the following equation:

$$n_0 = \frac{32\pi}{p_0} \cdot \left(\frac{k_B \cdot T \cdot m_e}{h^2} \right)^3 \cdot \left(\frac{m_e^*}{m_e} \cdot \frac{m_h^*}{m_e} \right)^{\frac{3}{2}} \cdot \exp\left(-q \cdot \frac{E_g(T)}{k_B \cdot T} \right) \tag{T2.7}$$

The values of m_e^* is given in T2.3. The values of k_B, m_e, q and h are $1.38 \cdot 10^{-23}$ J/K, $9.1 \cdot 10^{-31}$ kg, $1.6 \cdot 10^{-19}$ As and $6.626 \cdot 10^{-34}$ Js, respectively. The temperatures are 273.14 and 343.14 K.

The density of minority charge carriers is 8.8 cm^{-3} at 0°C and $6 \cdot 10^5$ cm^{-3} at 70°C.

T2.6: Temperature-dependent Fermi-energy

According to Equations (2.34) and (2.27), the Fermi-energy in p-type doped c-Si is calculated after the following equation:

$$E_F = E_V + \frac{k_B}{q} \cdot T \cdot \ln\left[\frac{2}{p_0} \cdot \left(2\pi \cdot \frac{k_B \cdot T \cdot m_h^*}{h^2} \right)^{\frac{3}{2}} \right] \tag{T2.8}$$

Values of parameters are given in T2.4. For p-type doped c-Si with $p_0 = 3 \cdot 10^{16}$ cm^{-3} the Fermi energy is 0.135 eV or 0.179 eV above the valence band edge at 0 and 70°C, respectively.

T2.7: Fermi-level splitting

According to Equations (2.38) and (2.29), the Fermi-level splitting is calculated after the following equation:

$$E_{Fn} - E_{Fp} = \frac{k_B}{q} \cdot T \cdot \ln \left(\frac{\left(\frac{n_i^2}{p_0} + \Delta n \right) \cdot (p_0 + \Delta p)}{n_i^2} \right) \tag{T2.9}$$

Parameters are given in the tasks before. The value of n_i^2 is calculated following Equations (2.30) and (2.27) and amounts to about 10^{20} cm^{-6}. The Fermi-level splitting in p-type doped c-Si with $p_0 = 3 \cdot 10^{16}$ cm^{-3} is 0.627, 0.687 and 0.754 eV under illumination with $\Delta n = 10^{14}$, 10^{15} and 10^{16} cm^{-3}, respectively.

T2.8: Temperature-dependent Fermi-level splitting

The Fermi-level splitting is calculated by using Equation (T2.9). The reduced band gap (see task T2.2) has to be taken into account to calculate n_i^2 at 65°C. At 65°C the Fermi-level splitting is reduced by about 0.1 V in comparison to 25°C, i.e. the Fermi-level splitting is reduced by about 14%, which is higher than the increase of I_{SC} by about 1%.

A.3 Solutions to Chapter 3

T3.1: Minimum lifetime and light trapping

The thickness of the absorber should be larger than three times the absorption length (α^{-1}) divided by the factor of the increase of the optical path (X_{path}).

$$d_{abs} \geq \frac{3 \cdot \alpha^{-1}}{X_{path}} \tag{T3.1}$$

The value of X_{path} is equal to 1 for a single pass, 2 for an absorber with back reflector and of the order of 10 for an absorber with a Lambertian back reflector for light trapping (Chapter 2). The corresponding values of d_{abs} are 300, 150 and 30 μm. The minimum lifetime condition is modified:

$$\tau_{min} = \frac{9 \cdot \alpha^{-2}}{X_{path} \cdot D} \tag{T3.2}$$

The diffusion constant of photogenerated charge carriers (D) and α^{-1} are $15\,\text{cm}^2/\text{s}$ and $100\,\mu\text{m}$, respectively, for a c-Si absorber. Values of $\tau_{\min}$ of 60, 15 and $0.6\,\mu\text{s}$ are obtained for a c-Si absorber with a single pass of light, with a back reflector and with a Lambertian back reflector for light trapping, respectively.

T3.2: Density of photogenerated charge carriers in a GaAs absorber

The density of photogenerated charge carriers can be estimated by combining Equations (3.8) and (T3.2):

$$\Delta n \approx \frac{3 \cdot \alpha^{-1}}{q \cdot X_{\text{path}} \cdot D} \cdot I_{\text{SC}}^{\max} \qquad (\text{T3.3})$$

The value of $I_{\text{SC}}^{\max}$ for $E_g = 1.42\,\text{eV}$ (GaAs) is about $0.03\,\text{A/cm}^2$ under illumination at AM1.5 and X_{path} is equal to 2 for an absorber with a back reflector. The resulting densities of photogenerated charge carriers are about 10^{12} and $5 \cdot 10^{11}\,\text{cm}^{-3}$ for a GaAs absorber without and with a back reflector, respectively. For comparison, the density of photogenerated charge carriers is about 3 orders of magnitude higher in a c-Si absorber.

T3.3: Density of majority charge carriers in a c-Si absorber

The lifetime of minority charge carriers is limited by Auger recombination in c-Si. The densities of majority charge carriers are obtained by using Equation (3.15). The maximum densities of majority charge carriers are $8.4 \cdot 10^{15}$ and $2.6 \cdot 10^{17}\,\text{cm}^{-3}$ for $\tau_{\min}$ equal to 7 ms and $7\,\mu\text{s}$, respectively. The diffusion lengths are calculated after Equation (3.5) by using a diffusion constant of $15\,\text{cm}^2/\text{s}$. The diffusion lengths are 3 mm and $95\,\mu\text{m}$ for $\tau_{\min}$ equal to 7 ms and $7\,\mu\text{s}$, respectively.

T3.4: Intrinsic carrier lifetime

The intrinsic densities of charge carriers are about 10^{10} and $10^6\,\text{cm}^{-3}$ for c-Si and GaAs, respectively (Chapter 2). The lifetime is limited by radiative recombination for very low densities of free charge carriers. The intrinsic carrier lifetimes are about $3 \cdot 10^4$ s for c-Si and $5 \cdot 10^3$ s for GaAs.

T3.5: Maximum limit of lifetime

A hypothetic lifetime of 157 s has to be related to densities of majority charge carriers and of defects in the bulk silicon crystal. Densities of majority charge carriers should be calculated for radiative and Auger recombination after Equations (3.12′) and (3.15), respectively. For a lifetime of 157 s densities of majority charge carriers of $2 \cdot 10^{12}$ and $6 \cdot 10^{13}$ cm^{-3} are obtained for radiative and Auger recombination, respectively. For comparison, the intrinsic density of charge carriers is about 10^{10} cm^{-3} in c-Si at room temperature, i.e. a related crystal is realistic from the point of view of doping. However, a lifetime of 157 s would require a density of defects as low as $6 \cdot 10^5$ cm^{-3} (after Equation (3.21)) what is not realistic. It is interesting to note that radiative recombination limits the density of majority charge carriers at very low densities.

T3.6: Lifetime at the solubility limit of doping

The lifetime is limited by Auger recombination at the solubility limit. Regarding to Equation (3.15) the lifetime is as low as 50 ps at the solubility limit. A density of defects of the order of $2 \cdot 10^{18}$ cm^{-3} can be tolerated at the solubility limit of doping (Equation (3.21)).

The diffusion constant is reduced at very high densities of free charge carriers. The diffusion length is of the order of 100 nm or less at the solubility limit of doping.

T3.7: Total lifetime

The total lifetime is calculated after Equation (3.31) by taking into account Equations (3.12″) for the radiative recombination lifetime, (3.15) for the Auger recombination lifetime and (3.21′) the Shockley–Read–Hall recombination lifetime.

$$\frac{1}{\tau_{total}} = B \cdot p_0 + C_A \cdot p_0^2 + \sigma_e \cdot v_{th} \cdot N_t + \frac{s_n}{d} \tag{T3.4}$$

The total lifetime is 72 μs for the given c-Si absorber.

A.4 Solutions to Chapter 4

T4.1: Diffusion potential

The diffusion current density is calculated after Equation (4.12). The density of intrinsic charge carriers in Equation (4.12) can be obtained after Equation (2.30). Values of n_i^2 are about 10^{20} and 10^{12} cm^{-6} for c-Si and GaAs, respectively, at room temperature according to

Chapter 2. The diffusion potentials are 0.9 and 1.38 V for *pn*-junctions with (a) $N_A = 10^{16}$ and $N_D = 10^{19}$ cm^{-3} for c-Si and (b) $N_A = 10^{18}$ and $N_D = 10^{17}$ cm^{-3} for GaAs, respectively.

T4.2: Space charge regions of a pn-junction

The widths of the space charge regions are given by Equations (4.5). The sum of the potential drops across the space charge regions at the *p*-type and type doped sides of the *pn*-junction is equal to the diffusion potential.

$$U_D = U_n + U_p \tag{T4.1}$$

The values of x_p and x_n are obtained by transforming Equations (4.5), (4.6) and (T4.1) to:

$$x_p = \sqrt{\frac{2 \cdot \varepsilon_r \cdot \varepsilon_0}{q \cdot N_A} \cdot \frac{U_D}{1 + \frac{N_A}{N_D}}} \tag{T4.2'}$$

$$x_n = \sqrt{\frac{2 \cdot \varepsilon_r \cdot \varepsilon_0}{q \cdot N_D} \cdot \frac{U_D}{1 + \frac{N_D}{N_A}}} \tag{T4.2''}$$

The diffusion potentials are calculated according to task T4.1. The relative dielectric constants of c-Si and GaAs are about 12 and 11, respectively. The values of x_p and x_n are about 350 and 0.35 nm, respectively, for a c-Si *pn*-junction with $N_A = 10^{16}$ and $N_D = 10^{19}$ cm^{-3} and about 12 and 120 nm, respectively, for a GaAs *pn*-junction with $N_A = 10^{18}$ and $N_D = 10^{17}$ cm^{-3}.

T4.3: Maximum electric field at a pn-junction

The maximum electrical field is reached at the *pn*-junction ($x = 0$). With respect to Equations (4.3') and (T4.2) the following equation is obtained for the maximum electric field at a *pn*-junction:

$$-\varepsilon_{max} = \sqrt{\frac{2 \cdot q \cdot N_A}{\varepsilon_r \cdot \varepsilon_0} \cdot \frac{U_D}{1 + \frac{N_A}{N_D}}} = \sqrt{\frac{2 \cdot q \cdot N_D}{\varepsilon_r \cdot \varepsilon_0} \cdot \frac{U_D}{1 + \frac{N_D}{N_A}}} \tag{T4.3}$$

The maximum electric field is -52 kV/cm at a *pn*-junction with $N_A = 10^{16}$ and $N_D = 10^{19}$ cm^{-3} for c-Si and -200 kV/cm $N_A = 10^{18}$ and $N_D = 10^{17}$ cm^{-3} for GaAs.

T4.4: Diode saturation current density of a semi-infinite pn-junction in the limit of Auger recombination

The diode saturation current density is calculated following Equation (4.34). The value of n_i^2 is given in task T4.1. The diffusion constants of electrons and holes can be approximated as 20 and 10 cm²/s for c-Si. The diffusion lengths can be calculated according to Equation (3.5) while the maximum lifetimes depend on Auger recombination after equation (3.15). The minimum diode saturation current density of the semi-infinite pn-junction is calculated for a c-Si solar cell following:

$$I_0 = q \cdot n_i^2 \cdot \sqrt{C_A} \cdot (\sqrt{D_n} + \sqrt{D_p}) \tag{T4.4}$$

It is interesting to remark that Equation (T4.4) does not depend on the densities of donors and acceptors. The Auger recombination rate constant (C_A) is about $2 \cdot 10^{-30}$ cm⁶/s for c-Si (Chapter 3).

The diode saturation current density of a semi-infinite pn-junction is $3.3 \cdot 10^{-13}$ A/cm² for c-Si solar cell in the Auger limit. As remark, this value is larger than the diode saturation current density of a record c-Si solar cell (Equation (1.22′)) and points to the importance of additional passivation measures.

T4.5: Diode saturation current density of a semi-infinite pn-junction in the limit of radiative recombination

The diode saturation current density is calculated following Equation (4.34). The value of n_i^2 is given in task T4.1. The diffusion constants of electrons and holes can be approximated as 60 and 10 cm²/s for GaAs. The diffusion lengths can be calculated according to Equation (3.5) while the maximum lifetimes depend on radiative recombination after Equation (3.12′). The diode saturation current density of the semi-infinite pn-junction is calculated for a GaAs solar cell following:

$$I_0 = q \cdot n_i^2 \cdot \sqrt{B} \cdot \left(\sqrt{\frac{D_n}{N_A}} + \sqrt{\frac{D_p}{N_D}} \right) \tag{T4.5}$$

The radiative recombination rate constant is about $2 \cdot 10^{-10}$ cm³/s for GaAs (Chapter 3).

The diode saturation current density of a semi-infinite *pn*-junction is $4 \cdot 10^{-20}$ A/cm^2 for a GaAs solar cell with $N_A = 10^{18}$ and $N_D = 10^{17}$ cm^{-3}. The found value of I_0 is lower than I$_0$ of a record GaAs solar cell and shows the optimization potential for the *pn*-junction.

T4.6: Temperature dependence of V_{OC} of c-Si solar cells

The general temperature dependence of V_{OC} is given by Equation (1.35) with $E_A = E_g$ (1.12 eV for c-Si at room temperature). The value of I_{00} in Equation (1.35) is obtained by combining Equations (2.30) and (T4.4).

$$V_{OC} = \frac{E_g}{q} - \frac{k_B}{q} \cdot \ln \left(\frac{q \cdot N_C \cdot N_V \cdot \sqrt{C_A} \cdot \left(\sqrt{D_n} + \sqrt{D_p} \right)}{I_{SC}} \right) \cdot T$$

(T4.6)

The temperature dependence of V_{OC} is given by

$$\frac{\Delta V_{OC}}{\Delta T} = -\frac{k_B}{q} \cdot \ln \left(\frac{q \cdot N_C \cdot N_V \cdot \sqrt{C_A} \cdot \left(\sqrt{D_n} + \sqrt{D_p} \right)}{I_{SC}} \right)$$

(T4.7)

The values of I_{SC} at AM1.5, N_C, N_V, C_A, D_n and D_p are about 0.04 A/cm^2, 10^{19} cm^{-3}, $3 \cdot 10^{19}$ cm^{-3}, $2 \cdot 10^{-30}$ cm^6/s, 20 cm^2/s and 10 cm^2/s, respectively. Therefore, the temperature dependence of V_{OC} for a c-Si solar cell is about $1.4 \cdot 10^{-3}$ V/K.

T4.7: Diode saturation current density for SRH recombination at the pn-junction

The diode saturation current density for Shockley–Read–Hall recombination is calculated after Equation (4.43) by taking into account the minimum SRH lifetime from Equation (3.21).

$$I_{0,SRH} = \frac{q}{2} \cdot n_i \cdot \delta_{SRH} \cdot v_{th} \cdot \sigma_{e(h)} \cdot N_t$$

(T4.8)

The value of n_i is 10^{10} for c-Si, the thermal velocity (v_{th}), the capture cross-section ($\sigma_{e(h)}$) and the characteristic length δ_{SRH} amount to about 10^7 cm/s, 10^{-15} cm^2 and 10^{-6} cm, respectively.

The diode saturation current density for SRH recombination in a c-Si solar cell is about 10^{-11} A/cm^2 for $N_t = 10^{12}$ cm^{-3} and 10^{-7} A/cm^2 for $N_t = 10^{16}$ cm^{-3}.

T4.8: Dependence of V_{OC} on SRH recombination at the pn-junction

According to the 2-diode model given by Equation (4.45), taking $R_S = 0$ and $R_p \rightarrow \infty$ and setting the current to 0 the following quadratic equation is obtained:

$$0 = \left[\exp\frac{q \cdot V_{OC}}{k_B \cdot T}\right]^2 + \frac{I_{0,SRH}}{I_0} \cdot \left[\exp\frac{q \cdot V_{OC}}{k_B \cdot T}\right] - \frac{I_{SC} + I_0 + I_{0,SRH}}{I_0}$$

$$(T4.9)$$

The solution of Equation (T4.9) with positive sign has physical meaning and V_{OC} can be calculated:

$$V_{OC} = \frac{2 \cdot k_B \cdot T}{q} \cdot \ln\left[\sqrt{\left(\frac{I_{0,SRH}}{I_0}\right)^2 + \frac{I_{SC} + I_0 + I_{0,SRH}}{I_0}} - \frac{I_{0,SRH}}{2 \cdot I_0}\right]$$

$$(T4.10)$$

For very low $I_{0,SRH}$ Equation (T4.10) leads to the known solution for V_{OC} given by Equation (1.19). The behavior of Equation (T4.10) is shown in Figure T4.1 for $I_0 = 10^{-26}$, 10^{-20}, 10^{-14} and 10^{-8} A/cm^2 ($I_{SC} = 0.04$ A/cm^2).

The plot shows that SRH recombination at the *pn*-junction limits completely V_{OC} for values of $I_{0,SRH}$ larger than the square root of I_0. On the other hand, the value of $I_{0,SRH}$ has to be about two orders of magnitude less than square root of I_0 for getting rid of the influence of SRH on V_{OC}.

T4.9: Geometry factor of solar cells with a pn-junction

The geometry factor is calculated after Equation (4.49′). The functions of $\sinh(x)$ and $\cosh(x)$ are defined as:

$$\sinh(x) = \frac{1}{2} \cdot (e^x - e^{-x}) \qquad (T4.11')$$

$$\cosh(x) = \frac{1}{2} \cdot (e^x + e^{-x}) \qquad (T4.11'')$$

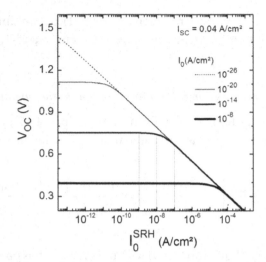

Figure T4.1.   Dependence of V_{OC} on the diode saturation current density limited by SRH recombination at the *pn*-junction for different values of the diode saturation current density limited by bulk recombination ($I_0 = 10^{-26}$, 10^{-20}, 10^{-14} and 10^{-8} A/cm², dotted, thin solid, medium solid and thick solid lines, respectively) for a short circuit current density of 0.04 A/cm².

The argument of the sinh and cosh is the ratio between the thickness of the base and the diffusion length. Regarding to Equations (3.5) and (3.21′) the argument of the sinh and cosh is obtained after the following equation:

$$\frac{H_p}{L_n} = H_p \cdot p_0 \cdot \sqrt{\frac{C_A}{D_n}} \qquad (T4.12)$$

The value of the ratio H_p/L_n is 0.047. Regarding to Equations (3.5) and (3.21′) the ratio $s_{\infty n}/s_n$ is obtained after the following equation:

$$\frac{s_{\infty n}}{s_n} = \frac{p_0}{s_n} \cdot \sqrt{C_A \cdot D_n} \qquad (T4.13)$$

The value of $s_{\infty n}/s_n$ is 6.3. According to Equations (4.49′) and (T4.11′) and (T4.11″) the geometry factor of the electron diode saturation current density is 0.2.

A.5 Solutions to Chapter 5

T5.1: Surface band bending in the presence of surface defects

The surface band bending can be calculated following Equations (5.9), (5.15′), (5.16), (5.13′) and (5.9′):

$$
\varphi_0 = \left[\frac{E_g - 2 \cdot \left(k_B \cdot T \cdot \ln(N_C/N_D)\right)}{2 \cdot q} - \frac{\varepsilon_r \cdot \varepsilon_0}{4 \cdot q} \cdot \frac{N_d}{N_{st}^2} \cdot \left(\frac{E_g}{q}\right)^2 \right]
$$

$$
- \sqrt{\left[\frac{E_g - 2 \cdot \left(k_B \cdot T \cdot \ln(N_C/N_D)\right)}{2 \cdot q} - \frac{\varepsilon_r \cdot \varepsilon_0}{4 \cdot q} \cdot \frac{N_d}{N_{st}^2} \cdot \left(\frac{E_g}{q}\right)^2 \right]^2 - \left[\frac{E_g - 2 \cdot \left(k_B \cdot T \cdot \ln(N_C/N_D)\right)}{2 \cdot q} \right]^2}
$$

(T5.1)

For the semiconductor with (a) $E_g = 1.1$ eV, the surface band bending is 0.33 ($N_{st} = 10^{13}$ cm^{-2}) or 0.025 V ($N_{st} = 10^{11}$ cm^{-2}). The values of φ_0 amount to 0.605 ($N_{st} = 10^{13}$ cm^{-2}) and 0.007 V ($N_{st} = 10^{11}$ cm^{-2}) for the semiconductor with $E_g = 1.42$ eV (b).

T5.2: Fermi-level pinning

The change of the charge in surface states (ΔQ_{st}) has to be compared with the change of the charge in the space charge region (ΔQ_{st}) at the semiconductor surface by assuming that

$$
\Delta E_{FS} = q \cdot \Delta \varphi_0 \tag{T5.2}
$$

A shift of E_{FS} is impossible (Fermi-level pinning) if ΔQ_{st} cannot be compensated by ΔQ_{SC}, i.e.

$$
\Delta Q_{SC} < \Delta Q_{st} \tag{T5.3}
$$

The value of ΔQ_{SC} can be calculated following Equation (5.9′):

$$\Delta Q_{sc} = \sqrt{\frac{2 \cdot \varepsilon_r \cdot \varepsilon_0}{q} \cdot N_D \cdot (\varphi_0 + \Delta\varphi_0)} - \sqrt{\frac{2 \cdot \varepsilon_r \cdot \varepsilon_0}{q} \cdot N_D \cdot \varphi_0} \quad (T5.4)$$

The value of ΔQ_{st} can be calculated following Equation (5.11) and by taking into account that $N_{st} = N_{st,a} = N_{st,d}$:

$$\Delta Q_{st} = 2 \cdot N_{st} \cdot \Delta E_{FS} \quad (T5.5)$$

The value of φ_0 can be set to the surface band bending in thermal equilibrium which has been calculated in task T5.1 (a) for the conditions of this task. The values of ΔQ_{st} are about $2 \cdot 10^{10}$ cm^{-2} (a) and $2 \cdot 10^{12}$ cm^{-2} (b). The corresponding values of ΔQ_{SC} are about $7 \cdot 10^{10}$ cm^{-2} (a) and $3 \cdot 10^{10}$ cm^{-2} (b). Therefore, a shift of E_{FS} by 0.1 eV is possible for $N_{st} = 10^{11}$ cm^{-2} (a) but impossible for $N_{st} = 10^{13}$ cm^{-2} (b).

T5.3: Engineering of surface band bending with interface defects

The surface band bending for n-type doped semiconductors can be calculated after Equation (5.16). For p-type doped semiconductors Equation (5.16) changes to

$$\varphi_0 = -\frac{2 \cdot B + A}{2} + \sqrt{\left(\frac{2 \cdot B + A}{2}\right)^2 - B^2} \quad (T5.6)$$

For calculating parameter A the value of N_D in Equation (5.15″) is replaced by N_A for p-type doped semiconductors. The differences $E_C - E_{F0}$ (n-type) and $E_{F0} - E_V$ (p-type) are zero for the given conditions. For calculating parameter B the value of $N_{st,a}$ in Equation (5.15‴) is replaced by $N_{st,d}$ for p-type doped semiconductors.

The values of φ_0 are (a) 0.249 V for n-type ($N_D = 10^{18}$ cm^{-3}) and -0.908 V for p-type ($N_A = 10^{19}$ cm^{-3}) and (b) 1.13 V for n-type ($N_D = 10^{18}$ cm^{-3}) and -0.22 V for p-type ($N_A = 10^{19}$ cm^{-3}) doped semiconductors.

A low band bending is desired for contacts with low tunneling resistance. A surface band bending of the order of 0.25 or -0.25 V can be easily reached at surfaces of moderately or highly doped n-type or p-type doped semiconductors, respectively, with high densities of surface defects while an excess of donor or acceptor defects by about 3–5 times is sufficient.

T5.4: Transmission probability of tunneling contacts with inorganic and organic semiconductors

Equation (5.26) is applied. For the organic semiconductors a value of m_e^* equal to m_e is assumed. The transmission probability is $9 \cdot 10^{-4}$ for the given organic and 0.027 for the given inorganic semiconductor.

T5.5: Recombination contact between highly doped semiconductors

The barrier heights for electron and hole tunneling ($\Phi_{B,e}$ and $\Phi_{B,h}$) are given by the drops of the potential across the space charge regions of the n-type and p-type doped regions ($\Phi_{B,e} = q \cdot U_n$ and $\Phi_{B,h} = q \cdot U_p$). The values of U_n and U_p follow from the condition of charge neutrality and from considering the band diagram (Figure 5.16).

$$U_n \cdot N_D^+ = U_p \cdot N_A^+ \tag{T5.7}$$

$$U_n = \frac{E_{g2} - \Delta E_C}{q} - U_p \tag{T5.8}$$

The values of U_n and U_p are:

$$U_n = \frac{E_{g2} - \Delta E_C}{q} \cdot \frac{1}{1 + \frac{N_D^+}{N_A^+}} \tag{T5.9$'$}$$

$$U_p = \frac{E_{g2} - \Delta E_C}{q} \cdot \frac{1}{1 + \frac{N_A^+}{N_D^+}} \tag{T5.9$''$}$$

The values of U_n and U_p are 1.283 and 0.257 V, respectively. The electron and hole tunneling resistances follow from Equation (2.29). The recombination resistance of the degenerated pn-heterojunction can be calculated:

$$R_{c,\text{tunn}} = R_{c,\text{tunn},e} + R_{c,\text{tunn},h}$$

$$= \frac{3 \cdot \hbar}{8 \cdot q \cdot v_{\text{th}}} \cdot \sqrt{\frac{1}{N_D^+ \cdot m_e^* \cdot \varepsilon_r \cdot \varepsilon_0}} \cdot \exp\left(\frac{8}{3} \cdot \sqrt{\frac{m_e^* \cdot \varepsilon_r \cdot \varepsilon_0}{N_D^+}} \cdot \frac{U_n}{\hbar}\right)$$

$$+ \frac{3 \cdot \hbar}{8 \cdot q \cdot v_{\text{th}}} \cdot \sqrt{\frac{1}{N_A^+ \cdot m_h^* \cdot \varepsilon_r \cdot \varepsilon_0}} \cdot \exp\left(\frac{8}{3} \cdot \sqrt{\frac{m_h^* \cdot \varepsilon_r \cdot \varepsilon_0}{N_A^+}} \cdot \frac{U_p}{\hbar}\right)$$

$$\tag{T5.10}$$

The values of $R_{c,\text{tunn},e}$ and $R_{c,\text{tunn},h}$ are about $6 \cdot 10^{-4}$ and $2 \cdot 10^{-7}$ Ωcm^2, respectively, i.e. tunneling of holes limits the given recombination contact. The tolerable series resistance of a triple-junction solar cell is about one order of magnitude larger than for a single junction solar cell due to the increase of V_{OC} and the reduction of I_{SC}. A tolerable series resistance of the order of 0.2 Ωcm^2 is obtained for illumination at AM1.5 if regarding to a V_{OC} of about 2.4 V and an I_{SC} of about 0.012 mA/cm^2 for a multijunction solar cell. The tolerable series resistance of the multijunction solar cell becomes equal to the resistance of one tunneling contact at concentrated sunlight with a concentration factor of 300. For illumination at higher concentrated sunlight a reduction of the given N_A^+ to a value closer to N_D^+ or an increase of N_D^+ will be beneficial.

T5.6: Schottky-barriers for charge-separation and ohmic contacts

A semiconductor at a Schottky-barrier is depleted, i.e. photogenerated minority charge carriers are separated towards the metal–semiconductor contact. In the case of Fermi-level pinning the surface recombination velocity at the metal–semiconductor contact is very high due to the high density of surface defects so that the contact behaves as an Ohmic contact for photogenerated minority charge carriers. This may be a technological advantage since the formation of a *pn*-junction can be omitted in the solar cell. However, (i) the diffusion potential across a *pn*-junction can be varied over a wider range than the surface band bending of a Schottky-barrier and (ii) the density of defects can be reduced to negligible values at a *pn*-homojunction what is not possible with a Schottky-barrier. Therefore, high solar energy conversion efficiencies cannot be obtained with solar cells based on Schottky-barriers.

A.6 Solutions to Chapter 6

T6.1: Shockley–Queisser limit

Equation (1.15) is used for calculating the solar energy conversion efficiency. The power of the sun is $0.1\,\text{W/cm}^2$ in the Shockley–Queisser limit. Following Figure T6.1 the values of I_{SC} are 0.069, 0.034, 0.013 and $0.003\,\text{A/cm}^2$ for band gaps of 0.3, 1.3, 2.0 and 2.7 eV, respectively.

The values of V_{OC} are calculated by using Equations (1.19) and (6.13) for getting the diode saturation current density. Equation (6.13) has to be written in the following way if the values of E_g are given in units of eV:

$$I_0 = \frac{2 \cdot q^3 \cdot \Omega_S \cdot k_B \cdot T_0}{h^3 \cdot c^2} \cdot \left(E_g[\text{eV}]\right)^2 \cdot \exp\left(-\frac{E_g}{k_B \cdot T_0}\right) \tag{T6.1}$$

The pre-factor (I_0') in Equation (T6.1) amounts to:

$$I_0' = \frac{2 \cdot q^3 \cdot \Omega_S \cdot k_B \cdot T_0}{h^3 \cdot c^2} = 1630\frac{\text{A}}{\text{cm}^2} \tag{T6.2}$$

The corresponding values of I_0 and V_{OC} are $1.5 \cdot 10^{-3}\,\text{A/cm}^2$ and 0.1 V, $5.2 \cdot 10^{-19}\,\text{A/cm}^2$ and 1.007 V, $2.3 \cdot 10^{-30}\,\text{A/cm}^2$ and 1.68 V, $8 \cdot 10^{-42}\,\text{A/cm}^2$ and 2.309 V for band gaps of 0.3, 1.3, 2.0 and 2.7 eV, respectively.

The fill factors (FFs) are obtained from Figure 2.4 by extrapolating. The values of FF are 0.6, 0.89, 0.92 and 0.95 for band gaps of 0.3, 1.3, 2.0 and 2.7 eV, respectively.

The solar energy conversion efficiencies of solar cells in the Shockley–Queisser limit are about 4, 30.5, 20 and 6.6% for band gaps of 0.3, 1.3, 2.0 and 2.7 eV, respectively. As remark, these values are accurate within the error caused by extrapolation.

T6.2: Maximum efficiency of tandem solar cells

The solar energy conversion efficiency of tandem solar cells is calculated with the following equation:

$$\eta_{\text{tandem}} = \frac{I_{SC}^{\text{bottom-single}}}{2} \cdot \frac{FF_{\text{bottom}} \cdot V_{OC}^{\text{bottom}} + FF_{\text{top}} \cdot V_{OC}^{\text{top}}}{P_{\text{sun}}} \qquad (T6.3)$$

The values of $I_{SC}^{\text{bottom-single}}/2$ are obtained from Figure T6.1 and amount to 0.033, 0.0235 and 0.0135 A/cm^2 for the band gaps of the bottom cells of 0.5, 1.0 and 1.5 eV, respectively.

The band gaps of the top solar cells correspond to the band gaps of single solar cells with the short circuit current densities equal to $I_{SC}^{\text{bottom-single}}/2$. The values of the band gaps of the top cells are 1.35, 1.65 and 2.00 eV for the band gaps of corresponding the bottom cells of 0.5, 1.0 and 1.5 eV, respectively.

The values of I_0 are calculated using Equations (T6.1) and (T6.2) and amount to $1.9 \cdot 10^{-6}$, $7.8 \cdot 10^{-20}$, $3.1 \cdot 10^{-14}$, $1.2 \cdot 10^{-24}$, $3 \cdot 10^{-23}$ and $2.3 \cdot 10^{-30}$ A/cm^2 for the band gaps of 0.5, 1.35, 1.0, 1.65, 1.5 and 2.0 eV, respectively.

Equation (1.19) is applied for calculating the values of V_{OC} which are equal to 0.254, 1.055, 0.711, 1.335, 1.236 and 1.666 V for the band gaps of 0.5, 1.35, 1.0, 1.65, 1.5 and 2.0 eV, respectively.

The fill factors are found from Figure 2.4 and amount to 0.735, 0.892, 0.86, 0.885, 0.88 and 0.925 for the band gaps of 0.5, 1.35, 1.0, 1.65, 1.5 and 2.0 eV, respectively.

The solar energy conversion efficiencies of ideal tandem solar cells are 37%, 42% and 36% for bottom cells with a band gap of 0.5, 1.0 and 1.5 eV, respectively.

T6.3: Band gaps of absorbers in multi-junction solar cells

The I_{SC} density of a multijunction solar cell with 5 sub-cells is calculated following Equation (6.18):

$$I_{SC}^{k=5} = \frac{I_{SC}^{\text{bottom-single}}}{5} \qquad (T6.4)$$

The ideal top cell has a band gap corresponding to a single solar cell with I_{SC}^{top} equal to $I_{SC}^{k=5}$. The band gaps of the bottom and top cells have the indexes E_{g1} and E_{g5}, respectively. The band gaps are obtained by using Equation (6.19).

$$(6-j) \cdot I_{SC}^{k=5}\Big|_{j \in (1,5)} = q \cdot \int_{E_{g(j)}}^{\infty} \Phi_{sun}(h\upsilon) \cdot d(h\upsilon) \tag{T6.5}$$

For the multijunction solar cell with the bottom cell having a band gap of 0.5 eV the values of I_{SC}^{single} are 0.0132, 0.0264, 0.0396, 0.0528 and 0.066 A/cm^2 corresponding to band gaps of $E_{g5} = 2.05$ eV, $E_{g4} = 1.53$ eV, $E_{g3} = 1.18$ eV, $E_{g2} = 0.8$ eV and $E_{g1} = 0.5$ eV, respectively.

For the multijunction solar cell with the bottom cell having a band gap of 1.0 eV, the values of I_{SC}^{single} are 0.0094, 0.0188, 0.0282, 0.0376 and 0.047 A/cm^2 corresponding to band gaps of $E_{g5} = 2.17$ eV, $E_{g4} = 1.75$ eV, $E_{g3} = 1.47$ eV, $E_{g2} = 1.22$ eV and $E_{g1} = 1.0$ eV, respectively.

T6.4: Concentrator multi-junction solar cells

The I_{SC} density of an ideal triple-junction solar cell is calculated using Equation (6.18).

$$I_{SC}^{k=3} = \frac{I_{SC}^{bottom-single}}{3} \tag{T6.6}$$

The value of $I_{SC}^{bottom-single}$ is 0.06 A/cm^2 ($E_{g1} = 0.7$ eV). Therefore the values of E_{g2} and E_{g3} are 1.17 and 1.75 eV corresponding to band gaps of single-junction solar cells with I_{SC} equal to 0.04 and 0.02 A/cm^2, respectively.

The solar energy conversion efficiency of the triple-junction solar cell is calculated after the following equation:

$$\eta_{triple} = \frac{I_{SC}^{E_{g1}-single}}{3} \cdot \frac{FF_{E_{g1}}^X \cdot V_{OC}^{E_{g1},X} + FF_{E_{g2}}^X \cdot V_{OC}^{E_{g2},X} + FF_{E_{g2}}^X \cdot V_{OC}^{E_{g2},X}}{P_{sun}(AM1.5)} \tag{T6.7}$$

As remark, the concentration factor in the current density and in the power of the sunlight cancelled out in Equation (T6.7). The procedures for finding I_0, V_{OC} and FF are identical to those in previous tasks. The values of η_{triple} are 53.5%, 56%, 57% and 58% for the concentration factors of 20, 200, 500 and 1000, respectively.

T6.5: Maximum concentration factor of sunlight

An absorber cannot emit photons at higher energy than receiving, i.e. the maximum temperature of an absorber cannot exceed the temperature of the sun. The energy flux of the concentrated sunlight is balanced by thermal irradiation of the absorber heated to the maximum temperature of $T_S \approx 5800$ K.

$$X_{\text{max}} \cdot P_{\text{sun}} = \sigma_S \cdot T_s^4 \tag{T6.8}$$

The energy flux received from the sun is the product of the solar constant $(1344\,\text{W/m}^2)$ and the maximum concentration factor (X_{max}). The Stefan–Boltzmann constant is $5.67 \cdot 10^{-8}\,\text{W/(m}^2\text{K}^4)$. The maximum concentration factor is about 46200.

T6.6: Thermodynamic limit of solar energy conversion

The maximum efficiency of thermal energy conversion is reached with the Carnot process (efficiency η_C).

$$\eta_C = 1 - \frac{T_0}{T_A} \tag{T6.9}$$

The temperatures T_A and T_0 correspond to the temperatures of the absorber or medium and to the ambient temperature, respectively. The absorber or medium is heated by the energy flux from the sun which is proportional to the temperature at the surface of the sun (T_S) to the power of four (Stefan–Boltzmann law). The heated absorber emits thermal irradiation which is lost for the conversion into work. The emission losses of the absorber are proportional to the absorber temperature to the power of four. The total efficiency of two subsequent processes is the product of the efficiencies of both processes. Therefore, the efficiency at the thermodynamic limit (η_{th}) is given by the following equation:

$$\eta_{\text{th}} = \left(1 - \left(\frac{T_A}{T_S}\right)^4\right) \cdot \left(1 - \frac{T_0}{T_A}\right) \tag{T6.10}$$

Equation (T6.10) has a maximum of 85% at a temperature of the absorber equal to 2478 K.

T6.7: Critical remark on solar cells with metal intermediate band

The photogeneration rate is increased due to absorption via the metal intermediate band (Figure T6.3). Ideal absorption is assumed for maximum solar energy conversion efficiency. Ideal absorption must be taken into account also for the thermal generation rate. As a consequence the diode saturation current density of a PV absorber increases drastically if implementing a metal intermediate band due to the spectrum of thermal irradiation. Consequently the Fermi-level splitting is drastically reduced and the solar energy conversion efficiency of a solar cell with a metal intermediate band cannot exceed the efficiency of a solar cell with a band gap which is equal to the higher of the differences between the valence or conduction band edges and the energy of the metal intermediate band. As remark, the nature of ideal charge-selective contacts has no influence on the Fermi-level splitting in an ideal PV absorber.

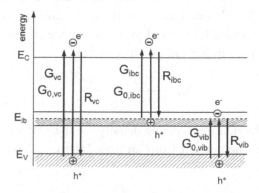

Figure T6.3. Generation and recombination processes in a PV absorber with a metal intermediate band meaning that there are occupied and unoccupied states available for excitation. G_{vc}, $G_{0,vc}$ G_{ibc}, $G_{0,ibc}$, G_{vib} and $G_{0,vib}$ denote photogeneration and thermal generation from the valence into the conduction band of the PV absorber, from the metal intermediate band into the conduction band of the absorber and from the valence band of the absorber into the intermediate band, respectively. Recombination takes place from the conduction band into the valence band of the absorber (R_{vc}), from the conduction band of the absorber into the metal intermediate band (R_{ibc}) and from the metal intermediate band into the conduction band of the absorber (R_{vib}).

A.7 Solutions to Chapter 7

T7.1: n-type doped base and p-type doped emitter

Equation (7.1′) is applied for calculating n_0 of the n-type doped base. The value of n_0 is $1 \cdot 10^{16}$ cm^{-3} if assuming a thickness of the base of 200 μm and a value of D_p which is 3.5 times smaller than 26 cm^2/s. The thickness of the p-type doped emitter is obtained by using the following equation:

$$d_{em} \leq \sqrt{\frac{D_n}{100 \cdot C_A \cdot p_0^2}} \qquad (T7.1)$$

The maximum useful density of acceptors is about the effective density of states at the valence band edge ($3 \cdot 10^{19}$ cm^{-3}). The value of d_{em} is about 200 nm.

T7.2: Silver required for c-Si based photovoltaics

A bus bar collecting the current from a square with an area of an L^2 is considered for the estimation of the amount of silver. The length of the bus bar and its cross-section are L and A_{cr}. The specific conductivity and the density of silver are denoted as σ_{Ag} and ρ_{Ag}. According to the definition of the $R_{s,tol}$ (Equation (1.29′)) and to the resistance (Equation (1.30)) the cross-section of the bus bar can be expressed as:

$$A_{cr} = \frac{100}{\sigma_{Ag}} \cdot L \cdot \frac{I_{SC}}{V_{OC}} \qquad (T7.2)$$

Following the definition of the density the mass of silver required for the collection of the current from an area $L^2 (m_{Ag-L^2})$ is calculated after equation:

$$m_{Ag-L^2} = 100 \cdot \frac{\rho_{Ag}}{\sigma_{Ag}} \cdot L^2 \cdot \frac{I_{SC}}{V_{OC}} \qquad (T7.3)$$

The mass of silver needed for 1 W of installed power is obtained from m_{Ag-L^2} by normalizing to the power of the solar cell with the area of L^2.

$$m_{Ag-W} = 100 \cdot \frac{\rho_{Ag}}{\sigma_{Ag}} \cdot \frac{L^2}{FF \cdot V_{OC}^2} \qquad (T7.4)$$

Bus bars cover a specific area of about $25\,cm^2$ in c-Si solar cells. If assuming a fill factor of 0.75 and a V_{OC} of $0.6\,V$ a mass of about $160\,mg$ of silver is required for one solar cell giving a power of $1\,W_p$. Therefore silver is an important cost factor for c-Si solar cells.

The electricity demand of mankind can be estimated if assuming $2\,kW$ per capita, a number of $7 \cdot 10^9$ people and a performance of 25% (solar cells produce electricity only during day time). The amount of silver which would be required for complete electricity production with c-Si based photovoltaic is about $9 \cdot 10^6$ t, i.e. about 400 times larger than the annual silver production. Therefore, silver has to be replaced in c-Si solar cells as a strategic issue for global electricity production with PV power plants.

T7.3: Boron doped c-Si(100) wafers

Large silicon crystals can be doped very homogeneously with boron since boron has a segregation coefficient close to one. Further, boron is an acceptor in c-Si and minority charge carriers have a longer diffusion length in p-type than in n-type doped c-Si. Therefore a base with a given thickness can be higher p-type than an n-type doped what leads to a higher diffusion potential and therefore to a higher V_{OC}.

T7.4: Diffusion of phosphorus in c-Si

The diffusion coefficient of phosphorus in c-Si is about $10^{-15}\,cm^2/s$ at 900°C. The thickness of the emitter can be calculated by using Equation (7.25) and amount to $56\,nm$ for a diffusion time of $1000\,s$.

T7.5: Diffusion of copper in c-Si

The diffusion from a localized Cu contamination corresponds to diffusion from an exhaustible source and therefore Equation (7.26) has to be applied for calculating the spread of Cu. The diffusion coefficient of Cu is $D_i = 5 \cdot 10^{-5}\,cm^2/s$ at 900°C. A contamination of one monolayer of Cu corresponds to a density of about $Q_{is} = 10^{15}\,cm^{-2}$. The density of Cu should be lower than $N_{i,max} = 10^{12}\,cm^{-3}$ with respect to the minimum lifetime condition in c-Si absorbers and Shockley–Read–Hall recombination (Chapter 3). The spread $x_{s,Cu}$ of the contamination can be defined as the depth at which $N_{i,\max}$ is reached. The value of $x_{s,Cu}$ can be calculated by using the following equation:

$$x_{s,Cu} = \sqrt{4 \cdot D_i \cdot t \cdot \ln\left(\frac{Q_{is}}{N_{i,\max} \cdot \sqrt{\pi \cdot D_i \cdot t}}\right)} \qquad (T7.5)$$

The spreads of Cu are 1.4 mm, 4.2 mm and 1.2 cm after diffusion from a Cu contamination of only one monolayer at 900°C for 10, 100 and 1000 s, respectively. This example demonstrates why any Cu contamination has to be avoided for the production of c-Si solar cells if the processes include conventional emitter diffusion and/or surface passivation by oxidation at high temperatures.

T7.6: V_{OC} of PERL and HIT c-Si solar cells

In PERL c-Si solar cells a conventional *pn*-junction is used as a charge-selective contact. Very high values of V_{OC} can only be achieved for a base with a diffusion length of minority charge carriers much larger than the thickness of the base concomitant with extremely low surface recombination velocities (Chapter 4). The overall surface recombination rate is still influenced by the Ohmic contacts despite the back surface field due to the relatively low barrier for minority charge carriers and due to photogeneration in barrier regions of the back surface field of PERL c-Si solar cells.

In HIT c-Si solar cells a-Si:H layers passivate the complete c-Si base and form charge-selective contacts with (i) additional barriers for minority charge carriers due to the band offsets at a-Si:H/c-Si heterojunctions. In addition (ii) the thin a-Si:H layers do not contribute to photogeneration. Therefore the density of minority charge carriers is stronger reduced along the Ohmic contacts for HIT than for PERL c-Si solar cells. Further (iii) the thickness of the base is only 98 μm in HIT c-Si solar cells while the thickness of the base is 260 μm in PERL c-Si solar cells. This improves additionally the ratio between the thickness of the base and the diffusion length of minority charge carriers. Therefore the geometry factor of the diode saturation current density can be stronger reduced for HIT than for PERL c-Si solar cells.

There is a disadvantage of HIT c-Si solar cells caused by absorption losses in the TCO front contact and in a-Si:H layers which limit I_{SC}.

T7.7: Potential of kerfless wafer transfer technologies

In conventional wafer technology about half of a large c-Si crystal is transformed to slick during wire sawing and following polishing of the wafers, i.e. half of the energy treated for the production of large c-Si crystals is wasted. Kerfless wafer transfer technologies avoid this waste in energy. In addition, the thickness of wafers can be strongly reduced to 25 μm by using exfoliation of c-Si absorbers. This means that the amount of c-Si can be reduced by up to 20 times in total. However, one has to keep in mind that the solar energy conversion efficiency of solar cells produced on exfoliated c-Si absorbers cannot reach the high efficiency of c-Si solar cells produced by advanced wafer based technologies. This problem one can overcome by applying beam induced wafering techniques, for example. It can be expected that kerfless wafer transfer technologies can lead to a reduction of the energy payback time of c-Si solar cells by about 30%. However, one has to keep in mind that dozens or hundreds of wafers can be cut by wire sawing from one large c-Si crystal at the same time. This parallel processing is not possible with kerfless wafer transfer technologies from one large c-Si crystal. Therefore parallel processing of many large c-Si crystals with big machines will be required for high through put in mass production of c-Si wafers by kerfless wafer transfer.

A.8 Solutions to Chapter 8

T8.1: Type I and type II semiconductor hetero-junctions

The band gaps are taken from Figure 8.2 and the natural valence band offsets in relation to GaAs are taken from Figure 8.6, the values of the band gaps. Then the natural conduction band offsets are calculated by using the following equation.

$$\Delta E_C = (E_{g2} - E_{g1}) - \Delta E_V \qquad (T8.1)$$

Natural GaP/InP, InN/AlAs and GaAs/InAs interfaces form type I, type II and type I heterojunctions, respectively.

T8.2: Influence of an interface dipole on the band offset

The charge density at the dipole layer can be estimated from the reciprocal area of one unit cell.

$$\Delta Q_{pol} = \frac{0.1 \cdot q}{a_{lc}^2} \tag{T8.2}$$

With respect to Equations (8.7) and (T8.2) the change of the band offset caused by the interfacial dipole can be estimated:

$$\Delta E_{dipol} = q \cdot \frac{0.1 \cdot q}{\varepsilon \cdot \varepsilon_0 \cdot a_{lc}} \tag{T8.3}$$

The value of ΔE_{dipol} is about 0.27 eV if assuming a lattice constant of 0.57 nm. According to Figure 8.6 and to the band gaps of the given semiconductors the natural valence band offset at the GaAs/InAs interface is about 0.06 meV and the natural conduction band offset is about 0.23 eV at the GaAs/AlAs interface. Therefore the type of the heterojunctions can change depending on the direction and the value of the interface dipole layer at the heterojunction.

T8.3: Tuning of band gaps in lattice-matched epitaxy

According to Figure 8.3 all band gaps of $Ga_xIn_{1-x}P$ with $x > 0.07$ are larger than all band gaps of $Ga_xIn_{1-x}As$. Therefore the maximum and minimum band gaps of $Ga_xIn_{1-x}As_yP_{1-y}$ are related to the band gaps of $Ga_xIn_{1-x}P$ and $Ga_xIn_{1-x}As$ with the $a_{lc} = 0.58$ nm. According to the lattice constants of the binary III–V semiconductors and to the Vegard's law (Equation (8.8)) stoichiometry parameters of $x = 0.16$ and $x = 0.64$ are obtained for $Ga_xIn_{1-x}P$ and $Ga_xIn_{1-x}As$, respectively. As remark, the stoichiometry parameter of $Ga_{0.16}In_{0.84}P$ is larger than 0.07, i.e. the assumption made before was correct. According to the dependence of the band gap on the stoichiometry and bowing parameters (Equation (8.2)) the band gaps of $Ga_{0.16}In_{0.84}P$ and $Ga_{0.64}In_{0.36}As$ are 1.51 and 0.93 eV, respectively.

T8.4: Absorber resistance of solar cells based on III–V semiconductors

The thickness of the absorber layer is about 3 μm in GaAs solar cells and a density of majority charge carriers of 10^{17} cm^{-3} can be assumed. The resistance of the GaAs absorber can be calculated according to Equation (7.3) and according to the values of the mobility of holes and electrons.

$$R_{\text{absorber}}^{\text{GaAs}} = \frac{d_{\text{abs}}}{q \cdot p(n)_0 \cdot \mu_{p(e)}} \approx 10^{-5} - 10^{-6} \, \Omega\text{cm}^2 \qquad \text{(T8.4)}$$

A GaAs absorber with a thickness of 3 μm and a density of majority charge carriers of 10^{17} cm^{-3} has a resistance of 10^{-5}–10^{-6} Ωcm^2. For comparison, the tolerable series resistance of a GaAs solar cell operated at AM1.5G is about 0.3 Ωcm^2 ($V_{\text{OC}} = 1.107$ V, $I_{\text{SC}} = 29.6$ mA/cm^2 (Kayes *et al.*, 2011). Therefore the bulk resistance of solar cells based on III–V semiconductors is not critical at all even for operation under highly concentrated sunlight.

T8.5: Epitaxial layers in tandem solar cells with III–V semiconductors

The layer system of the GaAs/Al$_{0.36}$Ga$_{0.64}$As tandem solar cell consists of 13 separate layers (layers 1–13). The epitaxy starts with the growth of a buffer layer (layer 1) for reducing the density of defects being present at the surface of the GaAs substrate. The buffer layer and the substrate are *n*-type doped. A barrier layer for holes (layer 2) is grown next. The *n*-type and *p*-type doped layers 3 and 4 absorb sunlight and form the *pn*-junction of the GaAs bottom cell. Layer 5 forms a barrier for electrons. The tunneling junction is formed by highly *p*-type and *n*-type doped GaAs layers (layers 6 and 7). A barrier layer for holes is deposited as next (layer 8). The *n*-type and *p*-type doped layers 9 and 10 absorb sunlight and form the *pn*-junction of the Al$_{0.36}$Ga$_{0.64}$As bottom cell. Layers 11 and 12 are responsible for the transport of holes towards the ohmic front contact while layer 12 forms an additional barrier layer for electrons. Layer 13 is made from highly *p*-type doped GaAs and forms the ohmic contact with a metal layer system.

T8.6: Maximum efficiency of a lattice-matched triple-junction solar cell

The short circuit current density of the triple-junction solar cell is limited by the top cell since the band gap of the top cell is significantly larger than the optimum band gap of the top cell for the given bottom cell (see Chapter 6). The value of I_{SC} is obtained from Figure T6.1. The values of the diode saturation current densities are calculated for the top, middle and bottom cells using Equation (6.12) and the corresponding fill factors can be taken from Figure 2.4 by extrapolating. The values of V_{OC} are calculated for the top, middle and bottom cells by using Equation (1.19).

$$\eta_{max}^{triple} = \frac{I_{SC}}{P_{sunlight}} \cdot \left[FF_{top} \cdot V_{OC,top} + FF_{middle} \cdot V_{OC,middle} \right.$$

$$\left. + FF_{bottom} \cdot V_{OC,bottom} \right] \tag{T8.5}$$

The maximum energy conversion efficiency of the lattice matched triple-junction solar cell is 47%. For comparison the maximum efficiency of a triple-junction solar cell with optimum band gaps is 50%.

T8.7: Lattice mismatch for lattice-matched epitaxy

First, the stoichiometry parameters are calculated by using Equation (8.2), the values of the corresponding band gaps of the binary III–V semiconductors for the Γ valleys (1.42, 0.42, 1.35 and 2.9 eV for GaAs, InAs, InP and GaP, respectively) and of the bowing parameters ($C = 0.477$ for $Ga_xIn_{1-x}As$ and $C = 0.65$ for $Ga_xIn_{1-x}P$ for $x < 0.64$).

$$x = \sqrt{\left[\frac{E_g(A) - E_g(B) - C}{2 \cdot C} \right]^2 + \frac{E_g(A_xB_{1-x}) - E_g(B)}{C}}$$

$$- \frac{E_g(A) - E_g(B) - C}{2 \cdot C} \tag{T8.6}$$

The stoichiometries $In_{0.97}Ga_{0.03}As$ and $In_{0.43}Ga_{0.57}P$ are calculated. According to the lattice constants of the binary III–V semiconductors and to the Vegard's law (Equation (8.8)) the lattice constants of $a_{lc} = 0.56653$ nm and $a_{lc} = 0.56894$ nm are found for $In_{0.97}Ga_{0.03}As$ and $In_{0.43}Ga_{0.57}P$, respectively. For comparison, the lattice constant of germanium is 0.56575 nm. The lattice mismatch is defined by Equation (8.13). The lattice mismatch is 0.14% between Ge and $Ga_{0.97}In_{0.03}As$ and 0.42% between $Ga_{0.97}In_{0.03}As$ and $In_{0.43}Ga_{0.57}P$.

T8.8: Lattice mismatch for lattice-mismatched epitaxy

The stoichiometry parameters are calculated regarding to Equation (T8.6). The stoichiometries $In_{0.91}Ga_{0.09}As$ and $In_{0.39}Ga_{0.61}P$ are calculated for the triple-junction solar cell grown on Ge and $Ga_{0.7}In_{0.3}As$ for the inverted metamorphic triple-junction solar cell. The lattice constants 0.56905, 0.57747 and 0.57062 nm are found for $In_{0.97}Ga_{0.03}As$, $Ga_{0.7}In_{0.3}As$ and

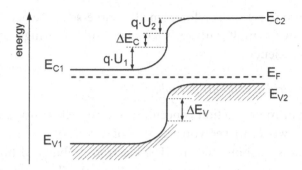

Figure T8.1. Band diagram of a type II *pn*-heterojunction.

$In_{0.43}Ga_{0.57}P$, respectively. The lattice mismatch is 0.58% between Ge and $Ga_{0.91}In_{0.09}As$ and 0.28% between $Ga_{0.91}In_{0.09}As$ and $In_{0.39}Ga_{0.61}P$. For comparison, the lattice mismatch is 2.1% between $Ga_{0.7}In_{0.3}As$ and GaAs.

T8.9: Type II hetero-junctions for charge separation

Figure T8.1 shows the band diagram of a type II *pn*-heterojunction. It is assumed that the semiconductor with index 1 is *n*-type (N_{D1}) and the semiconductor with index 2 *p*-type doped (N_{A2}). The diffusion potential can be expressed by the following equations.

$$q \cdot U_D = U_1 + U_2 = E_{g2} - (E_{C1} - E_F) - (E_F - E_{V2}) - \Delta E_C \quad (T8.7')$$

$$q \cdot U_D = E_{g1} - (E_{C1} - E_F) - (E_F - E_{V2}) - \Delta E_V \quad (T8.7'')$$

Equations (T8.7′) and (T8.7″) show that the maximum possible $q \cdot V_{OC}$ (ultimate efficiency, Chapter 2) is given by the differences between E_{g2} and ΔE_C or between E_{g1} and ΔE_V, i.e. $q \cdot V_{OC}$ is lower than each of the band gaps. As consequence, the solar energy conversion efficiency of a solar cell with a type II heterojunction as a charge selective contact is less than the solar energy conversion efficiency of a solar cell with a *pn*-homojunction. For avoiding losses caused by the band offsets one of the band offsets has to be reduced to zero. In this case the semiconductor with the larger band gap acts as a window layer and highly doped $p^{++}n^{++}$-tunneling homojunctions will be required for integrated series connection in related multijunction solar cells. However, it is practically impossible to reduce one of the band offsets to zero for a series of photovoltaic absorbers concomitant with the reduction of defects at the charge-selective heterojunction to zero. Therefore, type II charge-selective heterojunctions are not useful for multijunction solar cells with very high solar energy conversion efficiency.

T8.10: Finger grid of concentrator solar cells

The emitter of multijunction concentrator solar cells consists of a highly *n*-type doped window layer with a density of donors of about $5 \cdot 10^{18}$ cm^{-3} and a thickness of about 100 nm. The electron mobility of highly doped III–V semiconductors is of the order of 1000 cm^2/Vs. The emitter resistance given by Equation (7.11) should not be larger than the tolerable series resistance defined by Equation (1.29′). The specific resistance of the emitter is given by Equation (7.8). Therefore the distance between the grid fingers can be calculated using the following equation.

$$a \le \sqrt{\frac{1}{100} \cdot \frac{V_{OC}}{I_{SC}} \cdot 12 \cdot d_{em} \cdot q \cdot n_0 \cdot \mu_{n+}} \qquad \text{(T8.8)}$$

The value of V_{OC} of a triple-junction solar cell illuminated with highly concentrated sunlight is about 3 V. The value of I_{SC} of a triple-junction solar cell illuminated with sunlight concentrated by a factor of 2000 is about 30 A/cm^2. The resulting distance between grid fingers is about 100 μm. As remark, the width of the grid fingers should be about 5 μm or even less for minimizing losses by shading. Photolithography is required for producing related finger grids for concentrator solar cells.

A.9 Solutions to Chapter 9

T9.1: Bulk and tunneling resistances at grain boundaries in a TCO layer

The length, the height and the width of the TCO layer are denoted by L, H and B, respectively. The bulk resistance of the TCO layer can be calculated using Equation (9.2).

$$R_{bulk} = \frac{1}{q \cdot n_0 \cdot \mu_n} \cdot \frac{L}{B \cdot H} \qquad \text{(T9.1)}$$

The tunneling resistance through the barrier of one depletion region can be calculated by using Equation (5.30) and by keeping in mind the effective mass of the tunneling electron. Further the tunneling resistance has to be doubled since tunneling occurs via defect states at the grain boundary. The tunneling resistance of the TCO layer is obtained by taking into account the number of barriers across the length and the cross-section of tunneling barriers perpendicular to the direction of the current flow.

$$R_{\text{tunn}} = 2 \cdot R_{c,\text{tunn},e} \cdot \frac{L}{d_g} \cdot \frac{1}{BH} \tag{T9.2}$$

$$R_{c,\text{tunn},e} \approx \frac{\Omega \text{cm}^2}{1.26 \cdot 10^9} \cdot \sqrt{\frac{10^{19} \text{ cm}^{-3}}{N_D^+}}$$

$$\cdot \exp\left(79 \cdot \sqrt{\frac{10^{19} \text{ cm}^{-3}}{N_D^+}} \cdot \sqrt{\frac{m_e^*}{m_e} \cdot \frac{\Phi_B}{\text{eV}}}\right) \tag{T9.3}$$

The specific bulk resistance is $6 \cdot 10^{-4}$ Ωcm. The specific tunneling resistance is given by

$$\rho_{\text{tunn}} = \frac{2}{d_g} \cdot R_{c,\text{tunn},e} \tag{T9.4}$$

The specific tunneling resistances are 0.01 Ωcm for $\varphi_0 = 1.2$ V (a) and $6 \cdot 10^{-5}$ Ωcm for $\varphi_0 = 0.3$ V (b), i.e. the specific tunneling resistance is larger than the specific bulk resistance of the TCO layer for $\varphi_0 = 1.2$ V but lower for $\varphi_0 = 0.3$ V. This example demonstrates the importance of the control of defects at grain boundaries in TCO layers.

T9.2: TCO contact stripe of thin-film solar cells

Values of 8 mA/cm^2 and 1.7 V can be assumed for I_{SC} and V_{OC}, respectively, of a pin–pin a-Si:H solar cell with respect to the world record a-Si:H solar cell (Table 9.3). Transformation of Equations (9.3) and (9.4) and considering that the specific resistance is the reciprocal conductivity gives the following expression for the width of the TCO stripe:

$$L \leq \sqrt{\frac{1}{100 \cdot R_\square} \cdot \frac{V_{OC}}{I_{SC}}} \tag{T9.5}$$

The resulting value of L is 6.6 mm.

T9.3: Optical losses in TCO layers

The AM1.5d spectrum can be found in the internet. Optical losses in the ZnO:Al are caused by reflection and free carrier absorption. The refractive index of ZnO is calculated with respect to the Sellmeier equation (9.5) and the parameters given by Sun and Kwok (1999) for ZnO thin films. The reflection losses are considered in the transmission coefficient of the ZnO:Al layer which is given by Equation (9.6) and the parameter Equations (9.7') and (9.8).

The transmission losses due to free carrier absorption are calculated by using the Lambert–Beer law with an absorption coefficient which is proportional to the wavelength to the power of three and which is proportional to n_0 what is expressed in the pre-factor α_0.

$$\alpha(\lambda, n_0) = \alpha_0(n_0) \cdot \left(\frac{\lambda}{\lambda_0}\right)^3 \qquad (T9.6)$$

The value of α_0 is about 200 cm^{-1} at a wavelength $\lambda_0 = 1\,\mu$m and for n_0 equal to $5 \cdot 10^{19}$ cm^{-3} (Weiher, 1966).

The value of I_{SC} is obtained by integrating the product of the photon flux and the amount of light transmitted through the ZnO:Al layer from 330 nm to the wavelength corresponding to the band gap of the photovoltaic absorber (λ_g). The thickness of the TCO layer is denoted by d_{TCO}.

$$I_{SC} = q \cdot \int_0^{\lambda_g} \Phi_{sun}(\lambda) \cdot T_{tr} \cdot [1 - \exp(-\alpha(\lambda, n_0) \cdot d_{TCO})] \cdot d\lambda \qquad (T9.7)$$

Figure T9.1 shows a plot of I_{SC} as a function of λ_g for an ideal solar cell without losses and with optical losses due to reflection and free carrier absorption in the ZnO:Al surface layer.

The values of I_{SC} are 19.6 mA/cm^2 (a-Si:H, $\lambda_g = 730$ nm) and 42.3 mA/cm^2 (CuInSe$_2$, $\lambda_g = 1230$ nm) for AM1.5d without losses. Keep in mind that I_{SC} would be larger by a factor 1.1 for AM1.5G. The values of I_{SC} are reduced by about 14 and 15% if considering only reflection losses for a-Si:H and CuInSe$_2$, respectively, with a 0.5 μm thick ZnO surface layer. The reduction of I_{SC} increases to 19 (27) and 24 (44)% for a-Si:H and CuInSe$_2$, respectively, with a 0.5 (1.0) μm thick ZnO:Al surface layer (consideration of reflection and free carriers absorption). For comparison, I_{SC} is reduced by about 25 and 20% for the a-Si:H and CuIn$_{0.8}$Ga$_{0.2}$Se$_2$ record solar cells,

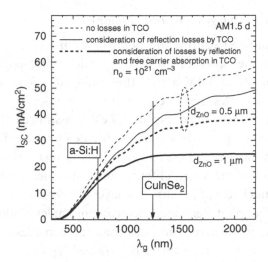

Figure T9.1. Dependence of the short circuit current density on the wavelength corresponding to the band gap of a PV absorber (λ_g) without consideration of losses in the TCO (thin dotted line), with consideration of optical losses due to reflection in a 0.5 μm thick TCO (ZnO) surface layer (thin solid line) and with consideration of optical losses due to reflection and free carrier absorption in a 0.5 and 1 μm thick ZnO:Al surface layer (thick dotted and thick solid lines, respectively). The values of λ_g are shown for a-Si:H and CuInSe$_2$ absorber layers.

respectively, in comparison to the maximum I_{SC} at the corresponding values of λ_g.

T9.4: Burstein shift in ZnO:Al

The density of states integrated from E_C to $E_C + \Delta E_g$ is equal to the density of free electrons in the degenerated semiconductor or metal oxide. Equation (2.16′) can be used while E_C is set to zero.

$$\Delta E_g = \left[\frac{cm^3}{6.7 \cdot 10^{21}} \cdot \left(\frac{m_e}{m_e^*} \right)^{\frac{3}{2}} \cdot N_D \right]^{\frac{2}{3}} eV \qquad (T9.8)$$

The resulting values of ΔE_g are 0.23 eV and 1.06 eV for densities of free electrons equal to 10^{20} and 10^{21} cm^{-3}, respectively.

T9.5: Open-circuit voltage in an a-Si:H solar cell

The densities of free electrons and holes in the n-type and p-type doped regions are of the order of 10^{18} cm^{-3} which correspond to a built-in potential of about 1.4 V across the undoped a-Si:H absorber layer (the densities of states at the conduction and valence band edges are about 10^{20} cm^{-2} and a band gap of 1.7 eV is assumed). The density of photogenerated charge carriers can be estimated by using Equation (3.8). The maximum short circuit current density of a solar cell with an a-Si:H absorber is about 16 mA/cm^2. A lifetime of photogenerated charge carriers of 10 ns and a thickness of the absorber layer of 1 μm are considered. Then the density of free photogenerated electrons and holes is about 10^{13} cm^{-3}, i.e. about 5 orders of magnitude less than the densities of the majority charge carriers in the p-type and n-type doped regions. A difference of 5 orders of magnitude in the densities of free charge carriers corresponds to a difference of 0.3 V in the chemical potentials. The value of V_{OC} can be estimated as the difference between the built-in potential and the differences of the chemical potentials for electrons and holes in the undoped absorber and the p-type and n-type doped regions, respectively. Therefore V_{OC} is about 0.8 V for a-Si:H solar cells. The remaining built-in potential under illumination of about 0.6 V is sufficient for realizing a drift length of the order of the thickness of the undoped a-Si:H absorber. Under light trapping the thickness of the undoped a-Si:H absorber layer can be reduced so that higher densities of photogenerated charge carriers can be obtained and a lower remaining built-in potential can be tolerated for sufficient drift. For comparison, V_{OC} is 0.886 V for the world record pin a-Si:H solar cell (Benagli *et al.*, 2009).

T9.6: Chalcopyrite tandem solar cell

The values of I_{SC}, V_{OC} and FF of suitable single junction bottom ($E_{g,bottom} = 1.17$ eV) and top ($E_{g,top} = 1.53$ eV) cells are 37.8 and 18.9 mA/cm^2, 0.741 and 0.960 V and 80.6 and 68.0%, respectively. The values of I_{SC} of the bottom and top cells are well matched if assuming that (i) the top cell and the recombination contact are transparent for all photons with energy below E_g of the top cell and (ii) a recombination contact can be realized. Therefore, the values of I_{SC}, V_{OC} and FF of the top cell remain unchanged in the tandem cell. In general, the values of V_{OC} and FF are slightly reduced for the bottom cell in comparison to the single-junction cell. A maximum η of the tandem solar cell can be estimated if considering the values of V_{OC} and FF of the single-junction solar cells. Under the given assumptions, the maximum η of the chalcopyrite tandem solar cell would be 23.6%, which is larger than η of the world record chalcopyrite solar cell.

T9.7: Advantages of CdTe thin-film technology

First, CdTe exists in only one phase and sublimates in correct stoichiometry. Therefore the stoichiometric CdTe phase forms in a relatively wide technological window. Second, CdTe has a large tolerance against incorporation of impurities from the targets and from atmosphere due to segregation. Third, $CdCl_2$ treatment at reduced temperature enables the formation of large grains. During this process the density of defects in the crystallites is reduced and additional segregation of impurity atoms towards grain boundaries takes place. Fourth, an intimate ohmic contact is formed at further reduced temperature by incorporation of acceptors into CdTe near the CdTe/contact layer interface.

T9.8: Strategic potential of CdTe, chalcopyrite and kesterite solar cells

The mass of an element (m_i) required for 1 W_p of installed power can be calculated by using the following equation:

$$\frac{m_i}{W_p} = \rho_{abs} \cdot \frac{M_i}{M_{abs}} \cdot \frac{d_{abs}}{P_{sun} \cdot \eta} \tag{T9.9}$$

The symbols ρ_{abs}, M_i, M_{abs}, d_{abs}, P_{sun} and d_{abs} denote the density of the absorber material, the molar mass the element i, the molar mass of the absorber material, the thickness of the absorber layer, the power of sunlight per cm^2 (0.1 W/cm^2) and the efficiency of the solar cells with the given absorber. Efficiencies of 12% for CdTe, 15% of chalcopyrite and 10% of kesterite PV modules are assumed. The densities of CdTe, chalcopyrite and kesterite absorbers are 5.85 and about 5.7 and 5 g/cm^2, respectively. Values of d_{abs} of 5, 3 and 3 μm are considered for solar cells based on CdTe, chalcopyrite and kesterite absorbers, respectively. The molar masses of Te, In and Sn are 127.6, 114.8 and 118.7 g/mol, respectively. The molar masses of CdTe, $CuIn_{0.75}Ga_{0.25}Se_2$ and $Cu_2ZnSn(Se_xS_{1-x})_4$ absorbers are 240 g/mol, 324.4 g/mol and about 500 g/mol, respectively.

(a) The masses needed for one W_p are about 0.13 g/W_p (CdTe), 0.03 g/W_p ($CuIn_{0.75}Ga_{0.25}Se_2$) and 0.03 g/W_p ($Cu_2ZnSn(Se_xS_{1-x})_4$).

The installed power of PV power plants should be of the order of 10^{14} W_p if all electricity will be produced only by PV, if considering a population of $7 \cdot 10^9$ individuals and if taking into account the fact that solar cells do not produce electricity during night.

(b) The required total masses of Te, In and Sn would be about 10^7 t (CdTe based PV), $3 \cdot 10^6$ t (chalcopyrite based PV) and $3 \cdot 10^6$ t (kesterite based PV). The required amounts of Te, In and Sn would exceed the annual production of Te, In and Se by more than 10^5 (!), 10^3 (!) and 10 times, respectively, i.e. conventional CdTe and chalcopyrite thin-film solar cells are not suitable for global electricity production.

A.10 Solutions to Chapter 10

T10.1: Solar cell with extremely thin absorber

The diffusion and drift lengths have to be compared (Equations (9.16) and (9.17)).

$$L = \sqrt{\frac{k_B \cdot T}{q} \cdot \mu \cdot \tau} \qquad (9.16)$$

$$L_{drift} = \mu \cdot \tau \cdot \varepsilon \qquad (9.17)$$

The diffusion length amounts to 50 nm. An electric field of about 10^4 V/cm can be assumed under illumination. Then the drift length is 100 nm. The effective thickness of the absorber layer should be larger than 3 times the absorption length of the absorber material. Therefore the surface area should be increased by a factor of 60 if taking into account the drift length.

T10.2: Surface atoms at colloidal quantum dots

The number of atoms in a quantum dot is obtained by using the definitions of the density and of the molar mass and by taking into account a sphere with a radius which is half of the diameter of the colloidal quantum dot (QD). The molar mass of PbS (M_{PbS}) is 239.3 g/mol. The Avogadro number (N_A) is $6.024 \cdot 10^{23}$ per mol. Lead and sulfur atoms are considered. The bond length is denoted by δ_{bond}. The number of bulk atoms of a quantum (N_{bulk}) dot is given by the radius reduced by the bond length. The number of surface atoms of a QD (N_{surf}) is equal to the difference between N_{PbS-QD} and N_{bulk}.

$$N_{PbS-QD} = 2 \cdot \frac{\rho_{PbS}}{M_{PbS}} \cdot N_A \cdot \frac{4\pi}{3} \cdot r_{QD}^3 = 160 \cdot r_{QD}^3 [nm^3] \quad \text{(T10.1)}$$

$$N_{bulk} = \frac{8\pi}{3} \cdot \frac{\rho_{PbS}}{M_{PbS}} \cdot N_A \cdot \left(r_{QD} - \delta_{bond}\right)^3 \quad \text{(T10.2)}$$

$$N_{surf} = N_{PbS-QD} - N_{bulk} \quad \text{(T10.3)}$$

The numbers of bulk atoms are 55, 786 and 3149 for PbS QDs with diameters of 2, 4 and 6 nm, respectively. At the same time the numbers of surface atoms are 105, 494 and 1171 for PbS QDs with diameters of 2, 4 and 6 nm, respectively, i.e. the numbers of bulk and surface atoms are very similar in colloidal quantum dots.

T10.3: QD absorber layer

The effective mass of excitons and the absorption edge the PbS QD absorber can be calculated after Equations (10.1) and (10.4). The increase of the absorption gap due to quantum confinement in PbS QDs can be expressed by the following equation:

$$\Delta E_{QC} = \frac{h^2}{8 \cdot r_{QD}^2 \cdot \mu_{e,h}} \approx 5.6 \cdot \frac{eV}{r_{PbS-QD}^2 [nm^2]} \quad \text{(T10.4)}$$

Reduction of the energy by the Coulomb potential in PbS QDs can be expressed by the following equation:

$$\Delta E_{Coul} = \left(\frac{1}{\varepsilon_{out}} + \frac{0.79}{\varepsilon_{in}}\right) \cdot \frac{q^2}{4\pi \cdot r_{QD} \cdot \varepsilon_0}$$

$$\approx 0.56 \cdot \frac{eV}{r_{PbS-QD} [nm]} \quad \text{(T10.5)}$$

The band gaps are 5.4, 1.5 and 0.8 eV for PbS-QDs with diameters of 2, 4 and 6 nm, respectively, which correspond to a maximum I_{SC} of less than

1 mA/cm^2 (2 nm), 28 mA/cm^2 (4 nm) and 53 mA/cm^2 (6 nm). In literature, for example, a value of the absorption edge of 1.3 eV is given for PbS-QDs with a diameter of 3.5 nm. Differences are caused by variations of effective electron and hole masses, by polarization phenomena at interface dipole layers, penetration of solvents, disordered distribution of distances between QD in a layer, etc.

T10.4: Strategic potential of PbS-QD solar cells and solar cells sensitized with CH$_3$NH$_3$PbI$_3$

Equation (T9.7) is applied for calculating the mass of Pb (m_{Pb}) required for 1 W_p of installed power. A solar energy conversion efficiency of 10% and an effective thickness of the PbS-QD absorber in terms of compact PbS of 100 nm are assumed. The molar masses of Pb and PbS are 207 and 239 g/mol, respectively. Less than 7 mg of Pb is needed for one W_p. The corresponding mass of Pb for complete electricity production for mankind would be about 7·10^5 t which is ten times less (!) than the annual lead production. Therefore solar cells based on PbS-QDs have a tremendous strategic potential.

T10.5: Absorption and transport gaps in nano-composite absorbers

The fundamental limitation of the solar energy conversion efficiency of nanocomposite solar cells cannot be described by the Shockley–Queisser limit due to the presence of an absorption gap limiting I_{SC} and a transport gap limiting V_{OC}. The difference between the absorption and transport gaps is generally needed in nanocomposite solar cells as the driving force for local charge separation in an effective bulk semiconductor.

In nanocomposite absorbers the densities of mobile charge carriers in the dark are not dominated by thermal generation as in ideal photovoltaic absorbers. In nanocomposite solar cells the densities of mobile charge carriers in the dark are dominated by electrical, chemical and electrochemical interactions at interfaces between materials transporting separately electrons or holes. The creation of a model describing fundamental limitation(s) of the efficiency of nanocomposite solar cells is still challenging.

T10.6: InP-nw solar cell

The effective lifetime related to surface recombination can be calculated following Equation (3.27). The density of defects at the InP/SiO$_2$ interface seems to be quite large since SiO$_2$ does not fit well with InP (passivation of A^{III}B^{V} semiconductors with oxides leads to non-stoichiometry and therefore to defect states). A surface recombination velocity of the order of 10^5–10^4 cm/s can be assumed what gives a lifetime of photogenerated charge carriers of 0.1–1 ns. The electric field is extended over about 1 μm and the electric field in the intrinsic InP layer is about 10^4 V/cm. Equation (9.17) can be used to calculate the drift time. The resulting drift time is about 0.04 ns, i.e. significantly shorter than the effective lifetime. Therefore surface passivation does not limit the short circuit current density of InP nanowire solar cells.

Bibliography

Aberle, A. G. (2000), Surface passivation of crystalline silicon solar cells: a review, *Prog. Photovolt.: Res. Appl.*, 8, pp. 473–487.

Aberle, A. G., Glunz, S., Warta, W. (1992), Impact of illumination level and poxide parameters on Shockley–Read–Hall recombination at the Si–SiO_2 interface, *J. Appl. Phys.*, 71, pp. 4422–4431.

Abou-Ras, D., Caballero, R., Kaufmann, C. A., Nichterwitz, M., Sakurai, K., Schorr, S., Unold, T., Schock, H. W. (2008), Impact of the Ga concentration on the microstructure of $CuIn_{1-x}Ga_xSe_2$, *Phys. Stat. Sol. (RRL)*, 3, pp. 135–137.

Adachi, S. (1989), Optical dispersion relations for GaP, GaAs, GaSb, InP, InAs, InSb, $Al_xGa_{1-x}As$, and $In_{1-x}Ga_xAs_yP_{1-y}$, *J. Appl. Phys.*, 66, pp. 6030–6040.

Adachi, S., Kimura, T., Suzuku, N. (1993), Optical properties of CdTe: experiment and modeling, *J. Appl. Phys.*, 74, pp. 3435–3441.

Albrecht, S., Saliba, M., Correa-Baena, J.-P., Jäger, K., Korte, L., Hagfeldt, A., Grätzel, M., Rech, B. (2016), Towards optical optimization of planar monolithic perovskite/silicon-heterojunction tandem solar cells, *J. Opt.*, 18, pp. 064012-1–10.

Alferov, Zh. I., Andreev, V. M., Korol'kov, V. I., Portnoi, E. L., Tret'yakov, D. N. (1969), Injection properties of n-$Al_xGa_{1-x}As$-p-GaAs heterojunctions, *Sov. Phys. Semicond.*, 2, pp. 843–844, translated from (1968), *Fiz. Tekh. Poluprovodn.*, 2, pp. 1016–1017.

Alferov, Zh. I., Andreev, Portnoi, E. L., Trukan, M. K. (1970), AlAs/GaAs heterojunction injection lasers with low room-temperature threshold, *Sov. Phys. Semicond.*, 3, pp. 1107–1110, translated from (1969), *Fiz. Tekh. Poluprovodn.*, 3, pp. 1328–1332.

Alferov, Zh. I., Andreev, Kagan, M. B., Protoasov, I. I., Trofim, V. G. (1971), Solar energy converters based on p-n $Al_xGa_{1-x}As$-GaAs heterojunctions, *Sov. Phys. Semicond.*, 4, pp. 2047–2048, translated from (1970), *Fiz. Tekh. Poluprovodn.*, 4, pp. 2378–2379.

Algora, C., Ortiz, E., Rey-Stolle, I., Diaz, V., Pena, R., Andreev, V. M., Khvostikov, V. P., Rumyantsev, V. D. (2001), A GaAs solar cell with an efficiency of 26.2% at 1000 suns and 25% at 2000 suns, *IEEE Trans. Electr. Dev.*, 48, pp. 840–844.

Allsop, N., Schönmann, A., Belaidi, A., Muffler, H.-J., Mertesacker, B., Bohne, W., Strub, E., Röhrich, J., Lux-Steiner, M. C., Fischer, C.-H. (2006), Indium sulfide thin films deposited by the spray ion layer gas reaction technique, *Thin Solid Films*, 513, pp. 52–56.

Alperovich, V. L., Paulish, A. G., Terekhov, A. S. (1994), Domination of adatom-induced over defect-induced surface states on p-type GaAs(Cs,O) at room temperature, *Phys. Rev. B.*, 50, pp. 5480–5483.

Anderson, P. W. (1958), Absence of diffusion in certain random lattices, *Phys. Rev.*, 109, pp. 1492–1515.

Andersson, B. A. (2000), Materials availability for large-scale thin-film photovoltaics, *Prog. Photovolt.: Res. Appl.*, 8, pp. 61–76.

Archer, J. (1960), Stain films on silicon, *J. Phys. Chem. Solids*, 14, pp. 104–110.

Ård, M. A., Granath, K., Stolt, L. (2000), Growth of $Cu(In,Ga)Se_2$ thin films by coevaporation using alkaline precursors, *Thin Solid Films*, 361–362, pp. 9–16.

Aspnes, D. E., Harbison, J. P., Studna, A. A., Florez, L. T. (1988), Reflectance-difference spectroscopy system for real-time measurements of crystal growth, *Appl. Phys. Lett.*, 52, pp. 957–959.

Aspnes, D. E., Kelso, S. M., Logan, R. A., Bhat, R. (1986), Optical properties of $Al_xGa_{1-x}As$, *J. Appl. Phys.*, 60, pp. 754–767.

Aspnes, D. E., Studna, A. A. (1983), Dielectric functions and optical parameters of Si, Ge, GaP, GaAs, GaSb, InP, InAs, and InSb from 1.5 to 6 eV, *Phys. Rev. B*, 27, pp. 985–1009.

Assael, M. J., Armyra, I. J., Brillo, J., Stankus, S. V., Wu. J., Wakeham, W. A. (2012), Reference data for the density and viscosity of liquid cadmium, cobalt, gallium, mercury, silicon, thallium and zinc, *J. Phys. Chem. Ref. Data*, 41, pp. 033101-1–16.

Atwater, H. A., Polman, A. (2010), Plasmonic for improved photovoltaic devices, *Nat. Mater.*, 9, pp. 205–213.

Auger, M. P. (1923), Sur les rayons β-secondaires produits dans un gas par des rayons X, *C. R. A. S.*, 177, pp. 169–171.

Bach, U., Lupo, D., Comte, P., Moser, J. E., Weissörtel, F., Salbeck, J., Spreitzer, H., Grätzel, M. (1998), Solid-state dye-sensitized mesoporous TiO_2 solar cells with high photon-to-electron conversion efficiencies, *Nature*, 395, pp. 583–585.

Bag, S., Gunawan, O., Gokman, T., Zhu, Y., Todorov, T. K., Mitzi, D. B. (2012), Low band gap liquid-processed CZTSe solar cell with 10.1% efficiency, *Energy Environ. Sci.*, 5, pp. 7060–7065.

Bardeen, J. (1947), Surface states and rectification at a metal semiconductor contact, *Phys. Rev.*, 71, pp. 717–727.

Barton, C. F. (1971), Electron effective mass in PbS single crystals by infrared reflectivity measurements, *J. Appl. Phys.*, 42, pp. 445–450.

Basch, A., Beck, F. J., Soderstrom, T., Varlamov, S., Catchpole, K. R. (2012), Combined plasmonic and dielectric rear reflectors for enhanced photocurrent in solar cells, *Appl. Phys. Lett.*, 112, pp. 243903-1–3.

Bean, K. E. (1978), Anisotropic etching of silicon, *IEEE Trans.*, ED-25, pp. 1185–1193.

Becker, P., De Bievre, P., Fujii, K., Glaser, M., Inglis, B., Luebbig, H., Mana, G. (2007), Consideration on future redefinition of the kilogram, the mole and of other units, *Metrologia*, 44, pp. 1–14.

Bedair, S. M., Lamorte, M. F., Hauser, J. R. (1979), A two-junction cascade solar-cell structure, *Appl. Phys. Lett.*, 34, pp. 38–39.

Belaidi, A., Dittrich, Th., Kieven, D., Tornow, J., Schwarzburg, K., Lux-Steiner, M. (2008), *Phys. Stat. Sol. (RRL)*, 2, pp.172–174.

Benagli, S., Borello, D., Vallat-Sauvain, E., Meier, J., Kroll, U., Hötzel, J., Spitznagel, J., Steinhauser, J., Castens, L., Djeridane, Y. (2009), High-efficiency amorphous silicon devices on LPCVD-ZNO TCO prepared in industrial KAI-M R&D reactor, *Proc. 24ᵗʰ European Photovoltaic Solar Energy Conference*, pp. 2293–2298.

Benick, J., Hoex, B., van de Sanden, M. C. M., Kessles, W. M. M., Schultz, O., Glunz, S. (2008), High efficiency n-type Si solar cells on Al_2O_3-passivated boron emitters, *Appl. Phys. Lett.*, 92, pp. 253504-1–3.

Bentzen, A., Holt, A., Christensen, J. S., Svensson, B. G. (2006), High concentration in-diffusion of phosphorus in Si from spray-on source, *J. Appl. Phys.*, 99, pp. 064502-1–8.

Bersch, E., Rangan, S., Bartynski, R. A. (2008), Band offsets of ultrathin high-k oxide films with Si, *Phys. Rev. B*, 78, pp. 085114-1–10.

Bett, A. W., Dimroth, F., Guter, W., Hoheisel, R., Oliva, E., Philipps, S. P., Schöne, J., Siefer, G., Steiner, M., Wekkeli, A., Welser, E., Meusel, M., Köstler, W., Strobl, G. (2009), Highest efficiency multi-junction solar cell for terrestrial and space application, *24ᵗʰ Europ. Photovolt. Sol. En. Conf. and Exhibition*, pp. 1–6.

Bhardwaj, A., Gupta, B. K., Raza, A., Sharma, A. K., Agnihotri, O. P. (1981), Fluorine-doped SnO_2 films for solar cell application, *Solar Cells*, 5, pp. 39–49.

Birkmire, R. W., Eser, E. (1997), Polycrystalline thin film solar cells: Present status and future potential, *Annu. Rev. Mater. Sci.*, 27, pp. 625–653.

Bischoff, F. (1954), Verfahren zum Herstellen von reinstem kristallinem Germanium, Verbindungen von Elementen der III. und V. oder II. und VI. Gruppe des periodischen Systems und von oxydischem Halbleitermaterial, *Deutsches Patentamt*, Auslegeschrift 1 140 549, pp. 1–3.

Bisquert, J., Vikhrenko, V. S. (2004), Interpenetration of time constants measured by kinetic techniques in nanostructured semiconductor electrodes and dye-sensitized solar cells, *J. Phys. Chem. B*, 108, pp. 2313–2322.

Bodnar, I. V. (2008), Study of single crystals of the $CuIn_3Se_5$ ternary compound, *Semiconductors*, 42, pp. 1030–1033.

Bolotin, K. I., Sikes, K. J., Hone, J., Stormer, H. L., Kim, P. (2008), Temperature-dependent transport in suspended graphene, *Phys. Rev. Lett.*, 101, pp. 096802-1–4.

Bordina, N. M., Golovner, T. M., Zadde, V. V., Zaitseva, A. K., Landsman, A. P., Streltsova, V. I. (1975), Operation of a thin silicon photo converter under illumination on both sides, *Appl. Sol. Energ.*, 6, pp. 81–86.

Brabec, C. J., Sariciftci, N. S., Hummelen, J. C. (2001), Plastic solar cells, *Adv. Funct. Mater.*, 11, pp. 15–26.

Brabec, C. J., Zerza, G., Cerullo, G., De Silvestri, S., Luzzati, S., Hummelen, J. C., Sariciftci, S. (2001), Tracing photoinduced electron transfer process in conjugated polymer/fullerene bulk heterojunctions in real time, *Chem. Phys. Lett.*, 340, pp. 232–236.

Braunstein, R., Moore, A. R., Herman, F. (1958), Intrinsic optical absorption in germanium-silicon alloys, *Phys. Rev.*, 109, pp. 695–710.

Brillson, L. J. (1978), Transition in Schottky barrier formation with chemical reactivity, *Phys. Rev. Lett.*, 40, pp. 260–263.

Britt, J., Ferekides, C. (1993), Thin-film CdS/CdTe solar cell with 15.8% efficiency, *Appl. Phys. Lett.*, 62, 2851–2852.

Brodsky, M. H., Title, R. S., Weiser, K., Pettit, G. D. (1970), Structural, optical, and electrical properties of amorphous silicon films, *Phys. Rev. B*, 1, pp. 2632–2641.

Brooks, W. E. (2011), Silver, *U.S. Geological Survey, Mineral Commodity Summaries*, January 2011, pp. 146–147.

Bruel, M. (1995), Silicon on insulator material technology, *Electron. Lett.*, 31, pp. 1201–1202.

Brus, L. (1984), Electron-electron and electron-hole interactions in small semiconductor crystallites: The size dependence of the lowest excited state, *J. Chem. Phys.*, 80, pp. 4403–4409.

Burschka, J., Pellet, N., Moon, S. J., Baker, R. H., Gao, P., Nazeruddin, M. K., Grätzel, M. (2013), Sequential deposition as route to high-performance perovskite-sensitized solar cells, *Nature*, 499, pp. 316–320.

Burstein, E. (1953), Anomalous optical absorption limit in InSb, *Phys. Rev. (Letters to the Editor)*, 93, pp. 632–633.

Burton, J. A., Prim, R. C., Slichter, W. P. (1953), The distribution of solute in crystals grown from the melt. Part I. Theoretical, *J. Chem. Phys.*, 21, pp. 1987–1991.

Capasso, F., Cho, A. Y., Mohammed, K., Foy, P. W. (1985), Doping interface dipoles: tunable heterojunction barrier heights and band-edge discontinuities by molecular beam epitaxy, *Appl. Phys. Lett.*, 46, pp. 664–666.

Caplan, P. J., Poindexter, E. H., Deal, B. E., Razouk, R. R. (1979), ESR centers, interface states, and oxide fixed charge in thermally oxidized silicon wafers, *J. Appl. Phys.*, 50, pp. 5847–5854.

Carlberg, T. (1986), Calculated solubilities of oxygen in liquid and solid silicon, *J. Electrochem. Soc.*, 139, pp. 1940–1942.

Carlin, J. F. (2013), Tin, *U.S. Geological Survey, Mineral Commodity Summaries*, January, pp. 170–171.

Carlson, D. E., Wronski, C. R. (1976), Amorphous silicon solar cell, *Appl. Phys. Lett.*, 28, pp. 671–673.

Casey, H. C., Miller, B. I., Pinkas, E. (1973), Variation of minority-carrier diffusion length with carrier concentration in GaAs liquid-phase epitaxial layers, *J. Appl. Phys.*, 44, pp. 1281–1287.

Chapin, D. M., Fuller, C. S., Pearson, G. L. (1954), A new silicon p-n junction photocell for converting solar radiation into electrical power, *J. Appl. Phys. (Letters to the Editor)*, 25, pp. 676–677.

Chelikowsky, J. R., Cohen, M. L. (1976), Nonlocal pseudopotential calculations for the electronic structure of eleven diamond and zinc-blende semiconductors, *Phys. Rev. B*, 14, pp. 556–582.

Chen, S., Liu, Y., Zhang, L., Chow, P. C. Y., Wang, Z., Zhang, G., Ma, W., Yan, H. (2017), A wide-bandgap donor polymer for highly efficient non-fullerene organic solar cells with a small voltage loss, *J. Am. Chem. Soc.*, 139, pp. 6298–6301.

Chirilă, A., Reinhard, P., Pianezzi, F., Bloesch, P., Uhl, A. R., Fella, C., Kranz, L., Keller, D., Gretener, C., Hagendorfer, H., Jaeger, D., Erni, R., Nishiwaki, S., Buecheler, S., Tiwari, A. N. (2013), Potassium-induced surface modification of $Cu(In,Ga)Se_2$ thin films for high-efficiency solar cells, *Nat. Mater.*, 12, pp. 1107–1111.

Chiu, P. T., Law, D. C., Woo., R. L., Singer, S. B., Bhusari, D., Hong, W. D., Zakaria, A., Boisvert, J., Mesropian, S., King, R. R., Karam, N. H. (2014), 35.8% space and 38.8 terrestrial 5J direct bonded cells, *Proc. 40th IEEE Photovolt. Spec. Conf.*, pp. 11–13.

Cho, A. Y., Arthur, J. R. (1975), Molecular beam epitaxy, *Prog. Solid State Chem.*, 10, pp. 157–191.

Cody, G. D., Tiedje, T., Abeles, B., Brooks, B., Goldstein, Y. (1981), Disorder and the optical-absorption edge of hydrogenated amorphous silicon, *Phys. Rev. Lett.*, 47, pp. 1480–1483.

Cojacaru-Mirédin, O., Choi, P., Wuerz, R., Raabe, D. (2011), Atomic-scale characterization of the $CdS/CuInSe_2$ interface in thin-film solar cells, *Appl. Phys. Lett.*, 98, pp. 103504-1–3.

Colville, F. (2009), Laser processing enables high-efficiency silicon-cell concepts, *Photovoltaics World*, March/April, pp. 46–49.

Contreras, M. A., Egaas, B., Dippo, P., Webb, J., Granata, J., Ramanathan, K., Asher, S., Noufi, R. (1997), On the role of Na and modifications to $Cu(In,Ga)Se_2$ absorber materials using thin-MF (M = Na, K, Cs) precursor layers, *Proc. 26th IEEE Photovolt. Spec. Conf.*, 335, pp. 359–362.

Coropceanu, V., Cornil, J., Da Silva Filho, D. A., Olivier, Y., Silbey, R., Brédas, J.-L. (2007), Charge transport in organic semiconductors, *Chem. Rev.*, 107, pp. 926–952.

Cowley, A. M., Sze, S. M. (1965), Surface states and barrier height of metal-semiconductor systems, *J. Appl. Phys.*, 36, pp. 3212–3220.

Cuevas, A., Stocks, M., Armand, S., Stuckings, M. (1997), High minority carrier lifetime in phosphorus-gettered multicrystalline silicon, *Appl. Phys. Lett.*, 70, pp. 1017–1019.

Cuff, K. F., Ellett, M. R., Kuglin, C. D., Williams, L. R. (1964), The band structure of PbTe, PbSe, and PbS, *Proc. 7th Int. Conf. de Physique des Semiconducteurs*, 7, pp. 677–684.

Czochralski, J. (1918), Ein neues Verfahren zur Messung der Kristallisations-geschwindigkeit der Metalle, *Z. Physik. Chemie*, 92, pp. 219–221.

Davis, J., R., Rohatgi, A., Hopkins, R. H., Blais, P. D., Rai-Choudhuri, P., McCormick, J. R., Mollenkopf, H. C. (1980), Impurities in silicon solar cells, *IEEE Transact. Eletcron. Dev.* ED-27, pp. 677–687.

Däweritz, L., Hey, R. (1990), Reconstruction and defect structure of vicinal GaAs(001) and Al$_x$Ga$_{1-x}$As(001) surfaces during MBE growth, *Surf. Sci.*, 236, pp. 15–22.

Deal, B. E., Grove, A. S. (1965), General relationship for the thermal oxidation of silicon, *J. Appl. Phys.*, 36, pp. 3770–3778.

Delbos, S. (2012), Kesterite thin films for photovoltaics: A review, *EPJ Photovolt.*, 3, pp. 35004-1–13.

de Nobel, D. (1959), Phase equilibria and semiconducting properties of CdTe, *Philips Res. Rep.*, 14, pp. 361–399.

Devlin, W. J., Wood, C. E. C., Stall, R., Eastman, L. F. (1980), A molybdenum source, gate and drain metallization system for GaAs MESFET layers grown by molecular beam epitaxy, *Solid-State Electron.*, 23, pp. 823–829.

Diebold, U. (2003), The surface science of titanium dioxide, *Surf. Sci. Rep.*, 48, pp. 53–229.

Dimmler, B., Dittrich, H., Menner, R., Schock, H. W. (1987), Performance and optimization of heterojunctions based on Cu(Ga,In)Se$_2$, *IEEE PVSC*, Proceedings (A88-34226 13-44), pp. 1454–1460.

Dimroth, F. (2006), Photovoltaic hydrogen generation process and device, *BRD Patent*, 2006/042650, pp. 1–24.

Dimroth, F., Grave, M., Beutel, P., Fiedeler, U., Karcher, C., Tibbits, T. N., Oliva, E., Siefer, G., Schachtner, M., Wekkeli, A., Bett, A. W., Krause, R., Piccin, M., Blanc, N., Drazek, C., Guiot, E., Ghyselen, B., Salvetat, T., Tauzin, A., Sgnamarcheix, T., Dobrich, A., Hannappel, T., Schwarzburg, K. (2014), Wafer-bonded four-junction GaInP/GaAs/GaInAsP/GaInAs concentrator solar cells with 44.7% efficiency, *Prog. Photovolt.: Res. Appl.*, 22, pp. 277–282.

Dimroth, F., Lanyi, P., Schubert, U., Bett, A. W. (2000), MOVPE grown $Ga_{1-x}In_xAs$ solar cells for GaInP/GaInAs tandem applications, *J. Electronic Mater.*, 29, pp. 42–46.

Dingemans, G., Seguin, R., Engelhart, P., van de Sanden, M. C. M., Kessels, W. M. M. (2010), Silicon surface passivation by ultrathin Al_2O_3 films synthesized by thermal and plasma atomic layer deposition, *Phys. Stat. Sol. (RRL)*, 4, pp. 10–12.

Dingemans, G., van de Sanden, M. C. M., Kessels, W. M. M. (2011), Excellent Si surface passivation by low temperature SiO_2 using an ultrathin Al_2O_3 capping film, *Phys. Stat. Sol. (RRL)*, 5, pp. 22–24.

Dismukes, J. P., Ekstrom, L., Paff, R. J. (1964), Lattice parameter and density in germanium-silicon alloys, *J. Phys. Chem.*, 68, pp. 3021–3027.

Dittrich, T., (2000), Porous TiO_2: Electron transport and application to dye sensitized injection solar cells, *Phys. Stat. Sol. (a)*, 182, pp. 447–455.

Dittrich, T., Awino, C., Prajongtat, P., Rech, B., Lux-Steiner, M. C. (2015), Temperature dependence of the band gap of $CH_3NH_3PbI_3$ stabilized with PMMA: A modulated surface photovoltage study, *J. Phys. Chem. C*, 119, pp. 23968–23972.

Dittrich, T., Belaidi, A., Ennaoui, A. (2011), Concepts of inorganic solid-state nanostructured solar cells, *Sol. Energy Mater. Sol. C.*, 95, pp. 1527–1536.

Dittrich, T., Bitzer, T., Rada, T., Timoshenko, V. Yu., Rappich, J. (2002), Non-radiative recombination at reconstructed Si surfaces, *Solid-State Electron.*, 46, pp. 1863–1872.

Dittrich, T., Burke, Th., Koch, F., Rappich, J. (2001), Passivation of an anodic oxide/p-Si interface stimulated by electron injection, *J. Appl. Phys.*, 89, pp. 4636–4642.

Dittrich, T., Lang, F., Shargaieva, O., Rappich, J., Nickel, N. H., Unger, E., Rech, B. (2016), Diffusion length of photo-generated charge carriers in layers and powders of $CH_3NH_3PbI_3$ perovskite, *Appl. Phys. Lett.*, 109, pp. 073901-1–4.

Dixon, J. R., Ellis, J. M. (1961), Optical properties of n-type indium arsenide in the fundamental absorption edge region, *Phys. Rev.*, 123, pp. 1560–1566.

Drechsel, J., Männig, B., Kozlowski, F., Pfeiffer, M., Leo, K., Hoppe, H. (2005), Efficient organic solar cells based on a double p-i-n architecture using doped wide-gap transport layers, *Appl. Phys. Lett.*, 86, pp. 244102-1–3.

Dullweber, T., Gatz, S., Hannebauer, H., Falcon, T., Hesse, R., Schmidt, J., Brendel, R. (2012), Towards 20% efficient large-area screen-printed rear-passivated silicon solar cells, *Prog. Photovolt.: Res. Appl.*, 20, pp. 630–638.

Duran, C., Buck, T., Kopecek, R., Libal, J., Traverso, F. (2010), Bifacial solar cells with boron back surface field, *25th Europ. Photovolt. Sol. En. Conf. and Exhibition*, pp. 2348–2352.

Eades, W., Swanson, R. M. (1985), Calculation of surface generation and recombination velocities at the $Si-SiO_2$ interface, *J. Appl. Phys.*, 58, pp. 4267–4276.

Eichberger, R., Willig, F. (1990), Ultra-fast electron injection from excited dye molecules into semiconductor electrodes, *Chem. Phys.*, 141, pp. 159–173.

Eisele, S. J., Röder, T. C., Köhler, J. R., Werner, J. H. (2009), 18.9% efficient full area laser doped silicon solar cells, *Appl. Phys. Lett.*, 95, pp. 133501-1-3.

Engelhardt, P., Herder, N.-P., Grischke, R., Merkle, A., Meyer, R., Brendel, R. (2007), Laser structuring for back junction silicon solar cells, *Prog. Photovolt.: Res. Appl.*, 15, pp. 237–243.

Fa. Carl Zeiss (1939), Verfahren zur Erhöhung der Lichtdurchlässigkeit optischer Teile durch Erniedrigung des Brechungsexponenten an den Grenzflächen dieser optischen Teile, *Patentschrift*, 685767, pp. 1–3.

Fahey, P. M., Griffin, P. B., Plummer, J. D. (1989), Point defects and dopant diffusion in silicon, *Rev. Mod. Phys.*, 61, pp. 289–384.

Fengler, S., Zillner, E., Dittrich, Th. (2013), Density of surface states at CdSe quantum dots by fitting temperature-dependent surface photovoltage transients with random walk simulations, *J. Phys. Chem. C.*, 117, pp. 6462–6468.

Ferry, V. E., Polman, A., Atwater, H. A. (2011a), Modelling light trapping in nanostructured solar cells, *ACS Nano*, 5, pp. 10055–10064.

Ferry, V. E., Verschuuren, M. A., van Lare, M. C., Schropp, R. U. I., Atwater, H. A., Polman, A. (2011b), Optimized spatial correlations for broadband light trapping nanopatterns in high efficiency ultrathin film a-Si:H solar cells, *Nano Lett.*, 11, 4239–4245.

FhG-ISE (2014), New world record for solar cell efficiency at 46%, *Press Release*, No. 26/14, pp. 1–4.

Filtvedt, W. O., Holt, A., Ramachandra, P. A., Melaaen, M. C. (2012), Chemical vapor deposition of silicon from silane: Review and growth mechanisms and modeling/scaleup of fluidized bed reactors, *Sol. Energ. Mater. Sol. C.*, 107, pp. 188–200.

Fischer, C.-H., Allsop, N. A., Gledhill, S. E., Köhler, T., Krüger, M., Sáez-Araoz, R., Fu, Yanpeng, Schwieger, R., Richter, J., Wohlfahrt, P., Bartsch, P., Lichtenberg, N., Lux-Steiner, M. C. (2011), The spray-ILGAR® (ion layer gas reaction) method for the deposition of thin semiconductor layers: process and applications for thin film solar cell, *Sol. Energ. Mater. Sol. C.*, 95, pp. 1518–1526.

Fisher, A. C., Peter, L. M., Ponomarev, E. A., Walker, A. B., Wijayantha, K. G. U. (2000), Intensity dependence of the Bach reaction and transport of electrons in dye-sensitized nanocrystalline TiO_2 solar cells, *J. Phys. Chem. B*, 104, pp. 949–958.

Friedlmeier, T. M., Jackson, P., Bauer, A., Hariskos, D., Kiowski, O., Wuerz, R., Powalla, M. (2015), Improved photocurrent in Cu(In,Ga)Se$_2$ solar cells: From

20.8% to 21.7% efficiency with CdS buffer and 21.0% Cd-free, *IEEE J. Photovolt.*, 5, pp. 1487–1491.

Fujishima, A., Honda, K. (1972), Electrochemical photolysis of water at a semiconductor surface, *Nature*, 238, pp. 37–38.

Gärtner, W. (1959), Depletion-layer photoeffects in semiconductors, *Phys. Rev.*, 116, pp. 84–87.

Gast, M., Köntges, M., Brendel, R. (2008), Lead-free on-laminate-laser-soldering: A new module assembling concept, *Prog. Photovolt.: Res. Appl.*, 16, pp. 151–157.

George, M. W. (2013), Tellurium, *U.S. Geological Survey, Mineral Commodity Summaries*, January, pp. 164–165.

Gloeckler, M., Sankin, I., Zhao, Z. (2013), CdTe solar cells at the threshold to 20% efficiency, *IEEE J. Photovolt.*, 3, pp. 1389–1393.

Glunz, S. W., Rein, S., Warta, W., Knobloch, J., Wettling, W. (2001), Degradation of carrier lifetime in Cz silicon solar cells, *Sol. Energ. Mater. Sol. C.*, 65, pp. 219–229.

Goetzberger, A., Voß, B., Knobloch, J. (1997), *Sonnenenergie: Photovoltaik*, B. G. Teubner, Stuttgart.

Goodman, A. M. (1961), A method for the measurement of short minority carrier diffusion lengths in semiconductors, *J. Appl. Phys.*, 32, pp. 2550–2552.

Grant, R. W., Waldrop, J. R., Kraut, E. A. (1978), Observation of the orientation dependence of interface dipole energies in Ge-GaAs, *Phys. Rev. Lett.*, 40, pp. 656–659.

Grätzel, M. (2012), Status and progress in dye sensitized solar cells, *EU-PVSEC Proc., 27th European Photovoltaic Solar Energy Conference and Exhibition*, pp. 2158–2162.

Green, M. A., Emery, K., Hishikawa, Y., Warta, W., Dunlop, E. D. (2016), Solar cell efficiency tables (version 48), *Prog. Photovolt.: Res. Appl.*, 24, pp. 905–913.

Griffith, D. J. (2004), *Introduction into Quantum Mechanics*, 2nd ed., Prentice Hall, Singapore.

Grimm, A., Kieven, D., Lauermann, I., Lux-Steiner, M. Ch., Hergert, F., Schwieger, R., Klenk, R. (2012), Zn(O,S) layers for chalcopyrite solar cells sputtered from a single target, *EPJ Photovolt.*, 3, pp. 30302-1–4.

Grimsehl, E. (founded), Schallreuter, W. (continued), Haferkorn, H. (revised) (1988), *Grimsehl Lehrbuch der Physik, Band 3: Optik*, 19th ed., BSB B. G. Teubner Verlagsgesellschaft, Leipzig.

Grimsehl, E. (founded), Schallreuter, W. (continued), Lösche, H. (new ed.) (1990), *Grimsehl Lehrbuch der Physik, Band 4: Struktur der Materie*, 18th corrected ed., BSB B. G. Teubner Verlagsgesellschaft, Leipzig.

Groenendaal, L., Jonas, F., Freitag, D., Pielartzik, H., Reynolds, J. R. (2000), Poly(3,4-ethylenedioxythiophene) and its derivatives: Past, present, and future, *Adv. Mater.*, 12, pp. 481–494.

Guberman, D. E. (2013), Lead, *U.S. Geological Survey, Mineral Commodity Summaries*, January, pp. 90–91.

Guillemoles, J. F. (2000), Stability of $Cu(In,Ga)Se_2$ solar cells: A thermodynamic approach, *Thin Solid Films*, 361–362, pp. 338–345.

Günes, S., Neugebauer, H., Sariciftci, N. S. (2007), Conjugated polymer-based organic solar cells, *Chem. Rev.*, 107, pp. 1324–1338.

Gutsche, H. (1958), Method for producing highest-purity silicon for electric semiconductor devices, *United States Patent*, 3,042,494, pp. 1–6.

Guyot-Sionnest, P. (2012), Electrical transport in colloidal quantum dot films, *J. Phys. Chem. Lett.*, 3, pp. 1169–1175.

Haase, F., Kajari-Schröder, S., Winter, R., Nese, M., Brendel, R. (2012), Back-contacted high-efficiency silicon solar cells — conversion efficiency dependence on cell thickness, *Photovoltaics International*, 18th edition, 4th quarter, pp. 50–57.

Haddon, R. C., Brus, L. E., Raghavachari, K. (1986), Electronic structure and bonding of icosahedral C_{60}, *Chem. Phys. Lett.*, 125, pp. 459–464.

Haines, W. G., Bube, R. H. (1978), Effects of heat treatment on the optical and electrical properties of indium-tin oxide films, *J. Appl. Phys.*, 49, pp. 304–307.

Hall, R. N. (1952), Electron-hole recombination in germanium, *Phys. Rev.*, 87, pp. 387.

Hammond, M. L. (2001), Silicon epitaxy by chemical vapor deposition, in *Handbook of Thin Film Deposition Processes and Techniques*, 2nd ed., eds. Seshan, K., William Andrew Inc., Norwich, pp. 45–110.

Han, L., Islam, A., Chen. H., Malapaka, C., Chiranjeevi, B., Zhang, S., Yang, X., Yanagida, M. (2012), *Energy Environ. Sci.*, 5, pp. 6057–6060.

Han, L., Mandlik, P., Cherenack, K. H., Wagner, S. (2009), Amorphous silicon thin-film transistors with field-effect mobilities of $2 \, cm^2/(Vs)$ for electrons and $0.1 \, cm^2/(Vs)$ for holes, *Appl. Phys. Lett.*, 94, pp. 162105-1–3.

Haschke, J., Amkreutz, D., Rech, B. (2016), Liquid phase crystallized silicon on glass: technology, material quality and back contacted heterojunction solar cells, *Jap. J. Appl. Phys.*, 55, pp. 04EA04-1–10.

Haugeneder, A., Negas, M., Kallinger, C., Spirkl, W., Lemmer, U., Feldmann, J., Scherf, U., Harth, E., Gügel, A., Müllen, K. (1999), Exciton diffusion and dissociation in conjugated polymer/fullerene blends and heterostructures, *Phys. Rev. B*, 59, pp. 15346–15351.

Heeger, A. J., Kivelson, S., Schrieffer, J. R., Su, W.-P. (1988), Solitons in conducting polymers, *Rev. Mod. Phys.*, 60, pp. 781–850.

Heliatek (2016), Heliatek sets new organic photovoltaik world record efficiency of 13.2%, *Pressemitteilung*, 8.2.2016, pp. 1–3.

Helms, C. R., Poindexter, E. H. (1994), The silicon-silicon-dioxide system: Its microstructure and imperfections, *Rep. Prog. Phys.*, 57, pp. 791–852.

Herberholz, R., (1997), Prospects of wide-gap chalcopyrites for thin film photo-voltaic modules, *Sol. Energ. Mater. Sol. C.*, 49, pp. 227–237.

Herron, J., Podewils, C. (2011), Im silbernen Käfig, *Photon*, September 2011, pp. 110–116.

Hieslmair, H., Latchford, I., Mandrell, L., Chun, M., Adibi, B. (2012), Ion implantation for silicon solar cells, *Photovoltaics International*, Fourth Quarter, November 2012, pp. 58–64.

Higuchi, M., Sugawa, S., Ikenaga, W., Ushio, J., Nohira, H., Maruizumi, T. (2007), Subnitride and valence band offset at Si_3N_4/Si interface formed using nitrogen-hydrogen radicals, *Appl. Phys. Lett.*, 90, pp. 123114-1–3.

Hill, I. G., Rajagopal, A., Kahn, A. (1998), Molecular level alignment at organic semiconductor-metal interfaces, *Appl. Phys. Lett.*, 73, pp. 662–664.

Himpsel, F. J., McFeely. F. R., Taleb-Ibrahimi, A., Yarmoff, J. A., Hollinger, G. (1988), Microscopic structure of the SiO_2/Si interface, *Phys. Rev. B*, 38, pp. 6084–6096.

Hiroi, H., Iwata, Y., Horiguchi, K., Sugimoto, H. (2016a), 960-mV open-circuit voltage chalcopyrite solar cell, *IEEE J. Photovolt.*, 6, pp. 309–312.

Hiroi, H., Iwata, Y., Adachi, S., Sugimoto, H., Yamada, A. (2016b), New world-record efficiency for pure-sulfide $Cu(In,Ga)S_2$ thin-film solar cell with Cd-free buffer layer via KCN-free process, *IEEE J. Photovolt.*, 6, pp. 760–762.

Hoex, B., Schmidt, J., Pohl, P., van de Sanden, M. C. M., Kessles, W. M. M. (2008), Silicon surface passivation by atomic layer deposition Al_2O_3, *J. Appl. Phys.*, 104, pp. 0449003-1–12.

Hotta, S., Rughooputh, S. D. D. V., Heeger, A. J. (1987), Conducting polymer composites of soluble polythiophenes in polystyrene, *Synt. Met.*, 22, pp. 79–87.

Hummelen, J. C., Knight, B. W., LePeq, F., Wudl, F. (1995), Preparation and characterization of fulleroid and methanofullerene derivatives, *J. Org. Chem.*, 60, pp. 532–538.

IEC (2008), Photovoltaic devices — part 3: Measurement principles for terrestrial photovoltaic (PV) devices with reference spectral irradiance data, *IEC 60904-3*, ed. 2, pp. 1–62.

Istratov, A. A., Flink, C., Hieslmair, H., Weber, E. R., Heiser, T. (1998), Intrinsic diffusion coefficient of interstitial copper in silicon, *Phys. Rev. Lett.*, 81, pp. 1243–1246.

Jackson, P., Hariskos, D., Lotter, E., Paetel, S., Wuerz, R., Menner, R., Wischmann, W., Powalla, M. (2011), New world record efficiency for $Cu(In,Ga)Se_2$ thin-film solar cells beyond 20%, *Prog. Photovolt.: Res. Appl.*, 19, pp. 894–897.

Jackson, P., Wuerz, R., Hariskos, D., Lotter, E., Witte, W., Powalla, M. (2016), Effects of alkali elements in $Cu(In,Ga)Se_2$ solar cells with efficiencies up to 22.6%, *Phys. Stat. Sol. RRL*, 10, pp. 583–586.

Jackson, W. B., Amer, N. M. (1982), Direct measurement of gap-state absorption in hydrogenated amorphous silicon by photothermal deflection spectroscopy, *Phys. Rev. B*, 25, pp. 5599–5562.

Jacoboni, C., Canali, C., Ottaviani, G., Alberigi Quarante, A. (1977), A review of some charge transport properties of silicon, *Solid-State Electron.*, 20, pp. 77–89.

Jaffe, J. E., Zunger, A. (1984), Theory of the band-gap anomaly in ABC_2 chalcopyrite semiconductors, *Phys. Rev. B*, 29, pp. 1882–1906.

Jasenek, A., Rau, U. (2001), Defect generation in $Cu(In,Ga)Se_2$ heterojunction solar cells by high-energy electron and proton irradiation, *J. Appl. Phys.*, 90, pp. 650–658.

Jeon, N. J., Noh, J. H., Kim, Y. C., Yang, W. S., Ryu, S., Seok, S. I. (2014), Solvent engineering for high-performance inorganic-organic hybrid perovskite solar cells, *Nat. Mater.*, 13, pp. 897–903.

Jeong, S., McGehee, M. D., Cui, Y. (2013), All-back-contacted ultra-thin silicon nanocone solar cells with 13.7% power conversion efficiency, *Nat. Commun.*, 4, pp. 2950-1–7.

Jordan, D. C., Kurtz, S. R., VanSant, K., Newmiller, J. (2016), Compendium of photovoltaic degradation rates, *Prog. Photovolt.: Res. Appl.*, 24, pp. 978–989.

Kahng, D. (1960), Electric field controlled semiconductor device, *US Patent*, 3,102,230.

Kapur, P., Moslehi, M. M., Deshpande, A., Rana, V., Krama, J., Seutter, S., Deshazer, H., Coutant, S., Calcaterra, A., Kommera, S., Su, Y., Grupp, D., Tamilmani, S., Dutton, D., Stalcup, T., Du, T., Wingert, M. (2013), A manufacturable, non-plated, non Ag metallization based 20.44% efficient, 242 cm^2area, back contacted solar cell on 40 µm thick mono-crystalline silicon, *28th European Photovoltaic Solar Energy Conference and Exhibition*, pp. 2228–2231.

Kayes, B. M., Nie, H., Twist, R., Sprytte, S. G., Reinhardt, F., Kizilyalli, I. C., Higashi, G. S. (2011), 27.6% conversion efficiency, a new record for single-junction solar cells under 1 sun illumination, *37th IEEE Photovoltaic Specialists Conf.*, pp. 4–8.

Kayes, B. M., Zhang, L., Twist, R., Ding, I.-K. (2014), Flexible thin-film tandem solar cells with >30% efficiency, *IEEE J. Photovolt.*, 4, pp. 729–733.

Keister, J. W., Rowe, J. E., Kolodziej, J. J., Niimi, H., Madey, T. E., Lucovski, G. (1999), Band offsets for ultrathin SiO_2 and Si_3N_4 films on Si(111) and Si(100) from photoemission spectroscopy, *J. Vac. Sci. Technol. B*, 17, pp. 1831–1835.

Keldysh, L. V. (1979), Coulomb interaction in thin films of semiconductors and semimetals, *Pisma Zh. Eksp., Teor. Fiz.*, 29, pp. 716–719.

Kepler, R. G., Hoestery, D. C. (1974), High-field mobility in anthracene crystals, *Phys. Rev. B*, 9, pp. 2743–2745.

Kern, W. (1984), Purification of Si and SiO_2 surfaces with hydrogen peroxide, *Semicond. Int.*, pp. 94–99.

Kern, W., Puotinen, D. A. (1970), Cleaning solutions based on hydrogen peroxide for use in silicon semiconductor technology, *RCA Rev.*, 31, pp. 187–206.

Kieven, D., Dittrich, Th., Belaidi, A., Tornow, J., Schwarzburg, K., Allsop, N., Lux-Steiner, M. (2008), *Appl. Phys. Lett.*, 92, pp. 153107-1-3.

Kim, J., Hiroi, H., Todorov, T. K., Gunawan, O., Kuwahara, M., Gokmen, T., Nair, D., Hopstaken, M., Shin, B., Lee, Y. S., Wang, W., Sugimoto, H., Mitzi, D. (2014), High efficiency $Cu_2ZnSn(S,Se)_3$ solar cells by applying a double $In_2S_3/CdTS$ emitter, *Adv. Mater.*, 26, pp. 7427–7431.

Kim, T.-J., Holloway, P. H. (1997), Ohmic contacts to GaAs epitaxial layers, *CRC Cr. Rev. Sol. State Mater. Sci.*, 22, pp. 239–273.

Kim, K. S., Lee, K. H., Cho, U.-I., Choi, J. S. (1986), A study of nonstoichiometric empirical formulas for semiconductive metal oxides, *Bull. Korean Chem. Soc.*, 7, pp. 29–35.

King, R. R., Haddad, M., Isshiki, T., Colter, P., Ermer, J., Yoon, H., Joslin, D. E., Karam, N. H. (2000), Metamorphic CaInP/GaInAs/Ge solar cells, *Proc. 28th IEEE Photovoltaic Specialists Conf.*, pp. 982–985.

King, R. R., Law, D. C., Edmondson, K. M., Fetzer, C. M., Kinsey, G. S., Yoon, H., Sherif, R. A., Karam, N. H. (2007), 40% efficient metamorphic GaInP/GaInAs/Ge multijunction solar cells, *Appl. Phys. Lett.*, 90, pp. 183516-1-3.

Klenk, R. (2001), Characterization and modeling of chalcopyrite solar cells, *Thin Solid Films*, 387, pp. 135–140.

Klenk, R., Klaer, J., Scheer, R., Lux-Steiner, M. Ch., Lick, I., Meyer, N., Rühle, U. (2005), Solar cells based on $CuInS_2$ — an overview, *Thin Solid Films*, 480–481, pp. 509–514.

Klenk, R., Steigert, A., Rissom, T., Greiner, D., Kaufmann, C. A., Unold, T., Lux-Steiner, M. C. (2014), Junction formation by Zn(O,S) sputtering yields CIGSe based cells with efficiencies exceeding 18%, *Prog. Photovolt.: Res. Appl.*, 22, pp. 161–165.

Knotter, D. M. (2000), Etching mechanism of vitreous silicon dioxide in HF based solutions, *J. Am. Chem. Soc.*, 122, pp. 4345–4351.

Kolodinski, S., Werner, J., Wittchen, T., Queisser, H. J. (1993), Quantum efficiencies exceeding unity due to impact ionization in silicon solar cells, *Appl. Phys. Lett.*, 63, pp. 2405–2507.

Konagai, M. (2011), Present status and future prospects of silicon thin-film solar cells, *Jpn. J. Appl. Phys.*, 50, pp. 039991-1-12.

Kraft, A., Wolf, C., Bartsch, J., Glatthaar, M., Glunz, S. (2015), Long term stability of copper front side contacts for crystalline silicon solar cells, *Sol. Energ. Mater. Sol. C.*, 136, pp. 25–31.

Králik, B., Chang, E. C., Louie, S. G. (1998), Structural properties and quasiparticle band structure of zirconia, *Phys. Rev. B*, 57, pp. 7027–7026.

Krämmer, C., Huber, C., Zimmermann, C., Lang, M., Schnabel, T., Abzieher, T., Ahlswede, E., Kalt, H., Hetterich, M. (2014), Reversible order-disorder related band gap changes in $Cu_2ZnSn(S,Se)_4$ via post-annealing of solar cells measured by electroreflectance, *Appl. Phys. Lett.*, 105, pp. 262104-1–3.

Kroto, H. W., Heath, J. R., O'Brien, Curl, R. F., Smalley, R. E. (1985), C_{60}: buckminsterfullerene, *Nature*, 318, 162–163.

Kuech, T. F., Tischler, M. A., Wang, P.-J., Scilla, G. (1988), Controlled carbon doping of GaAs by metalorganic vapor phase epitaxy, *Appl. Phys. Lett.*, 53, pp. 1317–1319.

Kumar, M., Dubey, A., Adhikari, N., Venkatesan, S., Qiao, Q. (2015), Strategic review of secondary phases, defects and defect-complexes in kesterite CZTS-Se solar cells, *Energy Environ. Sci.*, 8, pp. 3134–3159.

Kumar, N. S., Chyi, J. I., Peng, C. K., Morkoc, H. (1989), GaAs metal-semiconductor field-effect transistor with extremely low resistance non-alloyed ohmic contacts using InAs/GaAs superlattice, *Appl. Phys. Lett.*, 55, pp. 775–776.

Kupec, J., Stoop, R. L., Witzigmann, B. (2010), Light absorption and emission in nanowire array solar cells, *Optics Express*, 18, pp. 27589-1–17.

Kurtin, S., McGill, T. C., Mead, C. A. (1970), Fundamental transition in the electronic nature of solids, *Phys. Rev. Lett.*, 22, pp. 1433–1436.

Lan, X., Vozny, O., Garcia de Arquer, F. P., Liu, M., Xu, J., Proppe, A. H., Walters, G., Fan, F., Tan, H., Liu, M., Yang, Z., Hoogland, S., Sargent, E. H. (2016), 10.6% certified colloidal quantum dot solar cells via solvent-polarity-engineered halide passivation, *Nano Lett.*, 16, pp. 4630–4634.

Lauinger, T., Schmidt, J., Aberle, A. G., Hezel, R. (1996), Record low surface recombination velocities on 1 Ωcm p-silicon using remote plasma silicon nitride passivation, *Appl. Phys. Lett.*, 68, pp. 1232–1234.

Law, M., Beard, M. C., Choi, S., Luther, J. M., Hanna, M. C., Nozik, A. J. (2008), Determining the internal quantum efficiency of PbSe nanocrystal solar cells with the aid of an optical model, *Nano Lett.*, 8, pp. 3904–3910.

Lebedev, E., Dittrich, T., Petrova-Koch, V., Karg, S., Brütting, W. (1997), Charge carrier mobility in poly-(p-phenylenevinylene) studied by time-of-flight technique, *Appl. Phys. Lett.*, 71, pp. 2686–2688.

Lee, C. H., Yu, G., Moses, D., Pakbaz, K., Zhang, C., Sariciftci, N. S., Heeger, A. J., Wudl, F. (1993), Sensitization of the photoconductivity of conducting polymers by C_{60}: Photoinduced electron transfer, *Phys. Rev. B*, 48, pp. 15425–15433.

Lee, M. M., Teuscher, J., Miyasaka, T., Murakami, T. N., Snaith, H. J. (2012), Efficient hybrid solar cells based on meso-superstructured organometal halide perovskites, *Science*, 338, pp. 643–647.

Leguijt, C., Lölgen, P., Eikelboom, J. A., Weber, A. W., Schuurmans, F. M., Sinke, W. C., Alkemade, P. F. E., Sarro, P. M., Merée, C. H. M., Verhoef, L. A. (1996), Low temperature surface passivation for silicon solar cells, *Sol. Energ. Mater. Sol. C.*, 40, pp. 297–345.

Liu, H., Zhitomirsky, D., Hoogland, S., Tang, J., Kramer, I. J., Ning, Z., Sargent, E. H. (2012), Systematic optimization of quantum junction colloidal quantum dot solar cells, *Appl. Phys. Lett.*, 101, pp. 151112-1–3.

Liu, D., Kelly, T. L. (2014), Perovskite solar cells with a planar heterojunction structure prepared using room-temperature solution processing techniques, *Nat. Photonics*, 8, pp. 133–138.

Liu, M., Johnston, M. B., Snaith, H. J. (2013), Efficient planar heterojunction perovskite solar cells by vapor deposition, *Nature*, 501, pp. 395–398.

Löper, P., Stuckelberger, M., Niesen, B., Werner, J., Filipič, Moon, S.-J., Yum, J.-H., Topič, M., De Wolf, S., Ballif, C. (2015), Complex refractive index spectra of $CH_3NH_3PbI_3$ perovskite thin films determined by spectroscopic ellipsometry and spectrophotometry, *J. Phys. Chem. Lett.*, 6, pp. 66–71.

Lu, S., Ji, L., He, W., Dai, P., Yang, H., Arimochi, M., Yoshida, H., Uchida, A., Ikeda, M. (2011), High-efficiency GaAs and GaInP solar cells grown by all solid-state molecular-beam-epitaxy, *Nanoscale Res. Lett.*, 6, pp. 576-1–4.

Luther, J. M., Law, M., Beard, M. C., Song, Q., Reese, M. O., Ellingson, R. J., Nozik, A. J. (2008a), Schottky solar cells based on colloidal nanocrystal films, *Nano Lett.*, 8, pp. 3488–3492.

Luther, J. M., Law, M., Song, Q., Perkins, C. L., Beard, M. C., Nozik, A. J. (2008b), Structural, optical, and electrical properties of self-assembled films of PbSe nanocrystals treated with 1,2-ethanedithiol, *ACS Nano*, 2, pp. 271–280.

Ma, W., Luther, J. M., Zheng, H., Wu, Y., Alivisatos, A. P. (2009), *Nano Lett.*, 9, pp. 1699–1703.

Macor, L., Gervaldo, M., Fungo, F., Otero, L., Dittrich, Th., Lin, C.-Y, Chi, L.-C., Fang, F. C., Lii, S.-W., Wong, K.-T., Tsai, C.-H., Wu, C.-C. (2012), Photoinduced charge separation in donor–acceptor spiro compounds at metal and metal oxide surfaces: application to dye-sensitized solar cell, *RSC Adv.*, 2, pp. 4869–4878.

Mainz, R., Klenk, R. (2011), *In-situ* analysis of elemental depth distributions in thin films by combined evaluation of synchrotron X-ray fluorescence and diffraction, *J. Appl. Phys.*, 109, pp. 123515-1–13.

Malinkiewicz, O., Yella, A., Lee, Y. H., Minguez Espallargas, G., Graetzel, M., Nazeeruddin, M. K., Bolink, H. J. (2014), Perovskite solar cells employing organic charge-transport layers, *Nat. Photonics*, 8, pp. 128–132.

Markov, D. E., Amsterdam, A., Blom, P. W. M., Sieval, A. B., Hummelen, J. C. (2005), Accurate measurement of the exciton diffusion length in a conjugated polymer using a heterostructure with a side-chain cross-linked fullerene layer, *J. Phys. Chem. A*, 109, pp. 5266–5274.

Marshall, J. M., Street, R. A., Thompson, M. J. (1986), Electron drift mobility in amorphous Si:H, *Phil. Mag. B*, 54, 51–60.

Mårtensson, T., Carlberg, P., Borgström, M., Montelius, L., Seifert, W., Samuelson, L. (2004), Nanowire arrays defined by nanoimprint lithography, *Nano Lett.*, 4, pp. 699–702.

Massa, E., Mana, G., Kuetgens, U., Ferroglio, I. (2009), Measurement of the lattice parameter of a silicon crystal, *New J. Phys.*, 11, pp. 053013-1–12.

Masuko, K., Shigematsu, M., Hashiguchi, T., Fujishima, D., Kai, M., Yoshimura, N., Yamaguchi, T., Ichihashi, Y., Mishima, T., Matsubara, N., Yamanishi, T., Takahama, T., Taguchi., M., Maruyama, E., Okamoto, S. (2014), Achievement of more than 25% conversion efficiency with crystalline silicon heterojunction solar cell, *IEEE J. Photovolt.*, 4, pp. 1433–1435.

May, M. M., Lewerenz, H.-J., Lackner, D., Dimroth, F., Hannappel, T. (2015), Efficient direct solar-to-hydrogen conversion by *in-situ* interface transformation of a tandem structure, *Nat. Commun.*, 6, pp. 8286-1–7.

McCandless, B., Birkmire, R. (1991), Analysis of post deposition processing for CdTe/CdS thin film solar cells, *Sol. Cells*, 31, pp. 527–535.

McMeekin, D. P., Sadoughi, G., Rehmann, W., Eperon, G. E., Saliba, M., Hörantner, M. T., Haghighirad, A., Sakai, N., Korte, L., Rech., B., Johnston, M. B., Herz, L. M., Snaith, H. J. (2016), A mixed-cation lead mixed-halide perovskite absorber for tandem solar cells, *Science*, 351, pp. 151–155.

Mehdi, I., Reddy, U. K., Oh, J., East, J. R., Haddad, G. I. (1989), Nonalloyed and alloyed low-resistance ohmic contacts with good morphology for GaAs using a graded InGaAs cap layer, *J. Appl. Phys.*, 65, pp. 867–869.

Mei, A., Li, X., Liu, L., Ku, Z., Liu, T., Rong, Y., Xu, M., Hu. M., Chen, J., Yang, Y., Grätzel, M., Han, H. (2014), A hole-conductor-free, fully printable mesoscopic perovskite solar cell with high stability, *Science*, 345, pp. 295–298.

Meier, J., Flückiger, R., Keppner, H., Shah, A. (1994), Complete microcrystalline p-i-n solar cell — crystalline or amorphous cell behavior?, *Appl. Phys. Lett.*, 65, pp. 860–862.

Meyer, B. K., Polity, A., Farangis, B., He, Y., Hasselkamp, D., Krämer, Th., Wang, C. (2004), Structural properties and bandgap bowing of $ZnO_{1-x}S_x$ thin films deposited by reactive sputtering, *Appl. Phys. Lett.*, 85, pp. 4929–4931.

Miller, A., Abrahams, E. (1960), Impurity conduction at low concentrations, *Phys. Rev.*, 120, pp. 745–755.

Miller, O. D., Yablonovich, E., Kurtz, S. R. (2012), Strong internal and external luminescence as solar cells approach the Shockley–Queisser limit, *IEEE J. Photovolt.*, 2, pp. 303–311.

Mimura, T., Hiyamizu, S., Fuji, T., Nanbu, K. (1980), A new field-effect transistor with selectively doped GaAs/n-Al_xGa_{1-x}As heterojunctions, *Jpn. J. Appl. Phys.*, 19, pp. L225–L227.

Miyazaki, S., Narasaki, M., Suyama, A., Yamaoka, M., Murakami, H. (2003), Electronic structure and energy band offsets for ultrathin silicon nitride on Si(100), *Appl. Surf. Sci.*, 216, pp. 252–257.

Moors, M., Baert, K., Caremans, T., Duerinckx, F., Cacciato, A., Szlufcik, J. (2012), Industrial PERL-type solar cells exceeding 19% with screen-printed contacts and homogeneous emitter, *Sol. Energ. Mater. Sol. C.*, 106, pp. 84–88.

Moreels, I., Lambert, K., Smeets, D., de Muynck, D., Nollet, T., Martins, J. C., Vanhaecke, F., Vantomme, A., Delerue, C., Allan, G., Hynes, Z. (2009), Site-dependent optical properties of colloidal PbS quantum dots, *ACS Nano*, 3, pp. 3023–3030.

Morita, K., Miki, T. (2003), Thermodynamics of solar-grade-silicon refining, *Intermetallics*, 11, pp. 1111–1117.

Mott, N. F. (1985), The pre-exponential factor in the conductivity of amorphous silicon, *Phil. Mag. B*, 51, 19–25.

Muzillo, C. P., Mansfield, L. M., Ramanathan, K., Anderson T. J. (2016), Properties of $Cu_{1-x}K_xInSe_2$ alloys, *J. Mater. Sci.*, 51, pp. 6812–6823.

Nahory, R. E., Pollack, M. A., Johnston, Jr., W. D., Barns, R. L. (1978), Band gap versus composition and demonstration of Vegards law for $In_{1-x}Ga_xAs_yP_{1-y}$ lattice matched to InP, *Appl. Phys. Lett.*, 33, pp. 659–661.

Nakada, T., Mizutani, M. (2002), 18% efficiency Cd-free Cu(In,Ga)Se$_2$ thin-film solar cells fabricated using chemical bath deposition (CBD)-ZnS buffer layers, *Jpn. J. Appl. Phys.*, 41, L165–L167.

Nazeeruddin, M. K., Kay, A., Rodicio, I., Humphrey-Baker, R., Müller, E., Liska, P., Vlachopoulos, N., Grätzel, M. (1993), Conversion of light to electricity by cis-X$_2$Bis(2,2′-bipyridyl-4,4′-dicarboxylate)ruthenium(II) charge-transfer sensitizers (X = Cl$^-$, Br$^-$, I$^-$, CN$^-$, and SCN$^-$) on nanocrystalline TiO$_2$ electrodes, *J. Am. Chem. Soc.*, 115, pp. 6382–6390.

Nebel, C. E., Street, R. A., Johnson, N. M., Kocka, J. (1992), High electric field transport in a-Si:H. I. transient photoconductivity, *Phys. Rev. B*, 46, pp. 6789–6802.

Nickel, N. H., Lang, F., Brus, V. V., Shargaieva, O., Rappich, J. (2017), Unreveling the light-induced degradation mechanisms of CH$_3$NH$_3$PbI$_3$ perovskite, *Adv. Electron. Mater.*, DOI:10.1002/aelm.201700158.

Nie, W., Tsai, H., Asadpour, R., Blancon, J.-C., Neukirch, A. J., Gupta, G., Crochet, J. J., Chhowalla, M. , Tretiak, S., Alam, M. A., Wang, H.-L., Mohite, A. D. (2015), High-efficiency solution-processed perovskite solar cells with millimeter-scale grains, *Science*, 347, pp. 522–525.

Noh, J. H., Im, S. H., Heo, J. H., Mandal, T. N., Seok, S. I. (2013), Chemical management for colorful, efficient, and stable inorganic–organic hybrid structured solar cells, *Nano Lett.*, 13, pp. 1764–1769.

Noufi, R., Axton, R., Herrington, C., Deb, S. K. (1984), Electronic properties versus composition of thin films of CuInSe$_2$, *Appl. Phys. Lett.*, 45, pp. 668–670.

NREL (2014), NREL demonstrates 45.7% efficiency for concentrator solar cell, *Press Release*, NR-4514.

Nusbaumer, H., Moser, J.-E., Zakeeruddin, S. M., Nazeeruddin, M. K., Grätzel, M. (2001), CoII(dbbip)22+ complex rivals tri-iodide/iodide redox mediator in dye-sensitized photovoltaic cells, *J. Phys. Chem. B*, 105, pp. 10461–10464.

Nusbaumer, H., Zakeeruddin, Moser, J.-E., Grätzel, M. (2003), An alternative efficient redox couple for the dye-sensitized solar cell system, *Chem. Eur. J.*, 9, pp. 3756–3763.

Ohnesorge, B., Weigand, R., Bacher, G., Forchel, A., Riedl, W., Karg, F. H. (1998), Minority-carrier lifetime and efficiency of $Cu(In,Ga)Se_2$ solar cells, *Appl. Phys. Lett.*, 73, pp. 1224–1226.

O'Regan, B., Grätzel, M. (1991), A low-cost, high-efficiency solar cell based on dye-sensitized colloidal TiO_2 films, *Nature*, 353, pp. 737–740.

Otero, P., Quelle, I., Fonrodona, M., Santos, S., Rodríguez, J. A., Grande, E., Mata, C., Vetter, M., Andreu, J. (2012), Improvement of very large area (5.7 m²) a-Si:H PV module manufacturing by PECVD process control, *EU-PVSEC Proc.*, 27th *European Photovoltaic Solar Energy Conference and Exhibition*, Munich, Germany, pp. 2100–2112.

Pacios, R., Nelson, J., Bradley, J., Bradley, D. C., Brabec, C. J. (2003), Composition dependence of electron and hole transport in polyfluorene: [6,6]-phenyl C61-butyric acid methyl ester blend films, *Appl. Phys. Lett.*, 83, pp. 4764–4766.

Paire, M., Shams, A., Lombez, L., Péré-Leperne, N., Collin, S., Pelouard, J.-L., Guillemoles, J.-F., Lincot, D. (2012), Resistive and thermal scale effect for $Cu(In,Ga)Se_2$ polycrystalline thin film microcells under concentration, *Energy Environ. Sci.*, 4, pp. 4972–4977.

Parker, I. D. (1994), Carrier tunneling and device characteristics in polymer light-emitting diodes, *J. Appl. Phys.*, 75, pp. 1656–1666.

Pattantyus-Abraham, A. G., Kramer, I. J., Barkhouse, A. R., Wang, X., Konstantatos, G., Debnath, R., Levina, L., Raabe, I., Nazeeruddin, M. K., Grätzel, M., Sargent, E. H. (2010), Depleted-heterojunction colloidal quantum dot solar cell, *ACS Nano*, 4, pp. 3374–3380.

Patwa, R., Herfurth, H., Mueller, G., Bui, K. (2013), Laser drilling up to 15,000 holes/s in silicon wafer for PV solar cells, *Proc. SPIE*, pp. 88260G-1-7.

Paulson, P. D., Birkmire, R. W., Shafarman, W. N. (2003), Optical characterization of $CuIn_{1-x}Ga_xSe_2$ alloys by spectroscopic ellipsometry, *J. Appl. Phys.*, 94, pp. 879–888.

Pawlak, B. J., Janssens, T., Singh, S., Kuzma-Filipek, I., Robbelein, J., Posthuma, N. E., Poortmans, J., Christiano, F., Bazizi, E. M. (2012), Studies of implanted boron emitters for solar cell applications, *Prog. Phovolt.: Res. Appl.*, 20, pp. 106–110.

Perrin, J. (1995), Reactor design for a-Si:H deposition, in *Plasma Deposition of Amorphous Silicon-Based Materials*, eds. Capezzuto, P., Madan, A., Academic Press, Boston, pp. 177–203.

Persson, C. (2010), Electronic and optical properties of Cu_2ZnSnS_4 and $Cu_2ZnSnSe_4$, *J. Appl. Phys.*, 107, pp. 053710-1–8.

Persson, C., Platzer-Björkmann, C., Malmström, J., Törndahl, T., Edoff, M. (2006), Strong valence band offset bowing of $ZnO_{1-x}S_x$ enhances p-type nitrogen doping of ZnO-like alloys, *Phys. Rev. Lett.*, 97, pp. 146403-1–4.

Peter, L. (2009), "Sticky electrons" transport and interfacial transfer of electrons in the dye-sensitized solar cell, *Accounts Chem. Res.*, 42, pp. 1839–1847.

Pfann, W. G. (1957), Zone melting, *Metall. Rev.*, 2, pp. 29–76.

Philip, R. R., Pradeep, B., Shripathi, T. (2005), Photoconductivity in the ordered vacancy compound $CuIn_5Se_8$, *Appl. Surf. Sci.*, 250, pp. 216–222.

Picozzi, S., Asahi, R., Geller, C. B., Freeman, A. J. (2002), Accurate first-principles detailed-balance determination of Auger recombination and impact ionization rates in semiconductors, *Phys. Rev. Lett.*, 89, pp. 197601-1–4.

Platzer-Björkman, C., Törndahl, T., Abou-Ras, D., Malmström, J., Kessler, J., Stolt, L. (2006), Zn(O,S) buffer layers by atomic layer deposition in $Cu(In,Ga)Se_2$ based thin film solar cells: Band alignment and sulfur gradient, *J. Appl. Phys.*, 100, pp. 044506-1–9.

Poindexter, E. H., Caplan, P. J., Deal, B. E., Razouk, R. R. (1981), Interface states and electron spin resonance centers in thermally oxidized (111) and (100) silicon wafers, *J. Appl. Phys.*, 52, pp. 879–884.

Polman, A., Atwater, H. A. (2012), Photonic design principles for ultrahigh-efficiency photovoltaics, *Nat. Mater.*, 11, pp. 174–177.

Ponomarev, E. A., Levy-Clement, C. (1998), Macropore formation on p-type Si in fluoride containing organic electrolytes, *Electrochem. Solid-State Lett.*, 1, pp. 42–45.

Prins, P., Grozema, F. C., Schins, J. M., Patil, S., Scherf, U., Siebbeles, L. D. A. (2006), High intrachain hole mobility on molecular wires of ladder-type poly(p-phenylenes), *Phys. Rev. Lett.*, 96, 146601-1–4.

Radhakrishnan, H. S., Martini, R., Depauw, V., van Nieuwenhuysen, K., Bearda, T., Szlufscik, J., Poortmans, J. (2015), Kerfless layer-transfer of thin epitaxial silicon foils using noven multiple layer porous silicon stacks with near 100% detachment yield and large minority carrier diffusion lengths, *Sol. Energ. Mater. Sol. C.*, 135, pp. 113–123.

Rao, R. A., Mathew, L., Sarkar, D., Smith, S., Saha, S., Garcia, R., Stout, R., Gurmu, A., Ainom, M., Onyegam, E., Xu, D., Jawarani, D., Fossum, J., Banerjee, S., Das, U., Upadhyaya, A., Rohatgi, A., Wang, Q. (2012), A low cost kerfless thin crystalline Si solar cell technology, *38th IEEE Photovoltaic Specialists Conference* (PVSC2012), pp. 001837–001840.

Rauscher, S., Dittrich, T., Aggour, M., Rappich, J., Flietner, H., Lewerenz, H. J. (1995), Reduced interface state density after photocurrent oscillations and electrochemical hydrogenation of n-Si(111): A surface photovoltage investigation, *Appl. Phys. Lett.*, 66, pp. 3018–3020.

Ribeiro Jr., M., Fonseca, R. C., Ferreira, L. G. (2009), Accurate prediction of the Si/SiO_2 interface band offset using the self-consistent *ab initio* DFT/LDA-1/2 method, *Phys. Rev. B*, 79, pp. 241312-1–4.

Robbins, H., Schwartz, B. (1959), Chemical etching of silicon: Part I, The system HF, HNO_3, and H_2O, *J. Electrochem. Soc.*, 106, pp. 505–508.

Robbins, H., Schwartz, B. (1960), Chemical etching of silicon: Part II, The system HF, HNO_3, H_2O and $HC_2H_3O_2$, *J. Electrochem. Soc.*, 107, pp. 108–111.

RReDc (2013), ASTM G173-03 reference spectra derived from SMARTS v. 2.9.2. Available at: http://rredc.nrel.gov/solar/spectra/am1.5/ASTMG173/ASTMG173.html.

Ryvkin, S. M. (1964), *Photoelectric Effects in Semiconductors*, Consulting Bureau, New York.

Sáez-Araoz, R., Krammer, J., Harndt, S., Koehler, T., Krueger, M., Pistor, P., Jasenek, A., Hergert, F., Lux-Steiner, M. Ch., Fischer, C.-H. (2012), *Prog. Photovolt.: Res. Appl.*, 20, pp. 855–861.

Saliba, M., Orlandi, S., Matsui, T., Aghazeda, S., Cavaccini, M., Correa-Baena, J.-P., Gao, P., Scopelitti, R., Moscini, E., Dahmen, K.-H., De Angelis, F., Abate, A., Hagfeldt, A., Pozzi, G., Graetzel, M., Nazeeruddin, M. K. (2016), A molecularly engineered hole-transporting material for efficient perovskite solar cells, *Nature Energy*, 1, pp. 15017-1–7.

Saliba, M., Matsui, T., Seo, J.-Y., Domanski, K., Correa-Baena, J.-P., Zakeeruddin, S. M., Tress, W., Abate, A., Hagfeldt, A., Grätzel, M. (2016a), Cesium-containing triple cation perovskite solar cells: improved stability, reproducibility and high efficiency, *Energy Environ. Sci.*, 9, pp. 1989–1997.

Saliba, M., Matsui, T., Domanski, K., Seo, J.-Y., Ummadisingu, A., Zakeeruddin, S. M., Correa-Baena, J.-P., Tress, W., Abate, A., Hagfeldt, A., Grätzel, M. (2016b), Incorporation of rubidium cations into perovskite solar cells improves photovoltaic performance, *Science*, 354, pp. 206–209.

Samuelson, L., Junno, B., Paulsson, G., Fornell, J. O., Ledebo, L. (1992), CBE growth of (001) GaAs: RHEED and RD studies, *J. Cryst. Growth*, 124, pp. 23–29.

Sariciftci, N. S., Smilowitz, L., Heeger, A. J., Wudl, F. (1992), Photoinduced electron transfer from a conducting polymer to buckminsterfullerene, *Science*, 258, pp. 1474–1476.

Sariciftci, S., Braun, D., Zhang, C., Srdanov, V. I., Heeger, A. J., Stucky, G., Wudl, F. (1993), Semiconducting polymer-buckminsterfullerene heterojunctions: Diodes, photodiodes, and photovoltaic cells, *Appl. Phys. Lett.*, 62, pp. 585–587.

Sasaki, K., Agui, T., Nakaido, K., Takahashi, N., Onitsuka, R., Takamoto, T. (2013), Development of InGaP/GaAs/InGaAs inverted triple junction concentrator solar cells, *AIP Conf. Proc.*, 1556, pp. 22–25.

Scanlon, W. W. (1958), Intrinsic optical absorption and the radiative recombination lifetime in PbS, *Phys. Rev.*, 109, pp. 47–50.

Schaller, R. D., Klimov, V. (2004), High efficiency carrier multiplication in PbSe nanocrystals: Applications for solar energy conversion, *Phys. Rev. Lett.*, 92, pp. 186601-1–4.

Scharber, M. C., Mühlbacher, D., Koppe, M., Denk, P., Waldauf, C., Heeger, A. J., Brabec, C. J. (2006), Design rules for donors in bulk-heterojunction solar cells — towards 10% energy-conversion efficiency, *Adv. Mater.*, 18, pp. 789–794.

Schechtman, B. H., Spicer, W. E. (1970), Near infrared to vacuum ultraviolet absorption spectra and the optical constants of phthalocyanine ans porphyrin films, *J. Mol. Spec.*, 33, pp. 28–48.

Scheer, R. (1999), Qualitative and quantitative analysis of thin film heterostructures by electron beam induced current, *Sol. St. Phen.*, 67–68, pp. 57–68.

Scheer, R., Knieper, C., Stolt, L. (1995), Depth dependent collection functions in thin film chalcopyrite solar cells, *Appl. Phys. Lett.*, 67, pp. 3007–3009.

Scheer, R., Messmann-Vera, L., Klenk, R., Schock, H.-W. (2011), On the role of non-doped ZnO in CIGSe solar cells, *Prog. Photovolt.: Res. Appl.*, 20, pp. 619–624.

Scheer, R., Walter, T., Schock, H. W., Fearheiley, M. L., Lewerenz, H. J. (1993), $CuInS_2$ based thin film solar cell with 10.2% efficiency, *Appl. Phys. Lett.*, 63, pp. 3294–3296.

Schmidt, M., Korte, L., Laades, A., Stangl, R., Schubert, Ch., Angermann, H., Conrad, E., Maydell, K. v. (2007), Physical aspects of c-Si:H/c-Si hetero-junction solar cells, *Thin Solid Films*, 515, pp. 7475–7480.

Schmidt, J., Merkle, A., Brendel, R., Hoex, B., van de Sanden, M. C. M., Kessels, W. M. M. (2008), Surface passivation of high-efficiency silicon solar cells by atomic-layer-deposited Al_2O_3, *Prog. Photovolt.: Res. Appl.*, 16, pp. 461–466.

Schneiderlöchner, E., Preu, R., Lüdemann, R., Glunz, S. W. (2002), Laser-fired rear contacts for crystalline silicon solar cells, *Prog. Photovolt.: Res. Appl.*, 10, pp. 29–34.

Scholes, G. D., Rumles, G. (2006), Excitons in nanoscale systems, *Nat. Mater.*, 5, pp. 683–696.

Schottky, W. (1938), Halbleitertheorie der Sperrschicht, *Naturwissenschaften*, 26, p. 843.

Schroeder, D. J., Hernandez, J. L., Berry, G. D., Rockett, A. A. (1998), Hole transport and doping in epitaxial $CuIn_{1-x}Ga_xSe_2$, *J. Appl. Phys.*, 83, pp. 1519–1526.

Schubert, G., Huster, F., Fath, P. (2006), Physical understanding of printed thick-film front contacts of crystalline Si solar cells — review of existing models and recent developments, *Sol. Ener. Mater. Sol. C.*, 90, pp. 3399–3406.

Schumann, B., Neumann, H., Tempel, A., Kühn, G., Nowak, E. (1980), Structural and electrical properties of $CuIn_{0.7}Ga_{0.3}Se_2$ epitaxial layers on GaAs substrates, *Cryst. Res. Technol.*, 15, pp. 71–76.

Schüppel, R., Timmreck, R., Allinger, N., Mueller, T., Furno, M., Uhrich, C., Leo, K., Riede, M. (2010), Controlled current matching in small molecule organic tandem solar cells using doped spacer layers, *J. Appl. Phys.*, 107, pp. 044503-1–6.

Schweickert, H., Reuschel, K., Gutsche, H. (1956), Vorrichtung zur Gewinnung reinsten Halbleitermaterials für elektrotechnische Zwecke, *Deutsches Patentamt*, Auslegeschrift 1 061 593, pp. 1–5.

Seeger, K. (2001), *Semiconductor Physics*, Springer Berlin, New York.

Shanks, H. R., Maycock, P. D., Sidler, P. H., Danielson, G. C. (1963), Thermal conductivity of silicon from 300 to 1400°K, *Phys. Rev.*, 130, pp. 1743–1748.

Sharp (2013), Sharp develops concentrator solar cells with world's highest conversion efficiency of 44.4%, *Press Release*, June 14, pp. 1–2.

Shay, J. L., Tell, B., Kasper, H. M., Schiavone, L. M. (1972), p-d hybridization of the valence bands of I–III–VI$_2$ compounds, *Phys. Rev. B*, 5, pp. 5003–5005.

Shay, J. L., Tell, B., Kasper, H. M., Schiavone, L. M. (1973), Electronic structure of AgInSe$_2$ and CuInSe$_2$, *Phys. Rev. B*, 7, pp. 4485–4490.

Shay, J. L., Wagner, S., Kasper, H. M. (1975), Efficient CuInSe$_2$/CdS solar cells, *Appl. Phys. Lett.*, 27, pp. 89–90.

Shenai-Khatkhate, D. V., Goyette, R. J., DiCarlo, R. L., Dripps, G. (2004), Environment, health and safety for sources used in MOVPE growth of compound semiconductors, *J. Cryst. Growth*, 272, pp. 816–821.

Shin, Y. J., Kim, S. K., Park, B. H., Jeong, T. S., Shin, H. K., Kim, T. S., Yu, P. Y. (1991), Photocurrent study of the splitting of the valence band for a CdS single-crystal platelet, *Phys. Rev. B*, 44, pp. 5522–5526.

Shockley, W., Queisser, H. J. (1961), Detailed balance limit of *pn*-junction solar cells, *J. Appl. Phys.*, 32, pp. 510–519.

Shockley, W., Read, W. T. (1952), Statistics of the recombination of holes and electrons, *Phys. Rev.*, 87, pp. 835–842.

Smilowitz, L., Sariciftci, N. S., Wu, R., Gettinger, C., Heeger, A. J., Wudl, F. (1993), Photoexcitation spectroscopy of conducting-polymer — C$_{60}$ composites: Photoinduced electron transfer, *Phys. Rev. B*, 47, pp. 13835–13842.

Smith, R. L., Collin, S. D. (1992), Porous silicon formation mechanisms, *J. App. Phys.*, 71, pp. R1–R22.

Solar Junction (2012), Solar junction breaks its own world record — Silicon Valley based solar energy company achieves 44% cell efficiency, October 15, available at: http://www.sj-solar.com.

Sollmann, D. (2009), Sechs Neuner sind das Ziel, *Photon*, pp. 42–45.

Solmi, S., Parisini, A., Angelucci, R., Armigliato, A. (1996), Dopant and carrier concentration in Si in equilibrium with monoclinic SiP precipitates, *Phys. Rev. B*, 53, pp. 7836–7841.

Sonntag, P., Preissler, N., Bokalic, M., Trahms, M., Haschke, J., Schlatmann, R., Topič, M., Rech, B., Amkreutz, D. (2017), Silicon solar cells on glass with power conversion efficiency above 13% at thickness below 15 micrometer, *Sci. Rep.*, 7, pp. 873-1–12.

Spear, W. E., Le Comber, P. G. (1975), Substitutional doping of amorphous silicon, *Solid State Commun.*, 17, pp. 1193–1196.

Spicer, W. E., Chye, P. W., Skeath, P. R., Su, C. Y., Lindau, I. (1979), New and unified model for Schottky barrier and III–V insulator interface states formation, *J. Vac. Sci. Technol.*, 16, pp. 1422–1433.

Spinelli, P., Verschuuren, M. A., Polman, A. (2012), Broadband omnidirectional antireflection coating based on subwavelength surface Mie resonators, *Nat. Commun.*, 3, pp. 692-1–5.

Staebler, D. L., Wronski, C. R. (1977), Reversible conductivity changes in discharge-produced amorphous silicon, *Appl. Phys. Lett.*, 31, pp. 292–294.

Stall, R. A., Wood, C. E. C., Board, K., Dandekar, N., Eastman, L. F., Devlin, J. (1981), A study of Ge/GaAs interfaces grown by molecular beam epitaxy, *J. Appl. Phys.*, 52, pp. 4062–4069.

Stathis, J. H. (1995), Dissociation kinetics of hydrogen-passivated (100) Si/SiO_2 interface defects, *J. Appl. Phys.*, 77, pp. 6205–6207.

Steiner, M., Siefer, G., Schmidt, T., Wiesenfarth, M., Dimroth, F., Bett, A. W. (2016), 43% sunlight to electricity conversion efficiency using CPV, *IEEE J. Photovolt.*, 6, pp. 1020–1024.

Stephan, C., Schorr, S., Tovar, M., Schock, H.-W. (2011), Comprehensive insights into point defect and defect cluster formation in $CuInSe_2$, *Appl. Phys. Lett.*, 98, pp. 091906-1–3.

Stephens, A. W., Aberle, A. G., Green, M. A. (1994), Surface recombination velocity measurements at the silicon/silicon dioxide interface by microwave-detected photoconductance decay, *J. Appl. Phys.*, 76, pp. 363–370.

Strauss, U., Rühle, W. W., Köhler, K. (1993), Auger recombination in intrinsic GaAs, *Appl. Phys. Lett.*, 62, pp. 55–57.

Stutzmann, M., Biegelsen, D. K., Street, R. A. (1987), Detailed investigation of doping in hydrogenated amorphous silicon and germanium, *Phys. Rev. B*, 35, pp. 5666–5701.

Stutzmann, M., Jackson, W. B., Tsai, C. C. (1985), Light-induced metastable defects in hydrogenated amorphous silicon: A systematic study, *Phys. Rev. B*, 32, pp. 23–47.

Summit, R., Marley, J. A., Borelli, N. F. (1964), The ultraviolet absorption edge of stannic oxide (SnO_2), *J. Phys. Chem. Solids*, 25, pp. 1465–1469.

Sun, X. W., Kwok, H. S. (1999), Optical properties of epitaxially grown zinc oxide films on sapphire by pulsed laser deposition, *J. Appl. Phys.*, 86, pp. 408–411.

Supasai, T., Rujisamphan, N., Ullrich, K., Chemseddine, A., Dittrich, T. (2013), Formation of a passivating $CH_3NH_3PbI_3/PbI_2$ interface during moderate heating of $CH_3NH_3PbI_3$ layers, *Appl. Phys. Lett.*, 86, pp. 183906-1–3.

Szabo, N., Sagol, B. E., Seidel, U., Schwarzburg, K., Hannappel, T. (2008), InGaAsP/InGaAs tandem cells for a solar cell configuration with more than three junctions, *Phys. Stat. Sol. (RRL)*, 2, pp. 254–256.

Sze, S. M. (1981), *Physics of Semiconductor Devices*, 2nd ed., John Wiley & Sons, New York.

Sze, S. M., Irvine, J. C. (1968), Resistivity, mobility, and impurity levels in GaAs, Ge and Si at 300 K, *Solid State Electron.*, 11, pp. 599–602.

Taguchi, M., Terakawa, A., Maruyama, E., Tanaka, M. (2005), Obtaining a higher V_{OC} in HIT solar cells, *Prog. Photovolt.: Res. Appl.*, 13, pp. 481–488.

Taguchi, M., Yano, A., Tohoda, S., Matsuyama, K., Nakamura, Y., Nishiwaki, T., Fujita, K., Maruyama, E. (2014), 24.7 record efficiency HIT solar cell on thin silicon wafer, *IEEE J. Photovolt.*, 4, pp. 96–99.

Takahashi, K., Yamada, S., Unno, T. (2005), High-efficiency AlGaAs/GaAs tandem solar cells, *U.D.C.*, 621.383.51 : 523.9-7, pp. 1–6.

Talapin, D. V., Murray, C. B. (2005), PbSe nanocrystal solids for n- and p-channel thin film field-effect transistors, *Science*, 310, pp. 86–89.

Talapin, D. V., Rogach, A. L., Kornowski, A., Haase, M., Weller, H. (2001), Highly luminescent monodisperse CdSe and CdSe/ZnS nanocrystals synthesized in a hexadecylamine-trioctylphosphine oxide-trioctylphoshine mixture, *Nano Lett.*, 1, pp. 207–211.

Tanaka, M., Taguchi, M., Matsuyama, T., Sawada, T., Tsuda, S., Nakano, S., Hanafusa, H., Kuwano, Y. (1992), Development of new a-Si/c-Si heterojunction solar cells: ACJ-HIT (artificially constructed junction-heterojunction with intrinsic thin layer), *Jpn. J. Appl. Phys.*, 31, pp. 3518–3522.

Tang, C. W. (1986), Two-layer organic photovoltaic cell, *Appl. Phys. Lett.*, 48, pp. 183–185.

Tang, H., Prasad, K., Sanijes, R., Schmid, P. E., Lévy, F. (1994), Electrical and optical properties of TiO_2 thin films, *J. Appl. Phys.*, 75, pp. 2042–2047.

Tell, B., Shay, J. L., Kasper, H. M. (1971), Electrical properties, optical properties, and band structure of $CuGaS_2$ and $CuInS_2$, *Phys. Rev. B*, 4, pp. 2463–2471.

Teng, C. W., Muth, J. F., Özgür, Ü., Bergmann, M. J., Everitt, H. O., Sharma, A. K., Jin, C., Narayan, J. (2000), Refractive indices and absorption of coefficients of $Mg_xZn_{1-x}O$ alloys, *Appl. Phys. Lett.*, 76, pp. 979–981.

Thomas, D. G. (1960), The exciton spectrum of zinc oxide, *J. Phys. Chem. Solids*, 15, pp. 86–96.

Thornton, J. A. (1974), Influence of apparatus geometry and deposition conditions on the structure and topography of thick sputtered coatings, *J. Vac. Sci. Technol.*, 11, pp. 666–670.

Thornton, J. A., Lomasson, T. C., Talieh, H., Tseng, B.-H. (1988), Reactive sputtered $CuInSe_2$, *Solar Cells*, 24, pp. 1–9.

Tiedje, T., Cebulka, J. M., Morel, D. L., Abeles, B. (1981a), Evidence for exponential band tails in amorphous silicon hydride, *Phys. Rev. Lett.*, 46, pp. 1425–1428.

Tiedje, T., Moustakas, T. D., Cebulka, J. M. (1981b), Effect of hydrogen on the density of gap states in reactively sputtered amorphous silicon, *Phys. Rev. B*, 23, pp. 5634–5637.

Tolcin, A. C. (2012), Indium, *U.S. Geological Survey, Mineral Commodity Summaries*, January, pp. 74–75.

Tomasi, A., Paviet-Salomon, B., Jeangros, Q., Haschke, J., Christman, G., Barroud, L., Descoeudres, A., Seif, J. P., Nicolay, S., Despeisse, M., de Wolf, S., Ballif, C. (2017), Simple processing of back-contacted silicon heterojunction solar cells using selective-area crystalline growth, *Nat. Energy*, 2, pp. 17062-1–8.

Toor, F. (2012), Turning lemons into limonade: opportunities in the turbulent PV equipment market, *Photovoltaics International*, 18th ed., 4th quarter, pp. 12–16.

Toyoda, T., Sano, T., Nakajima, J., Doi, S., Fukumoto, S., Ito, A., Tohyama, T., Yoshida, M., Kanagawa, T., Motohiro, T., Shiga, T., Higuchi, K., Tanaka, H., Takeda, Y., Fukano, T., Katoh, N., Takeichi, A., Takechi, K., Shiozawa, M. (2004), Outdoor performance of large scale DSC modules, *J. Photochem. Photobiol. A: Chemistry*, 164, pp. 203–207.

Triboulet, R., Marfaing, Y., Cornet, A., Siffert, P. (1974), Undoped high-resistivity cadmium telluride for nuclear radiation detectors, *J. Appl. Phys.*, 45, pp. 2759–2765.

Trumbore, F. A. (1960), Solid solubilities of impurity elements in germanium and silicon, *Bell. Syst. Tech. J.*, 39, pp. 205–233.

Tsai, H., Nie, W., Blancon, J.-C., Stoumpos, C. C., Asadpour, R., Harutyunyan, B., Neukirch, A. J., Verduzco, R., Crocher, J. J., Tretiak, S., Pedesseau, L., Even, J., Alam, M. A., Gupta, G., Lou, J., Ajayan, P. M., Bedzyk, M. J., Kanatzidis, M. G., Mohite, A. D. (2016), High-efficiency two-dimensional Ruddlesden-Popper perovskite solar cells, *Nature*, 536, pp. 312–317.

Tsunomura, Y., Yoshimine, Y., Taguchi, M., Baba, T., Kinoshita, T., Kanno, H., Sakata, H., Maruyama, E., Tanaka, M. (2009), Twenty-two percent efficiency HIT solar cell, *Sol. Energ. Mater. Sol. C.*, 93, pp. 670–673.

Turcu, M., Pakma, O., Rau, U. (2002), Interdependence of absorber composition and recombination mechanism in $Cu(In,Ga)(Se,S)_2$ heterojunction solar cells, *Appl. Phys. Lett.*, 80, pp. 2598–2600.

Turner, W. J., Reese, W. E., Pettit, G. D. (1964), Exciton absorption and emission in InP, *Phys. Rev.*, 136, pp. A1467–A1470.

Uhrich, C. (2017), private communication about I_{SC}, V_{OC} and FF of the organic record solar cell of Heliatek, data to be published.

Ulzhöfer, C., Hermann, S., Harder, N.-P., Altermatt, P. P., Brendel, R. (2008), The origin of reduced fill factors of emitter-wrap-through-solar cells, *Phys. Stat. Sol. (RRL)*, 2, pp. 251–253.

Valle Rios, L. E., Neldner, K., Gurieva, G., Schorr, S. (2016), Existence of off-stoichiometric single phase kesterite, *J. Alloys Comp.*, 657, pp. 408–413.

Van de Walle, C. G. (1994), Energies of various configurations of hydrogen in silicon, *Phys. Rev. B*, 49, pp. 4579–4585.

Van Roesbroek, W., Shockley, W. (1954), Photon-radiative recombination of electrons and holes in germanium, *Phys. Rev.*, 94, pp. 1558–1560.

Van Sark, W. G. J. H. M. (2002), Methods of deposition of hydrogenated amorphous silicon for device applications, in *Handbook of Thin-Film Materials, Vol. 1,*

Deposition and Processing of Thin Films, ed. Nalwa, H. S., Academic Press, San Diego, pp. 1–102.

Varshni, Y. P. (1967a), Band-to-band radiative recombination in groups IV, VI and III–V semiconductors (II), *Phys. Stat. Sol.*, 20, pp. 9–36.

Varshni, Y. P. (1967b), Temperature dependence of the energy gap in semiconductors, *Physica*, 34, pp. 149–154.

Vautier, M., Guillard, C., Herrmann, J.-M. (2001), Photocatalytic degradation of dyes in water: Case study of indigo and indigo carmine, *J. Catal.*, 201, pp. 46–59.

Vegard, L. (1921), Die Konstitution der Mischkristalle und die Raumfüllung der Atome, *Z. Phys.*, 5, pp. 17–26.

Vetterl, O., Finger, F., Carius, R., Hapke, P., Houben, L., Kluth, O., Lambertz, A., Mück, A., Rech, B., Wagner, H. (2000), Intrinsic microcrystalline silicon: A new material for photovoltaics, *Sol. Energ. Mater. Sol. C*, 62, pp. 97–108.

Vogel, H. (1995), *Gerthsen Physik*, Springer-Verlag, Berlin Heidelberg.

Voncken, M. M. A. J., Schermer, J. J., Bauhuis, G. J., Mulder, P., Larsen, P. K. (2004), Multiple release layer study of the intrinsic lateral etch rate of the epitaxial lift-off process, *Appl. Phys. A*, 79, pp. 1801–1807.

Vurgaftman, I., Meyer, J. R., Ram-Mohan, L. R. (2001), Band parameters for III–V compound semiconductors and their alloys, *J. Appl. Phys. — Appl. Physics Rev.*, 89, pp. 5815–5873.

Wadia, C., Alivisatos, A. P., Kammen, D. M. (2009), Materials availability expands the opportunity for large-scale photovoltaics deployment, *Environ. Sci. Technol.*, 43, pp. 2072–2077.

Wagner, P., Helbig, R. (1974), Halleffekt und Anisotropie der der Beweglichkeit der Elektronen in ZnO, *J. Phys. Chem. Solids*, 35, pp. 327–335.

Wagner, S., Shay, J. L., Migliorato, P., Kasper, H. M. (1974), $CuInSe_2/CdS$ heterojunction photovoltaic detectors, *Appl. Phys. Lett.*, 25, pp. 434–435.

Wallentin, J., Anttu, N., Asoli, D., Huffmann, M., Åberg, I., Magnusson, M. H., Siefer, G., Fuss-Kailuweit, P., Dimroth, F., Witzigmann, B., Xu, H. Q., Samuelson, L., Deppert, K., Borgström, M. T. (2013), InP nanowire array solar cells achieving 13.8% efficiency by exceeding the ray optics limit, *Science*, 339, pp. 1057–1060.

Walter, M. G., Warren, E. L., McKone, J. R., Boettcher, S. W., Mi, Q., Santori, E. A., Lewis, N. S. (2010), Solar water splitting cells, *Chem. Rev.*, 110, pp. 6446–6473.

Walukiewicz, W. (1988), Mechanism of Fermi-level stabilization in semiconductors, *Phys. Rev. B*, 37, pp. 4760–4763.

Wang, P., Klein, C., Humphrey-Baker, R., Zakeeruddin, S. M., Grätzel, M. (2005), Stable $\geq$ 8% efficient nanocrystalline dye-sensitized solar cell based on an electrolyte of low volatility, *Appl. Phys. Lett.*, 86, pp. 123508-1–3.

Wang, C.-L., Lan, C.-M., Hong, S. H., Wang, Y.-F., Pan, T.-Y., Chang, C.-W., Kuo, H.-H., Kuo, M.-Y., Diau, E. W.-G., Lin, C.-Y. (2012), Enveloping porphyrins for efficient dye-sensitized solar cells, *Energy Environ. Sci.*, pp. 6933–6940.

Wang, M., Liu, J., Cevey-Ha, N.-L., Moon, S.-J., Liska, P., Humphrey-Baker, R., Moser, J. E., Grätzel, C., Wang, P., Zakeeruddin, S. M., Grätzel, M. (2010), High efficiency solid-state sensitized heterojunction photovoltaic device, *Nano Today*, 5, 169–174.

Wang, S., Macdonald, D. (2012), Temperature dependence of Auger recombination in highly injected crystalline silicon, *J. Appl. Phys.*, 112, pp. 113708-1–4.

Wang, Z., White, T. P., Catchpole, K. (2013), Plasmonic near-field enhancement for planar ultra-thin photovoltaics, *IEEE Photonics J.*, 5, pp. 8400608-1–8.

Wang, W., Winkler, M. T., Gunawan, O., Gokmen, T., Todorov, T. K., Zhu, Y., Mitzi, D. B. (2014), Device characteristics of CZTSSe thin-film-solar cells with 12.6% efficiency, *Adv. Energy Mater.*, 4, pp. 1301465-1–5.

Wang, A., Zhao, J., Green, M. A. (1990), 24% efficient silicon solar cells, *Appl. Phys. Lett.*, 57, pp. 602–604.

Wang, A., Zhao, J., Wenham, S. R., Green, M. A. (1996), 21.5% efficient thin silicon solar cell, *Prog. Photovolt.: Res. Appl.*, 4, pp. 55–58.

Watanabe, N., Kondo, Y., Ide, D., Matsuki, T., Takato, H., Sakata, I. (2010), *Prog. Photovolt.: Res. Appl.*, 18, pp. 485–490.

Wei, H. S., Zhang, S. B., Zunger, A. (1999), Effects of Na on the electrical and structural properties of $CuInSe_2$, *J. Appl. Phys.*, 85, pp. 7214–7218.

Wei, H. S., Zunger, A. (1993), Band offsets at the $CdS/CuInSe_2$ heterojunction, *Appl. Phys. Lett.*, 63, pp. 2549–2551.

Wei, S.-H., Zunger, A. (1998), Calculated natural band offsets of all II–VI and III–V semiconductors: Chemical trends and the role of cation d orbitals, *Appl. Phys. Lett.*, 72, pp. 2011–2013.

Weiher, R. L. (1966), Optical properties of free electrons in ZnO, *Phys. Rev.*, 152, pp. 736–739.

Weinberg, Z. A., Rubloff, G. W., Bassous, E. (1979), Transmission, photoconductivity, and the experimental band gap of thermally grown SiO_2 films, *Phys. Rev. B*, 19, pp. 3107–3116.

Wenham, S. R., Honsberg, C. B., Green, M. A. (1994), Buried contact silicon solar cells, *Sol. Energ. Mater. Sol. C.*, 34, pp. 101–110.

Werner, F., Cosceev, A., Schmidt, J. (2012), Silicon surface passivation by Al_2O_3: Recombination parameters and inversion layer solar cells, *Energy Procedia*, 27, pp. 319–324.

Werner, J., Kolodinski, S., Rau, U., Arch, J. K., Bauser, E. (1993), Silicon solar cell of 16.8 μ thickness and 14.7% efficiency, *Appl. Phys. Lett.*, 62, pp. 2998–3000.

Wild-Scholten, M. de, Sturm, M., Butturi, M. A., Noack, M., Narec, K. H., Timò, G. (2010), Environmental sustainability of concentrator PV systems: Preliminary LCA results of the Apollon project, *EU-PVSEC Proc.*, 25th *European Photovoltaic Solar Energy Conference and Exhibition/5th World Conference on Photovoltaic Energy Conversion*, 6–10 September, Valencia, Spain, pp. 3908–3911.

Wilke, A., Endres, J., Hörmann, U., Niederhaiúsen, J., Schlesinger, R., Frisch, J., Amsalem, P., Wagner, J., Gruber, M., Opitz, A., Vollmer, A., Brütting, W., Kahn, A., Koch, N. (2012), Correlation between interface energetics and open circuit voltage in inorganic photovoltaic cells, *Appl. Phys. Lett.*, 101, pp. 233301-1–4.

Wu, X. (2004), High-efficiency polycrystalline CdTe thin-film solar cells, *Solar Energy*, 77, pp. 803–814.

Würfel, P. (2005), *Physics of Solar Cells. From Principles to New Concepts*, WILEY-VCH, Weinheim.

Xiao, Z., Dong, Q., Bi, C., Shao, Y., Yuan, Y., Huang, J. (2014), Solvent annealing of perovskite-induced crystal growth for photovoltaic-device efficiency enhancement, *Adv. Mats.*, 26, pp. 6503–6509.

Yablonovich, E. (1982), Statistical ray optics, *J. Opt. Soc. Am.*, 72, pp. 899–907.

Yablonovich, E., Allara, D. L., Chang, C. C., Gmitter, T., Bright, T. B. (1986a), Unusually low surface recombination velocity on silicon and germanium surfaces, *Phys. Rev. Lett.*, 57, pp. 249–252.

Yablonovich, E., Gmitter, T. (1986b), Auger recombination in silicon at low carrier densities, *Appl. Phys. Lett.*, 49, pp. 587–589.

Yablonovich, E., Gmitter, T., Harbison, J. P., Bhat, R. (1987), Extreme selectivity in the lift-off of epitaxial GaAs films, *Appl. Phys. Lett.*, 51, pp. 2222–2224.

Yamada, S. (1960), On the electrical and optical properties of p-type cadmium telluride crystals, *J. Phys. Soc. Jpn.*, 15, pp. 1940–1944.

Yatsurugi, Y., Akiyama, N., Endo, Y., Nozaki, T. (1973), Concentration, solubility, and equilibrium distribution coefficient of nitrogen and oxygen in semiconductor silicon, *J. Electrochem. Soc.*, 120, pp. 975–979.

Yella, A., Lee, H.-W., Tsao, H. N., Yi, C., Chandiran, A. K., Nezeeruddin, M. K., Diau, E. W.-G., Yeh, C. Y., Zakeeruddin, S. M., Grätzel, M. (2011), Porphyrin-sensitized solar cells with cobalt (II/III)-based redox electrolyte exceed 12 percent efficiency, *Science*, 334, pp. 629–634.

Yoon, J., Jo, S., Chun, I. S., Jung, I., Kim, H.-S., Meitl, M., Menard, E., Li, X., Coleman, J. J., Paik, U., Rogers, J. A. (2010), GaAs photovoltaics and optoelectronics using releasable multilayer epitaxial assemblies, *Nature Lett.*, 465, pp. 329–334.

Yoshikawa, K., Kawasaki, H., Yoshida, W., Irie, T., Konishi, K., Nakano, K., Uto, T., Adachi, D., Kanematsu, M., Uzu, H., Yamamato, K. (2017), Silicon heterojunction solar cell with interdigitated back contacts for a photoconversion efficiency over 26%, *Nature Energy*, 2, pp. 17032-1–8.

You., J., Meng, L., Song, T.-B., Guo, T.-F., Yang, Y., Chang, W.-H., Hong, Z., Chen, H., Zhou, H., Chen, Q., Liu, Y., De Marco, N., Yang, Y. (2016), Improved air stability of perovskite solar cells via solution-processed metal oxide transport layers, *Nature Nanotechnol.*, 11, pp. 75–81.

Yu, A. Y. (1970), Electron tunneling and contact resistance of metal-silicon contact barriers, *Solid-State Electron.*, 13, pp. 239–247.

Yu, G., Gao, J., Hummelen, J. C., Wudl, F., Heeger, A. J. (1995), Polymer photovoltaic cells: Enhanced efficiency via a network of internal donor–acceptor heterojunctions, *Science*, 270, pp. 1789–1791.

Zhang, B., Tian, Y., Zhang, J. X., Cai, W. (2011), The studies of the role of fluorine in SnO_2:F films prepared by spray pyrolysis with $SnCl_4$, *J. Optoel. Adv. Mats.*, 13, pp. 89–93.

Zhang, S. B., Wie, S.-H., Zunger, A., Katayama-Yoshida, H. (1998), Defect physics of the $CuInSe_2$ chalcopyrite semiconductor, *Phys. Rev. B*, 57, pp. 9642–9656.

Zhao, J., Li, Y., Yang, G., Jiang, K., Lin, H., Ade, H., Ma, W., Yan, H. (2016b) Efficient organic solar cells processed from hydrocarbon solvents, *Nat. Energy*, 1, pp. 15027-1–7.

Zhao, J., Wang, A., Altermatt, P., Green, M. A. (1995), Twenty-four percent efficient silicon solar cells with double layer antireflection coatings and reduced resistance loss, *Appl. Phys. Lett.*, 66, pp. 3636–3638.

Zhao, J., Wang, A., Green, M. A. (1999), 24.5% efficiency silicon PERT cells on MCZ substrates and 24.7% efficiency PERL cells on FZ substrates, *Prog. Photovolt.: Res. Appl.*, 7, pp. 471–474.

Zhao, W., Qian, D., Zhang, S., Li, S., Inganäs, O., Gao, F., Hou, J. (2016), Fullerene-free polymer solar cells with over 11% efficiency and excellent thermal stability, *Adv. Mater.*, 28, pp. 4734–4739.

Zhou, H., Chen, Q., Li, G., Luo, S., Song, T., Duan, H.-S., Hong, Z., You, J., Liu, Y., Yang, Y. (2014), Interface engineering of highly efficient perovskite solar cells, *Science*, 345, pp. 542–546.

Zhu, L., Yu, B., Schoen, S., Li, X., Aldighathir, M., Richardson, B. J., Alamer, A., Yu, Q. (2016), Solvent-molecule-mediated manipulation of crystalline grains for efficient planar binary lead and tin triiodide perovskite solar cells, *Nanoscale*, 8, pp. 7621–7630.

Zide, J. M. O., Kleiman-Schwarzstein, A., Strandwitz, N. C., Zimmerman, J. D., Steenblock-Smith, T., Gossard, A. C., Forman, A., Ivanovskaya, A., Stucky, G. D. (2006), Increased efficiency in multijunction solar cells through the incorporation of semimetallic ErAs nanoparticles into the tunnel junction, *Appl. Phys. Lett.*, 88, pp. 162103-1–3.

Zillner, E., Fengler, S., Niyamakom, P., Rauscher, F., Köhler, K., Dittrich, Th. (2012), Role of ligand exchange at CdSe quantum dot layers for charge separation, *J. Phys. Chem. C*, 116, pp. 16747–16754.

Zulehner, W. (2000), Historical overview of silicon crystal pulling development, *Mater. Sci. Eng. B*, 73, pp. 7–15.

Index

Printed in the United States
by Bookmasters.

Printed in the United States
By Bookmasters